Springer Tracts in Modern Physics
Volume 137

Springer
Berlin
Heidelberg
New York
Barcelona
Budapest
Hong Kong
London
Milan
Paris
Santa Clara
Singapore
Tokyo

Springer Tracts in Modern Physics

Volumes 121–138 are listed at the end of the book

Covering reviews with emphasis on the fields of Elementary Particle Physics, Solid-State Physics, Complex Systems, and Fundamental Astrophysics

Manuscripts for publication should be addressed to the editor mainly responsible for the field concerned:

Florian Gebhard

The Mott Metal-Insulator Transition

Models and Methods

With 38 Figures

Springer

Dr. Florian Gebhard
Institut Max von Laue-Paul Langevin
J. P. 158
F-38042 Grenoble Cedex 9
France
E-mail: florian@ill.fr

Library of Congress Cataloging-in-Publication Data

Gebhard, Florian, 1962-
The mott metal-insulator transition : models and methods / Florian Gebhard.
p. cm. -- (Springer tracts in modern physics, ISSN 0081-3869 ; v. 137)
Includes bibliographical references and index.
ISBN 3-540-61481-8 (Berlin : alk. paper)
1. Metal-insulator transitions. 2. Electron-electron interactions. 3. Condensed matter. 4. Many-body problem. 5. Mathematical physics. I. Title. II. Series: Springer tracts in modern physics ; 137.
QC1.S797 no. 137
[QC176.8.M48]
530 s--dc21
[530.4'1] 96-29969
CIP

Physics and Astronomy Classification Scheme (PACS):
71.30.+h, 71.10.-w, 05.70.Fh, 71.27.+a, 71.15.-m

ISSN 0081-3869
ISBN 3-540-61481-8 Springer-Verlag Berlin Heidelberg New York

Typesetting: Camera-ready copy by the author using a Springer $\mathrm{T_EX}$ macro-package
SPIN: 10515683 56/3144-5 4 3 2 1 0 – Printed on acid-free paper

To My Family

Preface

Little do we reliably know about the Mott transition, and we are far from a complete understanding of the metal–insulator transition due to electron–electron interactions. Mott summarized his basic ideas on the subject in his wonderful book *Metal–Insulator Transitions* that first appeared in 1974 [1.1]. In his view, a Mott insulator displays a gap for charge-carrying excitations due to *electron correlations*, whose importance is expressed by the presence of local magnetic moments regardless of whether or not they are ordered. Since the subject is far from being settled, different opinions on specific aspects of the Mott transition still persist. This book naturally embodies my own understanding of the phenomenon, inspired by the work of the late Sir Nevill Mott.

The purpose of this book is twofold: first, to give a detailed presentation of the basic theoretical concepts for Mott insulators and, second, to test these ideas against the results from model calculations. For this purpose the Hubbard model and some of its derivatives are best suited. The Hubbard model describes a Mott transition with a mere minimum of tunable parameters, and various exact statements and even exact solutions exist in certain limiting cases. Exact solutions not only allow us to test our basic ideas, but also help to assess the quality of approximate theories for correlated electron systems. As I describe in this book, the Mott transition is indeed a consequence of electron correlations, and Mott's ideas are found to be essentially correct – not surprising for a physicist of his greatness.

The Hubbard model serves as our paradigm for correlated electron systems. This many-electron Hamiltonian is conceptually simple at first sight, but already poses a very difficult theoretical problem. Thus, we do not attempt to describe the phase diagram of *real* materials where the full band structure, band degeneracies, lattice distortions, impurity scattering, and long-range Coulomb interactions must be taken seriously. Since these effects tend to obscure the physics of the Mott transition, and are not expected to change the overall picture qualitatively, a quantitative comparison to experiments is not the primary goal of this book. Numerical results for the Hubbard model have also been left out for the most part of the book, for good reasons: (i) I am not an expert in this vastly growing field which deserves a book of its own, (ii) numerical approaches have their own limitations (finite size effects, minus sign problem, analytical continuation, etc.), and in general it is very difficult to obtain sufficient energy resolution to allow for definitive statements on the Mott–Hubbard metal–insulator transition. Finally, I have left out those analytical approaches to correlated electron systems which are not intended to describe the Mott transition. I apologize to the many workers on correlated electron systems who will find their contributions inadequately considered.

Despite these omissions, I hope that the book will be profitable for both researchers and graduate students working on correlated electron problems. The book certainly requires a working knowledge of solid-state physics, and a basic familiarity with many-particle concepts would also prove helpful. In any case, I have tried to provide simple derivations, although I have deferred excessively detailed calculations to the appendices or primary literature.

This project grew out of my Habilitation thesis at the Philipps University Marburg, and I am most grateful to Philippe Nozières for the opportunity to turn it into a book during my stay at the Institute Laue–Langevin. I profited a lot from his opinions and from our discussions on the Mott transition. I am also greatly indebted to my dear friend David Logan who helped me to tie up many loose ends, and to clarify my thoughts on insulators, charge gaps, magnetic moments, and the like. I also owe him my thanks for critical proof-reading of substantial parts of the book, as did my friends Peter Thomas, Fabian Eßler, and Kurt Schönhammer. Of course, the responsibility for errors and flaws rests entirely with me.

My special thanks go to authors and publishers who granted me permission to reproduce their figures, in particular to Marcelo Rozenberg and Thomas Pruschke for their POSTSCRIPT originals. This book could not have been accomplished without the help and support of many people, and I am grateful to my family and friends for their constant support and encouragement.

I appreciate the faithful collaboration with Dr. Kölsch and the production team from Springer-Verlag, and thank them for their help with the layout, their patience, and their flexibility.

Grenoble, January 1997 *Florian Gebhard*

Contents

1. Metal–Insulator Transitions

In the first section we specify an insulator as a substance with a vanishing electrical conductivity in a (weak) static electrical field at zero temperature. The transport of electrical charge in the solid state is provided by electrons that are subjected to the Coulomb interaction with the ions and the other electrons. Correspondingly, in the first category of insulators we find band, Peierls, and Anderson insulators, which can be understood in terms of single electrons that interact with the electrostatic field of the ions. Mott insulators constitute the second category, where the insulating behavior is understood as a cooperative many-electron phenomenon. In the absence of electron pairing an insulator may be characterized by a gap for charge excitations into states in which the wave functions of the excitations spatially extend over the whole specimen (gap criterion). This zero-temperature gap sets an energy scale that allows us to distinguish in practical terms between good conductors ("metals") and bad conductors ("insulators") at finite temperature.

In the second section we describe quantum and thermodynamic phase transitions as possible scenarios for the formation of a gap. In quantum phase transitions the gap opens as a consequence of the competition between the carriers' kinetic and interaction energy. The notion of a gap that separates two bands remains valid at all temperatures ("robust gap"). In thermodynamic phase transitions the gap opens as a consequence of the formation of long-range order (symmetry breaking) at some finite temperature ("soft gap").

In the third section we briefly describe band, Peierls, and Anderson insulators and their corresponding transitions. Pressure-induced metal–band insulator transitions and the metal–Anderson insulator transition upon doping provide examples for quantum phase transitions in which the formed charge gap is robust. The Peierls transition provides an example for a thermodynamic phase transition (symmetry breaking) in which the gap is the consequence of the formation of a charge-density wave.

In the fourth section we discuss Mott's correlated electron view of the metal–insulator transition. Mott emphasized the importance of magnetic moments as a signature of electron correlations, regardless of whether or not they are ordered. Correspondingly, we propose to distinguish between Mott–Hubbard and Mott–Heisenberg insulators. In the quantum phase transition from the metal into the Mott–Hubbard insulator the moments need not order.

They order in a thermodynamic phase transition into the Mott-Heisenberg insulator. The concept of the Mott–Heisenberg insulator resembles Slater's views on the metal-to-insulator transition as an ordering phenomenon. His approach is conceptually different since the magnetic moments form and order at the same temperature in Slater's one-electron picture, whereas pre-formed moments order at the transition into the Mott–Heisenberg insulator.

Some experimentally observed types of Mott insulators are listed in the fifth section, namely Wigner and Verwey insulators, multi-band Mott–Hubbard insulators, and charge-transfer insulators. We discuss some of their properties and their metal–insulator transitions to show that these materials are correlated electron systems. For other classification schemes on metal–insulator transitions we refer the reader to Mott's book [1.1] and the work by Ramakrishnan [1.2] and Imada [1.3].

1.1 Classification of Metals and Insulators

In this section we first provide the definition for metal and insulator for weak external fields at zero temperature, $T = 0$. Next, we use the dominant interaction that causes the insulating behavior as the criterion to specify four basic classes of insulators. In the third subsection we argue that an insulating ground state results from a gap for charge excitations from the ground state into states in which the wave functions of the excitations are extended. The conductivity at low temperature displays an activated behavior. As shown in the fourth subsection this allows us to identify insulators experimentally from the behavior of the resistivity for $T \to 0$. Lastly, we discuss the notion of the gap at finite temperatures.

1.1.1 Definition of Metal and Insulator

In solid-state physics the electrical conductivity of a material is based on the transport of electrons. Transport, i.e., the directed motion of charge, energy, entropy, etc., does not appear in thermodynamic equilibrium. Instead this process requires a non-equilibrium situation to occur. In addition, the electron transport in the bulk of a solid is basically influenced by three different effects:

1. the potential of the ions,
2. the electron–electron (Coulomb) interaction, and
3. externally applied (e.g., electrical and magnetic) fields.

For interacting many-particle systems far from thermal equilibrium, e.g., in optically excited semiconductors, a very complex formalism has to be employed to calculate transport properties [1.4, Chap. 3.7]. In this book, however, we are interested in the conducting properties near thermal equilibrium, i.e., the presence of external fields causes only small perturbations in our systems. Hence the first of our restrictions for a classification of insulators is to demand "weak" external fields such that Ohm's law between the current density j_α and the applied electrical field E_β holds in momentum/frequency space ($\alpha, \beta = 1, \ldots, d$; d is the dimension of the system):

$$j_\alpha(\boldsymbol{q},\omega) = \sum_\beta \sigma_{\alpha\beta}(\boldsymbol{q},\omega) E_\beta(\boldsymbol{q},\omega) \,. \tag{1.1}$$

Since we may now apply linear response theory, the conductivity tensor $\sigma_{\alpha\beta}(\boldsymbol{q},\omega)$ defined by (1.1) can be derived from an *equilibrium* current-current correlation function [1.4, Chap. 3.8]. This fluctuation-dissipation theorem implies that the conductivity as non-equilibrium property can be expressed in terms of a correlation function of the system in thermal equilibrium. The assumption of the validity of (1.1) rules out metal–insulator transitions caused by strong external fields, e.g., electrical breakdown. However, we have to keep in mind that the calculation of the conductivity tensor still requires the evaluation of a *two-electron* correlation function. Leaving aside the particular case of independent, free electrons, single-electron properties alone such as the density of states, will not be sufficient to explain electron transport in the solid state.

The second restriction is that the distinction between metal and insulator is strictly meaningful only at zero temperature ($T = 0$). This may appear to be a matter of theoretical convenience since, e.g., we may exclude the scattering of electrons by lattice vibrations from our considerations. In fact, the limit $T \to 0$ is actually of fundamental importance to distinguish between metal and insulator and the quantum phase transition between them.

With the restrictions to weak external fields and zero temperature we may define an insulator as follows (Re{...}: real part).

The static electrical conductivity of an insulator vanishes at zero temperature,

$$\sigma_{\alpha\beta}^{\mathrm{DC}}(T=0) \equiv \lim_{T\to 0} \lim_{\omega\to 0} \lim_{|\boldsymbol{q}|\to 0} \mathrm{Re}\left\{\sigma_{\alpha\beta}(\boldsymbol{q},\omega)\right\} = 0 \,. \tag{1.2}$$

In the case of a finite metallic conductivity we typically observe a Drude behavior at small frequencies,

$$\mathrm{Re}\left\{\sigma_{\alpha\beta}(T=0,\omega\to 0)\right\} = \left(\mathcal{D}_{\mathrm{c}}\right)_{\alpha\beta} \frac{\tau}{\pi(1+\omega^2\tau^2)} \;:\; \text{Drude metal,} \tag{1.3a}$$

where $\left(\mathcal{D}_{\mathrm{c}}\right)_{\alpha\beta}$ denotes the Drude weight and τ is a typical scattering time for uncorrelated collisions of electrons with, e.g., impurities. In a simple Drude theory we have $\left(\mathcal{D}_{\mathrm{c}}\right)_{\alpha\beta} = \pi e^2 (n/m^*)\delta_{\alpha\beta}$ with e as the elementary charge and n/m^* as the ratio between the charge carrier concentration and the effective mass [1.5, Chap. 1]. If electron scattering is absent, $\tau^{-1} \to 0$, translational invariance is restored, which leads to ideal metallic behavior:

$$\mathrm{Re}\left\{\sigma_{\alpha\beta}(T=0,\omega\to 0)\right\} = \left(\mathcal{D}_{\mathrm{c}}\right)_{\alpha\beta} \delta(\omega) \;:\; \text{ideal metal.} \tag{1.3b}$$

At zero temperature the conductivity alone is not sufficient to distinguish an ideal metal from a superconductor. The superconductor, however, also shows the Meissner effect [1.5, Chap. 34], and its ideal conductivity persists for temperatures up to the superconducting transition; see Sect. 1.2.

1.1.2 Classes of Insulators

The restriction to zero temperature and small external fields allows us to identify two basic categories of insulators, namely,

I. insulators due to the electron–ion interaction, and
II. insulators due to the electron–electron interaction.

To understand the insulating behavior of a material at zero temperature we may consider only static ion configurations. Then, the first category subsumes all insulators, which can be understood in terms of a *single-electron* theory. The following three classes of insulators are prominent representatives:

1. Bloch–Wilson or band insulators due to the electrons' interaction with the periodic potential of the ions;
2. Peierls insulators due to the electrons' interaction with static lattice deformations;
3. Anderson insulators due to the presence of disorder, e.g., the electrons' interaction with impurities and other lattice imperfections.

The insulators due to the electrons' mutual interaction are conceptually different. Here, we always have to solve a *many-electron* problem. The fourth class of insulators constitute the

4. Mott insulators due to the electron–electron interaction.

In Sect. 1.4 we will propose to distinguish further between Mott–Heisenberg and Mott–Hubbard insulators, depending on whether or not there is long-range order in the charge and/or spin density of the electrons.

In real materials the interactions used in the above classification scheme are always simultaneously present such that the classification of experimentally observed insulators is not always that straightforward; see, e.g., a recent discussion on the classification of vanadium dioxide (VO_2) [1.6]. In fact, an observed metal–insulator transition need not be dominated by one interaction alone to the exclusion of the others. Although a single interaction may drive the transition, its nature may be (and often is) strongly affected by the others. In the vicinity of the transition they can become relevant perturbations. Examples for this effect are the appearance of an "exciton insulator" near a metal–band insulator transition (see Sect. 1.3.1) and modifications of the disorder-induced metal–Anderson insulator transition (see Sect. 1.3.3) by the Coulomb interaction between electrons. Hence, these metal–insulator transitions are called band–Mott transitions and Anderson–Mott transitions, respectively.

As a result of Mott's intensive search for metal–insulator transitions dominantly driven by electron–electron interactions [1.1, 1.7] various insulating states such as the "exciton insulator" were and still are called "Mott insulators". Later in this book we will address only "pure" cases of Mott transitions, anticipating that the Coulomb interaction dominates the others throughout the transition.

1.1.3 Gap Criterion for the Insulator

The simple-minded Drude approach (Boltzmann transport theory) to metallic conductivity [1.5, Chap. 1] that leads to (1.3) can formally be derived from a Kubo formula for the current–current correlation function [1.4, Chap. 7]. This correlation function probes the (collective) propagation of an electron–hole excitation of the system at a fixed electron number. The analysis uncovers two fundamental requirements for electron transport in Fermi systems:

i. quantum-mechanical states for electron–hole excitations must be available at energies immediately above the energy of the ground state since the external field provides vanishingly small energy ($\omega \to 0$);
ii. these excitations must describe delocalized charges that can contribute to transport over the macroscopical sample size.

If both conditions are met, we shall have a metallic ground state.

In general, the electron and the hole interact with each other such that it is very difficult to describe the collective propagation of the electron–hole excitation. We would prefer to relate the conducting properties of a system to its *single-electron excitation spectrum.* In single-electron excitations we add an electron to or take it away from the system, i.e., we study energy differences between the spectra of the systems with N and $N \pm 1$ electrons, respectively. *If* there is no correlated motion between the electron and the hole of the electron–hole excitation, we may disentangle it into a superposition of two individual single-particle excitations. Then, the conducting properties directly relate to the single-electron excitation spectrum. This implies that in an insulator there exists a *gap for charge-carrying single-electron excitations* (gap criterion).

It is immediately apparent that the gap criterion must be taken with care. After all, we draw conclusions about the DC conductivity, a two-electron quantity, from the one-electron spectrum. Consequently, the criterion completely fails if there is pairing between electrons. For example, if the electrons form bound pairs, the pairs are mobile and contribute to a finite DC conductivity although a single-electron excitation requires a finite pair-breaking energy. The same applies to an ordinary BCS superconductor. One-particle excitations are gaped but the system does have a finite Drude weight even at finite temperature: it is a superconductor, not an insulator. Even in the case of a semiconductor the one-particle gap does not give the energy required for the optical excitation of an electron–hole pair. Instead, excitons appear in the absorption spectrum below the band absorption edge [1.8].

Despite this caveat the gap criterion is very useful in practical terms. On this basis we easily understand Bloch–Wilson (band) insulators (see Sect. 1.3.1) and Peierls insulators (see Sect. 1.3.2). In both cases there is a gap between the highest occupied and the lowest unoccupied single-electron level. A gap in this narrow sense is not present for an Anderson insulator (see Sect. 1.3.3). In this case condition (ii) is violated since the single-electron states near the Fermi energy are localized. To cover this case we define the gap as energy necessary to create a *charge-carrying*, i.e., *macroscopically extended* charge excitation. As discussed in more detail in Sect. 1.3.3 only single-electron states beyond the mobility edge can transport current. Hence, the gap can be defined as the distance between the Fermi energy and the mobility edge. It should be kept in mind, though, that the gap criterion for the Anderson insulator is only justified a posteriori because the information on the position of the mobility edge(s) is actually contained in the current–current correlation function.

For band, Peierls, and Anderson insulators the insulating behavior can be understood in terms of a gap for charge excitations between the Fermi energy and spatially extended states. The situation is less clear for many-particle systems than for these single-electron problems. When we restrict ourselves to the Mott transition later in this book, we may assume that disorder and pairing are absent. Then the properties of the conductivity tensor should be determined by the propagation of two *individual* charge excitations, and their mutual interactions (vertex corrections) should not qualitatively change their individual behavior. Given these assumptions, a gap for single charge excitations will again cause the DC conductivity to vanish. Technically speaking, we factorize the two-electron Green function into its one-electron components, whereby we assume that anomalous expectation values for Cooper pairs (superconductivity) do not occur, and vertex corrections are not important.

Under these precautions we will use the zero-temperature gap for single charge excitations (to extended states) as criterion for an insulator for all our systems, including many-electron systems with repulsive interactions. The zero-temperature chemical potentials

$$\mu^+(\lambda) = E_0(N+1,\lambda) - E_0(N,\lambda) \, , \tag{1.4a}$$

$$\mu^-(\lambda) = E_0(N,\lambda) - E_0(N-1,\lambda) \tag{1.4b}$$

are the minimum energies needed to add an electron to the ground state of the systems with N and $(N-1)$ electrons, respectively. λ denotes the set of interaction parameters that characterize our system, e.g., pressure or electron concentration. The gap to charge-carrying states is then given by

$$\Delta(\lambda) = (\mu^+(\lambda) - \mu^-(\lambda))_{\rm ext} \, , \tag{1.5a}$$

where the subscript "ext" indicates that the electron and/or the hole state is extended such that it can carry current. Finally, the gap criterion becomes

$$\Delta(\lambda) > 0 \quad : \quad \text{insulator} \, . \tag{1.5b}$$

Later in this book we will only consider pure Mott insulators where disorder and pairing effects are absent. In this case the jump in the zero-temperature chemical potential defines the Mott gap.

1.1.4 Electrical Conductivity at Low Temperature

The notions of a metal and an insulator are strictly limited to zero temperature. In fact, an "insulator" displays a finite DC conductivity even at $\Delta(\lambda) \gg k_{\mathrm{B}}T > 0$ since, by definition, thermal excitations above the gap are charge-carrying states. Of course, at room temperature ($T = 300\,\mathrm{K}$) the absolute values for the DC conductivity for various materials differ by many orders of magnitude. Hence, "metal" and "insulator" can normally be distinguished for all practical, i.e., technical purposes, but what we really mean is a "good" and a "bad" conductor at a given temperature. That such a distinction is not sufficient to define metal and insulator at finite temperature becomes apparent from the difference between semimetals and semiconductors. The specific resistivity $\rho(T) = \left[\sigma^{\mathrm{DC}}(T)\right]^{-1}$ of semiconductors at room temperature with $\rho_{\mathrm{SC}} = 10^{-3} \ldots\ 10^{9}$ Ohm-cm is close to that of semimetals with $\rho_{\mathrm{SM}} = 10^{-5} \ldots\ 10^{-4}$ Ohm-cm [1.5, Chaps. 1,28]. At $T = 0$, however, semimetals keep their (small) conductivity, whereas semiconductors display a diverging specific resistivity. It is only the limit of low temperature that allows us to identify semiconductors as insulators and semimetals as metals.

An observed strong increase or even divergence in the resistivity on lowering the temperature T may serve as a heuristic way to distinguish between metal and insulator. For insulators we expect an exponential increase on lowering the temperature since the zero-temperature gap for charge excitations $\Delta(\lambda)$ sets the energy scale for thermal activation processes:

$$\mathrm{Re}\left\{\sigma_{\alpha\beta}(k_{\mathrm{B}}T \ll \Delta(\lambda), \omega \to 0)\right\} \sim \exp\left(-\frac{\Delta(\lambda)}{2k_{\mathrm{B}}T}\right) \text{ for an insulator.} \quad (1.6)$$

A sharp transition between metal and insulator exists only at zero temperature where these two phases are actually defined. The experimental temperature often is a small energy scale compared to electronic energies ($\Delta(\lambda) \gg k_{\mathrm{B}}T$) such that the system behaves as if the zero-temperature limit has already been reached. Hence, in real experiments a very sharp change in the resistivity within a small range of temperatures can often be observed. In this sense we may also use the phrase "metal–insulator transition" at finite temperature.

1.1.5 Gap at Finite Temperature

According to (1.5) the gap $\Delta(\lambda)$ is defined only at zero temperature. It provides a natural energy scale for small temperatures, $k_{\mathrm{B}}T \ll \Delta(\lambda)$. For this reason we could use it in (1.6) to identify the conducting behavior of insulators at temperatures small compared to the zero-temperature gap. For finite temperature, $k_{\mathrm{B}}T = \mathcal{O}(\Delta(\lambda))$, it becomes more difficult to assign a proper meaning to the notion of "the" gap, as we now discuss.

Since the gap is a one-electron property it is natural to investigate the one-electron density of states, which can be measured with the help of (inverse) photoemission spectroscopy. Like the chemical potential $\mu(T)$ the density of states is continuous for finite temperature, and it is only in the limit $k_{\mathrm{B}}T \ll \Delta(\lambda)$ that the (zero-temperature) gap can be identified unambiguously. For $k_{\mathrm{B}}T = \mathcal{O}(\Delta(\lambda))$ the size of the gap can only be empirically determined because the sharp signatures of the zero-temperature gap are thermally smeared out.

In addition to the thermal smearing one usually also finds that the size of the gap varies with temperature. This "gap-shift" effect is well known in semiconductor optics [1.8]. For illustration let us consider a direct semiconductor (band insulator). If no excitations are present, it has an (optical) gap $\Delta(\lambda)$. Then, let us consider an optically excited state with a finite density $n_{\mathrm{e-h}}$ of electron–hole excitations. The creation of an *additional* electron–hole pair in the presence of the other excitations requires an energy $\Delta_{\mathrm{exc}}(\lambda, n_{\mathrm{e-h}})$ *different* from $\Delta(\lambda)$ for two reasons. First, there is Pauli blocking in the conduction and in the valence band, which increases $\Delta_{\mathrm{exc}}(\lambda, n_{\mathrm{e-h}})$ over $\Delta(\lambda)$. Second, the electrons in the conduction band and the holes in the valence band experience a mutually screened Coulomb interaction that depends on the density of electron–hole excitations. Pauli blocking and screening effects change $\Delta(\lambda)$ to $\Delta_{\mathrm{exc}}(\lambda, n_{\mathrm{e-h}})$ ("gap shift").

The question remains as to whether the gap stays finite or vanishes at some temperature T_{c}. In general, both scenarios are conceivable.

"Robust" Gap. In the first case we can classify the whole single-electron spectrum with the help of the (good quantum) number of electron–hole excitations "beyond the gap". We may not be able to define a unique gap but temperature does not destroy the notion of two different bands for electrons and holes. Two examples are band and Anderson insulators for which, by definition, the spectrum is *additive* ("rigid bands"). The zero-temperature gap $\Delta(\lambda)$ remains a characteristic energy scale for *all* temperatures.

"Soft" Gap. In the second case the band structure is "soft" as for Peierls insulators, i.e., the notion of electron–hole excitations between two separate bands depends on the presence of symmetry breaking. Within a simple mean-field description the order parameter $\Xi(\lambda) > 0$ uniquely determines the gap for electron–hole excitations, $\Delta(\lambda) \equiv \Delta(\Xi(\lambda))$. The size of the order parameter, however, depends on temperature, $\Xi(\lambda) \equiv \Xi(\lambda, T)$, such that we find a temperature-dependent gap $\Delta(\Xi(\lambda, T))$. This gap agrees with the zero-temperature gap for $T \to 0$ and ceases to exist at the critical temperature T_{c} when the long-range order breaks down and the symmetry is restored.

We will discuss the two mechanisms for the gap formation in more detail in the next section.

1.2 Types of Metal–Insulator Transitions

In this section we outline two different scenarios for the gap formation in metal–insulator transitions, both of which are realized in nature. The first route is to understand it as a quantum phase transition that results from the competition between the electrons' kinetic and potential energies at zero temperature. Symmetry breaking need not occur and the gap for electron–hole excitations is robust, i.e., it remains finite for all temperatures. Thermodynamic phase transitions offer a second way to explain the formation of a gap for charge excitations. The competition between the internal energy and the entropy favors an ordered state at low enough temperatures. The symmetry-broken state may display a gap for charge-carrying excitations that causes an insulating behavior below the critical temperature.

1.2.1 Quantum Phase Transition

We start our considerations with a toy-model description of the ionization problem; see [1.9] for a recent review on real systems. Metal–insulator transitions in the liquid phase are briefly reviewed in [1.1, Chap. 10], [1.10], and [1.11]. In this book we restrict ourselves to the solid state.

We consider hydrogen atoms of density ρ in a box. An electron with coordinates $\boldsymbol{x}_l$ shall interact with square-well model potentials produced by the protons at positions $\boldsymbol{R}_j$

$$V(\boldsymbol{x}_l) = \begin{cases} -V_0 & \text{for} \quad |\boldsymbol{x}_l - \boldsymbol{R}_j| \leq a \\ 0 & \text{for} \quad |\boldsymbol{x}_l - \boldsymbol{R}_j| > a \end{cases} . \tag{1.7}$$

The bare depth of a square-well potential V_0 is supposed to support only one bound state. Our quantum-mechanical system is described by two non-commuting operators, namely the kinetic energy of the electrons, which prefers to spread the electrons over the box, and the square-well potentials, which tend to localize them near the protons. The ground state may dominantly display the physics of either one of the operators: either the electrons are bound to the protons ("gas phase") or they are unbound ("plasma phase").

To make the problem tractable we assume that the presence of the other electrons and protons leads only to a density-dependent reduction of the potential produced by a single proton. Therefore, we analyze the corresponding single-electron problem for an electron in the screened potential of a proton,

$$\widetilde{V}(\boldsymbol{x}) = \begin{cases} -V(\rho) & \text{for} \quad |\boldsymbol{x}| \leq a \\ 0 & \text{for} \quad |\boldsymbol{x}| > a \end{cases} . \tag{1.8}$$

In the dilute limit we have $V(\rho \to 0) = V_0$, the potential is attractive, and all electrons are bound to their protons (gas phase). At densities above the critical density, $V(\rho_c) = \pi^2\hbar^2/(8m_e a^2)$ [1.12], the attractive potential of a single proton is too strongly screened by the other particles in the system to sustain a bound state. The electrons unbind and the system goes into the plasma phase.

The transition between the two phases is a "quantum phase transition". The plasma phase is the ground state of the electrons' kinetic energy operator, whereas the potential energy operator favors the gas phase. It is a macroscopic cooperative phenomenon since all atoms simultaneously unbind due to their mutual interaction (ρ dependence of $\widetilde{V}(\boldsymbol{x})$). In reality the situation is much more complex since the strength of the screening itself depends on the ionization degree of the system and thus also on temperature. Within our toy-model description, however, the effect is independent of temperature. As long as we are below the critical density the gas will be only partly ionized even for temperatures $k_{\mathrm{B}}T \gg V(\rho < \rho_{\mathrm{c}})$, and some bound hydrogen atoms are still present in the system. The notion of the gas phase remains valid at all temperatures. The robust gap in this toy-model description is proportional to the depth of the potential, $\Delta_{\mathrm{QPT}}(\rho) \sim V(\rho < \rho_{\mathrm{c}})$.

In general, a quantum phase transition can be induced by varying the relative strength of two non-commuting operators. In principle, none of the operators needs to favor a broken-symmetry ground state such that a quantum phase transition will not necessarily imply the formation of long-range order of any kind. However, our definition does not exclude this possibility since some of the operators may display an ordered ground state. In our toy-model example we find that the gas phase displays order in the joint positions of protons and electrons. The essential criterion for the quantum phase transition remains, however, that it is induced by a change in the coupling strengths between the respective operators.

In the case of solid-state physics the electrons, on the one hand, prefer to spread over the entire crystal to optimize their kinetic energy. On the other hand their various interactions tend to localize them in order to minimize their potential energy. It is this competition between the electrons' kinetic and potential energies that determines their conducting properties at zero temperature. The ground state can display dominantly either the electrons' itinerant (metallic) or localized (insulating) features. We may externally shift this energy balance by, e.g., applying pressure to our specimen to induce a metal–insulator transition. The gap for charge excitations (into extended states) is robust and does not vanish in the presence of other excitations. The metal–insulator transition is understood as a quantum phase transition.

The concept of quantum phase transitions is strictly limited to zero temperature and so are the notions of a metal and an insulator. In reality, both a "metal" and an "insulator" display a finite DC conductivity at $T > 0$ and we should discriminate between "good" and "bad" conductors at finite, e.g., room temperature. The generic situation for a system with a quantum phase transition as a function of an external parameter λ (e.g., pressure) is sketched in Fig. 1.1.

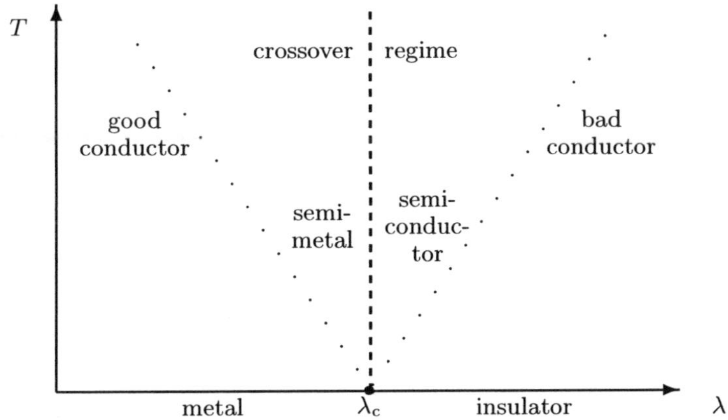

Fig. 1.1. Sketch of the metal–insulator transition as quantum phase transition: the gap for one-electron excitations opens at the critical interaction λ_c and separates the semi-metallic and semiconducting phases. Whereas the metal–insulator transition is sharp as a function of interaction strength at zero temperature, only a crossover region connects good and bad conductors at finite temperatures.

A true phase transition does not exist at any finite temperature. We observe instead a crossover regime from good to bad conductivity. In Sect. 1.3.1 we will show that the pressure-induced transition in ytterbium provides an example for this type of metal–insulator transition. In this metal–band insulator transition the crossover regime can be roughly defined as the range of pressures for which the gap (for the semiconductor) and the band overlap (for the semimetal) are small compared to temperature.

In the quantum phase transition scenario a sharp transition between metal and insulator exists only at zero temperature, where these two phases are actually defined. In real experiments a very sharp increase in the resistivity can still be observed in a metal-to-insulator transition since temperature often is a small energy scale compared to electronic energies such that the system behaves as if the zero-temperature limit was reached. In this sense we may also use the phrase "metal–insulator transition" at finite temperatures even if it has its origin in a quantum phase transition.

1.2.2 Thermodynamic Phase Transition

Thermodynamic phase transitions are the consequence of the competition between the internal energy and the entropy of a system such that a low-entropy (ordered) state is formed at low enough temperatures, which breaks a (continuous) symmetry of the Hamiltonian. For example, the formation of a charge-density wave in a Peierls transition (see Sect. 1.3.2) breaks the translational symmetry of the underlying lattice in favor of a reduced internal energy of the combined electron–lattice system.

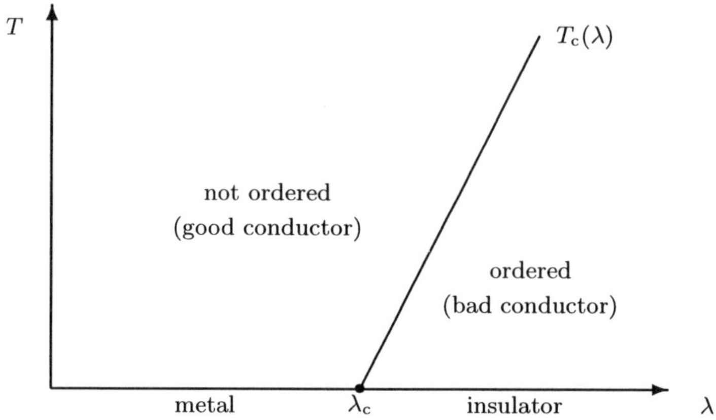

Fig. 1.2. Sketch of the metal–insulator transition as a thermodynamic phase transition: the gap for one-electron excitations opens at the critical temperature $T(\lambda)$ and separates the non-ordered metal from the ordered insulating phase.

The formation of an ordered state also influences the conducting properties. The most celebrated example is superconductivity. A superconductor is characterized by a finite Drude peak even at non-zero temperatures,

$$\sigma_{\alpha\beta}(T \leq T_c, \omega \to 0) = (\mathcal{D}_c(T))_{\alpha\beta}\,\delta(\omega) \;:\; \text{superconductor}, \tag{1.9}$$

where T_c is the thermodynamic phase transition temperature. The Drude peak can be taken as an order parameter that characterizes the superconducting state.

The formation of an ordered state in a thermodynamic phase transition may also result in a gap for charge excitations. For example, the Peierls gap is a function of the order parameter for the charge-density wave, $\Xi(T, \lambda)$. The value of the order parameter depends on the interaction parameter(s) λ, e.g., on the strength of the electron–phonon coupling. The gap is "soft" and vanishes at T_c. We sketch the situation in Fig. 1.2 where we have assumed that the ordered ground state is an insulator. We point out, however, that the occurrence of ordering need not imply a metal-to-insulator transition as seen from the example of metallic (anti-)ferromagnets.

In a thermodynamic phase transition the conducting properties change because of a shift in the energy-entropy balance. Hence, the control parameter for the thermodynamic transition is the temperature, whereas for the quantum phase transition we had, e.g., pressure that controlled the balance between the electrons' kinetic and potential energy. The essential difference from the quantum phase transition lies in the fact that an increase in *temperature* allows the gap for charge excitations to close, i.e., it allows us to transform the insulator at $T = 0$ into a "gapless", i.e., "true" metallic state at $T > T_c$.

Along the line $T = T_c(\lambda)$ we observe a transition from the symmetry-unbroken into the ordered phase. At finite temperatures there is always a finite region around the transition line within which the values for the critical exponents can be obtained from classical statistical mechanics [1.13, 1.14]. This region shrinks to zero at the so-called "quantum critical point" ($\lambda = \lambda_c$, $T = 0$) where the critical exponents in the whole transition region must be obtained from a quantum-mechanical approach. Crtical phenomena at zero temperature are reviewed in [1.15, 1.16].

In this context the $T = 0$ phase transition as a function of λ is often denoted as a "quantum phase transition" although it still is a transition between an ordered and a non-ordered phase. We suggest that this transition should still be called a "thermodynamic" phase transition since the quantum critical point is on the critical line for ordering transitions that occur for finite temperatures. Only the critical exponents, not the nature of the transition, are different for $T = 0$. As outlined in Sect. 1.2.1 we reserve the notion "quantum phase transition" for those transitions that occur *only* at zero temperature and do not necessarily imply symmetry breaking.

In the last section we pointed out that the metal–insulator transition in real materials will be influenced by all four basic electron interactions even if it is dominantly driven by one of them. Analogously, we must be aware of the fact that the signatures of both quantum and thermodynamic phase transitions will be present in experimentally determined phase diagrams.

1.3 Band, Peierls, and Anderson Insulators

In this section we discuss those three classes of insulators due to the electron–ion interaction. As outlined in the first section, the insulators in this category can be understood in the framework of *single-electron theory.* The Mott insulators for which many-body effects have to be included are the subject of the next two sections. Band and Anderson insulators display a robust gap and their metal-to-insulator transitions are quantum phase transitions. The metal–Peierls insulator transition provides an example for a thermodynamic phase transition in which the gap for charge excitations ceases to exist together with the periodic lattice distortion.

1.3.1 Band Insulators

The simplest case of an insulator is the Bloch–Wilson or band insulator [1.17]. To understand the insulating state we assume that the lattice is ideal, i.e., static deformations and disorder are absent. In addition, we neglect the electron–electron interaction or assume that it can be treated within an effective single-electron model [1.5, Chaps. 8–11,17]. The electrons' motion is then dominated by their interaction with the ions and the problem is reduced to an effective Hamiltonian for a single electron in a periodic potential.

Bloch's theorem then allows us to classify the one-electron states according to their crystal momentum $\boldsymbol{k}$ of the first Brillouin zone, their band index b, and their band energy $\epsilon(\boldsymbol{k}, b)$. These bands are "rigid" in the sense that the single electron states and their respective energy are totally independent of the band filling. This picture changes drastically when electron–electron interactions are taken into account. For simplicity we further neglect spin–orbit coupling. As a good approximation each band is then two-fold spin-degenerate ($\sigma = \uparrow, \downarrow$), i.e., each band may be occupied with two electrons per quantum number $\boldsymbol{k}$. In the ground state ($T = 0$) each state up to the Fermi energy E_{F}, $\epsilon(\boldsymbol{k}, b) \leq E_{\mathrm{F}}$, is occupied such that the electron density determines E_{F} as the energy of the last occupied state. The equation $\epsilon(\boldsymbol{k}, b) = E_{\mathrm{F}}$ defines the wave vectors $\boldsymbol{k}_{\mathrm{F}}$ that form the Fermi surface. The inverse effective mass $(m^*)^{-1}$ is related to the curvature (second derivative) of the dispersion relation on the Fermi surface.

The one-electron dispersion relation $\epsilon(\boldsymbol{k}, b)$ and the corresponding one-electron density of states,

$$D_0(E) = \sum_{b,\sigma} \int_{\substack{\text{Brillouin}\\ \text{zone}}} \frac{\mathrm{d}^d k}{(2\pi)^d} \delta\left(E - \epsilon(\boldsymbol{k}, b)\right) , \tag{1.10}$$

determine not only the thermodynamics but also the *transport* properties of the system [1.5, Chaps. 12,13], a feature specific to non-interacting systems. In general, however, one-electron properties alone do not determine the conductivity.

A review on the present-day band structure and quantum-chemical techniques can be found, e.g., in [1.18, Chaps. 2–5]. Band structure theory provides a very simple understanding for the existence of metals and insulators since completely filled bands cannot contribute to transport [1.17]. With each state with crystal momentum $\boldsymbol{k}$ the state with momentum $-\boldsymbol{k}$ is also occupied so that both contributions to transport properties will compensate each other [1.1, Chap. 1.3]. Four typical cases are schematically displayed in Fig. 1.3, which we now discuss.

The energy bands can be separated completely from each other due to the influence of the periodic potential, i.e., there can be an energy gap between bands; see Fig. 1.3a,b. If the chemical potential lies between two such bands all bands are either completely filled ("valence bands") or empty ("conduction bands"). The energy needed to take an electron away from the system is $\mu^- = E_{\mathrm{F}}$, and the energy needed to add an extra electron is $\mu^+ = E_{\mathrm{F}} + \Delta$ such that

$$\Delta = \mu^+ - \mu^- \tag{1.11}$$

is the definition for the energy necessary for a one-electron excitation.

The material is an insulator since there are no mobile charge carriers available at temperature $T = 0$. This situation holds both for wide-gap insulators (e.g., noble element crystals, diamond) and for intrinsic semiconductors

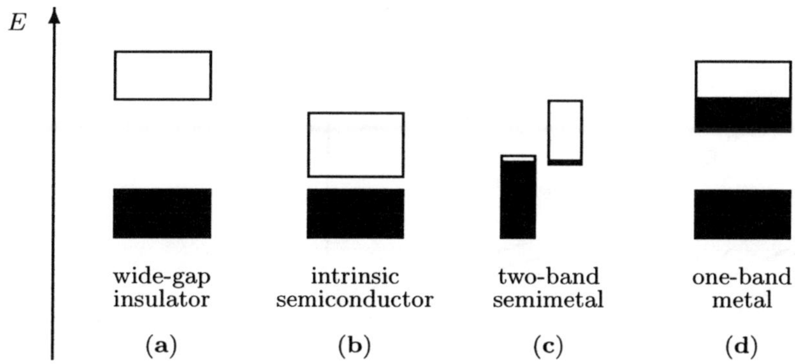

Fig. 1.3. Idealistic sketch of the density of states and band fillings for insulator, semiconductor, semimetal, and metal according to band theory; occupied energy levels at zero temperature are filled in.

(e.g., silicon); see Fig. 1.3a,b. In the latter case, however, the energy gap is only of the order of $0.1 \ldots 1\,\mathrm{eV}$ such that there is a considerable number of thermally activated carriers at room temperatures which give rise to a finite conductivity at $T = 300\,\mathrm{K}$. As discussed in Sect. 1.1, (1.6), their specific resistivity increases exponentially with decreasing temperature such that their insulating nature becomes apparent.

If the Fermi energy comes to lie in a band (conduction band) the state is metallic and the chemical potentials are equal, $\mu^+ = \mu^- \equiv \mu \equiv E_\mathrm{F}$. As sketched schematically in Fig. 1.3d, the density of free charge carriers at zero temperature can be high as in alkali metals like sodium (Na) and potassium (K). They have one conduction electron per lattice unit cell and the conduction band is half-filled since each $\boldsymbol{k}$ state is two-fold spin-degenerate. A system with an odd number of electrons per unit cell cannot be a band insulator. The carrier number can also be very low as in semimetals like arsenic (As), antimony (Sb), and bismuth (Bi), where there are an even number of electrons per unit cell. The small but finite overlap of the two bands means there is no band gap. A small density of holes in the "valence band" and electrons in the "conduction band" provide charge carriers for electrical conductivity even at low temperatures; see Fig. 1.3c.

These considerations lead to the following definition of a band insulator.

For a Bloch–Wilson or band insulator the interaction between electrons and the periodic ion potential gives rise to an energy gap between the lowest conduction band and the highest valence band. Consequently, there are no free carriers available for the transport of charge. A band insulator is possible only for an even number of valence electrons per lattice unit cell.

Owing to the small overlap between the conduction and the valence band in semimetals, these materials are suitable for the realization of a metal–band

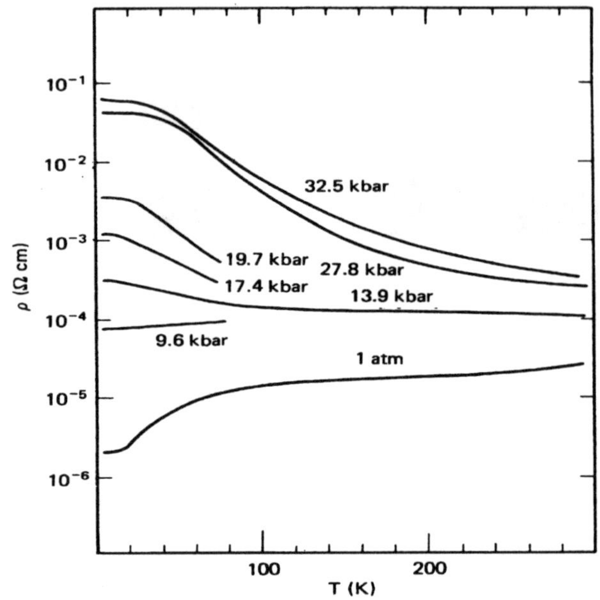

Fig. 1.4. Specific resistivity of ytterbium as a function of temperature at various pressures (after [1.19]).

insulator transition. For example, semimetallic ytterbium (Yb) can be transmuted into a semiconductor at a pressure $p_c \approx 10\,\text{kbar}$. Early experimental results are displayed in Fig. 1.4. The behavior of bismuth (Bi) under pressure shows similar signatures for a metal–band insulator transition, in particular, the measured Fermi surface vanishes with increasing pressure [1.19, 1.20]. Further examples are briefly reviewed in [1.21]. At first sight it is puzzling that the increase in the pressure *opens* a gap. One would naively expect that an increase in pressure also increases the overlap of wave functions such that the bandwidth increases. In contrast, the pressure increases the mixing of the $4f$ and $5d$ shells, which finally leads to the opening of a hybridization gap in these isostructural band crossing transitions.

The metal–band insulator transition shown in ytterbium also provides an instructive illustration for the concept of a quantum phase transition as discussed in the preceding section; see Fig. 1.1. Figure 1.4 shows that the zero-temperature gap is robust, i.e., it does not vanish when the temperature is increased. Nevertheless, the isostructural semimetallic and semiconducting phases can be identified clearly only on lowering the temperature because the signatures of the gap are smeared out for $k_B T = \mathcal{O}(\Delta(p))$. For the semimetallic state the resistivity decreases because scattering by thermal fluctuations is reduced. This reduction is overcome by the exponential vanishing of the charge carriers in the semiconducting state such that the resistivity increases on lowering the temperature; see (1.6). At room temperature ($k_B T \approx 25\,\text{meV}$) we observe a broad crossover near the critical pressure of 10 kbar since the zero-temperature gap is small, of the order of thermal energies. In a perfect ytterbium crystal and in the absence of ordering there would be a sharp quantum phase transition at zero temperature as a function of the pressure between the metallic and insulating phase. At finite temperatures this results in the observed crossover behavior due to the robust gap that opens at the quantum phase transition at $p = p_c$.

At very low temperature and close to the critical pressure another phenomenon may in general occur since the electron–ion interaction no longer dominates all other interactions. In fact, when the gap is small enough electron–hole pairs may spontaneously form and bind into excitons since their binding energy overcomes the single-electron gap. This "excitonic insulator" forms in the regime of the band insulator and persists to a region for which the single electron band theory would predict a metallic behavior [1.22]–[1.24].

The transition of a metallic phase into an exciton insulator is frequently referred to as "the Mott transition" since it was first predicted by Mott [1.22]; see also [1.1, Chap. 4.2] and [1.23] for an early review on the theory. In our classification scheme it is a band–Mott transition since the driving force behind the metal–insulator transition is the closing of the band gap, whereas the actual metal–band insulator transition is modified by the electron–electron interaction. In fact, the thermodynamic phase transition into an exciton insulator occurs at a finite temperature where the excitons undergo a "Bose condensation". We emphasize that this transition is quite distinct from the (genuine) Mott transition by (true) electron–electron interactions, which is the sense in which throughout this book we use the term "Mott transition".

Experimentally it is very difficult to prove the existence of the exciton insulator. One route is the optical excitation of semiconductors to show that the excitons actually Bose condense. See [1.24, 1.25] for an experimental and theoretical introduction to exciton condensation in optically excited semiconductors. Unfortunately, one is probing highly excited states in this case such that one has to cope with problems of finite life-times (e.g., thermalization, various decay mechanisms) in the interpretation of the experimental results.

The second route follows the ideas outlined above to find the exciton insulator near a metal–band insulator transition. Recently, experimental evidence for the existence of an exciton insulator in the thulium-selenium-tellurium compound $TmSe_{0.45}Te_{0.55}$, has been provided [1.26]. In this system a pressure increase gradually closes the one-electron gap but near the critical pressure the resistivity drastically increases again. Hall effect measurements show that the number of free carriers goes to zero in this phase, which suggests that excitons have been formed.

In the band–Mott transition the quantum metal–band insulator transition at zero temperature is hidden behind the thermodynamic phase transition into the exciton insulator. For the genuine Mott transition we will also find that the quantum phase transition from the metal into the Mott–Hubbard insulator is concealed by the thermodynamic phase transition into the Mott–Heisenberg insulator, at least in higher dimensions (see Chap. 5). We may even state that it is not an uncommon feature for real materials that zero-temperature quantum phase transitions in their phase diagram are hidden under an ordered phase. For example, we may speculate that in the high-temperature superconductors the thermodynamic phase transition into the superconducting state conceals a quantum phase transition that causes the crossover from non-Fermi-liquid to Fermi-liquid behavior as a function of doping in the normal state.

1.3.2 Peierls Insulators

In the previous section we focused on the influence of the *static* periodic ion potential on the motion of independent electrons. This approach is not quite correct even at zero temperature and in the absence of disorder and electron–electron interactions, since the periodic ion array may (dynamically) respond to the presence of the electrons. In fact, the electron–ion interaction can induce static lattice deformations that result in a new potential of a different periodicity. At low enough temperature the elastic energy necessary for the deformation can be overcome by the gain in energy for the electrons (Peierls effect). In turn, the formation of the new periodic potential will influence the electron transport and can give rise to a metal-to-insulator transition. In the Peierls insulator the symmetry of the original lattice is broken, which shows that the Peierls transition provides an example for a thermodynamic phase transition, where the static lattice deformation serves as the order parameter.

A simple example for this effect in electron systems coupled to a distortable lattice was independently given by Peierls [1.27, Chap. 5.3] and Fröhlich [1.28]. Let us consider a chain of atoms at distance a at zero temperature over which independent electrons with Fermi wave number k_{F} can move freely. If we adopt a mean-field picture a small static lattice deformation $\Delta\cos(2k_{\mathrm{F}}x)$ will cost the average elastic energy $E_{\mathrm{elastic}} \sim \Delta^2$. As a result of this deformation, the electrons experience a periodic potential $U(x) \sim \Delta\cos(2k_{\mathrm{F}}x)$, which opens a gap at $k = \pm k_{\mathrm{F}}$. The one-electron energies at $E_{\mathrm{F}} - \epsilon$ below the gap are lowered to $E_{\mathrm{F}} - \sqrt{\epsilon^2 + \Delta^2}$. The shift in the total electronic energy is essentially given by $E_{\mathrm{electron}} \sim \Delta^2 \ln \Delta$, as in BCS theory, such that the equilibrium deformation that results from the minimization of $E_{\mathrm{elastic}} + E_{\mathrm{electron}}$ is always finite, $\Delta \sim \exp\left[-1/\left(\alpha D_0(E_{\mathrm{F}})\right)\right]$ at zero temperature, for arbitrarily small coupling α between the electrons and the lattice. Both the lattice and the electrons form a charge-density wave with wave number $2k_{\mathrm{F}}$.

The above argument treats the lattice deformations and the Peierls transition at a critical temperature $k_{\mathrm{B}}T_{\mathrm{P}} \sim \Delta$ only within a (semi-)classical mean-field approximation. In low dimensions, however, the fluctuations of the order parameter at finite temperatures are so strong that a stable Peierls distortion is prohibited in one and two dimensions. A real system, however, is always three-dimensional, even if highly anisotropic, such that the results of the mean-field calculation are at least qualitatively applicable. The existence of the Peierls gap for charge excitations is linked to the presence of the charge-density wave, which vanishes at the Peierls transition temperature. Hence, the metal–Peierls insulator transition is the result of symmetry breaking (thermodynamic phase transition). A derivation of the Peierls distortion in polyacetylene can be found in [1.29, Chap. 2]; for calculations beyond the Hartree–Fock level, see [1.30].

As in the case of BCS superconductivity the existence of a gap in the one-electron spectrum does not immediately imply that the system is an insulator. If both the lattice deformation and the electron charge-density wave could move freely, we would have a Fröhlich superconductor [1.28]. In real materials, however, impurities and lattice defects fix the charge-density waves in position space ("pinning"). In addition, if the average separation of the electrons is comparable to the lattice spacing, the original diameter of the Brillouin zone $2\pi/a$ can be an integer multiple of $2k_F$ ("commensurate band filling") such that the charge-density wave is locked into a fixed superlattice structure [1.27, Chap. 5.3]. For example, this is the case in polyacetylene [1.29, Chap. 2]. In this simplest $(CH)_x$ polymer the conduction band of the carbon π electrons would be half-filled ($2k_F = \pi/a$) in the absence of the Peierls distortion. Thus, it would be predicted to be a (wide-band) metal. Instead, the polymer is a Peierls insulator with a new unit cell twice as big as the original one. The electron charge-density wave is commensurate with the new lattice structure of alternating short and long bonds. This effect has been proven to persist even in the presence of short-range electron–electron interactions [1.31].

Lattice deformations and changes in the crystal structure can also be induced by the interaction of electrostatic (crystal) fields with localized electrons. In crystals the angular momentum at the position of an atom is not a good quantum number and corresponding atomic degeneracies are lifted by electrical fields of the surrounding atoms (crystal field effect). In some cases, typically for transition-metal atoms surrounded by oxygen octahedra, the cubic lattice symmetry would formally lead to some remaining orbital degeneracies. As in the case of the Peierls distortion it may be energetically favorable at low temperatures to reduce the ideal symmetry of the oxygen octahedra by a (static) deformation. This results in a splitting of electron energy levels, a lowering of the occupied levels and thus to an energy gain that outweighs the lattice deformation energy (Jahn–Teller effect [1.32]).

The Jahn–Teller effect can result in a change in the lattice structure, e.g., from cubic to tetragonal, below a critical temperature T_{JT}, which changes the electronic band structure and thus the conductivity. In principle, we might observe a thermodynamic transition of a metallic state into a "Jahn–Teller" insulator. The Jahn–Teller distortion reduces the symmetry of the crystal structure and vanishes above T_{JT}. Since both Peierls and (conceivable) Jahn–Teller insulators are the consequence of thermodynamic phase transitions, we subsume them under the class of insulators due to the electrons' interaction with periodic lattice deformations.

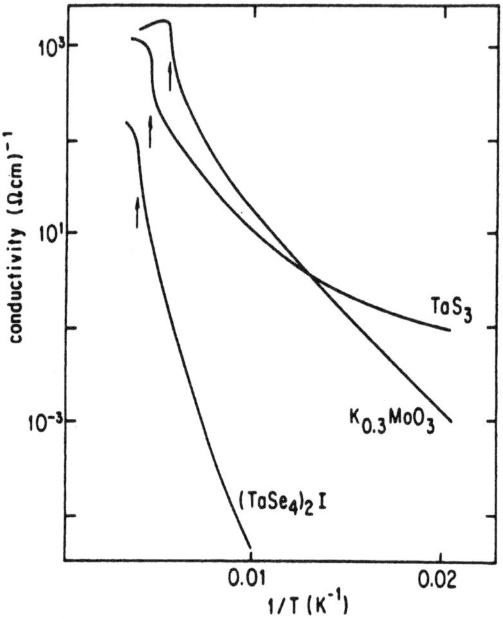

Fig. 1.5. Metal–Peierls insulator transition: electrical conductivity as a function of inverse temperature for o-TaS_3, $(TaSe_4)_2I$, and $K_{0.3}MoO_3$; the arrows indicate the Peierls transition found from the analysis of the temperature gradient of the resistivity (from [1.38]).

Below we concentrate on Peierls insulators for which we give the following definition.

For a Peierls insulator the interaction between electrons and the lattice results in the formation of a static periodic lattice deformation and a (commensurate) charge-density wave of the electrons.

The occurrence of the Peierls instability requires that the electronic structure (Fermi surface) and the lattice structure merge in such a way that a (small) lattice deformation results in only filled or empty bands. This condition can most easily be met in one-dimensional chains, e.g., in polymers [1.29, 1.33], or in other strongly anisotropic structures like in KCP ($K_2Pt(CN)_4Br_{0.30} \cdot x\ H_2O$) [1.34] or in the blue bronze $K_{0.3}MoO_3$ [1.35]–[1.37]. Since many other effects may compete with the Peierls distortion in low dimensions (localization effects due to disorder, influence of the electron–electron interaction) the existence of a Peierls transitions does not immediately follow from an observed anomaly of the conductivity at the transition. Instead, the expected unit-cell doubling has to be confirmed by X-ray and neutron scattering experiments.

The almost classical example for an observed Peierls transition is the blue bronze $K_{0.3}MoO_3$. It has a monoclinic face-centered crystal structure of ten chemical units with the shortest distance between the molybdenum atoms being in the *b* direction. The potassium atoms dope the conduction band of the Mo atoms which results in a strong metallic conductivity at room temperature mainly into the *b* direction [1.38, 1.39]. At 181 K a Peierls transition occurs [1.38]–[1.40] and the electrical resistivity exponentially rises with decreasing temperature. This behavior is shown in Fig. 1.5 which also displays the Peierls transitions in the two tantalum compounds o-TaS_3 and $(TaSe_4)_2I$. Recent X-ray and neutron scattering experiments have clearly proven the existence of a commensurate charge-density wave and the corresponding lattice deformations. They also allowed other possible mechanisms for the transition to be ruled out [1.41]. Further details on the Peierls transition can be found in the excellent review by Grüner [1.38].

In the Peierls effect the electron–ion interaction leads to a static distortion of the whole lattice structure. In general, this interaction always leads to a *local* distortion of the lattice around an electron, i.e., an electron is always surrounded by lattice distortions (phonons). The resulting fermionic quasi-particle is called a "polaron" [1.4, Chap. 6]. In the case of a weak electron–phonon coupling the range of the lattice deformation around the electron is large compared to the lattice constant. The polaron behaves like an essentially free electron with an increased effective mass ("large" or "Fröhlich polaron"). If, on the contrary, the polaron diameter and the lattice constant are of the same size ("small polaron"), the polaron need not always be free but can trap itself. Hence, a quantum phase transition between a metal and a "polaron insulator" is conceivable.

The idea of a quantum phase transition of a metal into a "polaron insulator" as a function of the carrier concentration dates back to Landau and Fröhlich [1.42]. At a large polaron density the polaron clouds will overlap and trapping becomes unlikely. At low carrier densities, however, the small polarons could be localized in non-overlapping regions of the lattice. Experimentally the self-trapping of optically excited excitons in metal halides has been confirmed [1.43]; for a recent review, see [1.44, Chap. 6]. Variational approaches provide an intuitive understanding of the trapping mechanism [1.44, Chap. 4] but it turns out to be very difficult to find a rigorous proof that this localization effect is indeed contained in standard electron–phonon models. For example, it has been shown [1.45] that Galilean invariance of the electrons' kinetic energy operator cannot be broken by the interaction with optical phonons. Furthermore [1.46], acoustic phonons with dispersion $\omega(\boldsymbol{k}) \sim |\boldsymbol{k}|^{\nu}$ that are coupled to electrons with the potential $V(\boldsymbol{k}) = \alpha|\boldsymbol{k}|^{-\lambda}$ lead to extended electron wave functions in d dimensions independent of the coupling strength α if $(d+2-2\lambda)/\nu > 2$. Finally, a rigorous theorem states that in three dimensions a localized electron state becomes asymptotically degenerate with the exact ground state *only* in the limit of infinitely strong electron–phonon interactions ($\alpha \to \infty$, $\nu = 0$, $\lambda = 1$) [1.47]. It is not surprising that only a few exact statements on the polaron problem are available because the description of a (heavy) particle moving in a bath of bosonic excitations addresses the fundamental problem of dissipation and quasi-particle damping in many-particle physics; for a recent treatment, see [1.48].

The polaronic localization effect is particularly pronounced in low dimensions [1.46] since there the lattice may easily be deformed. Strictly speaking, a long-range order of the atoms cannot exist in one and two dimensions for $T > 0$ because the translational symmetry as a continuous symmetry cannot be broken spontaneously (Mermin–Wagner theorem [1.49]). All real materials are three dimensional, even if highly anisotropic, and no convincing experimental evidence for the existence of a "polaron insulator" has been provided thus far. One reason for this lies in the fact that at low electron densities and in low dimensions other interaction effects compete and also lead to particle localization. A celebrated example is the Anderson localization due to the presence of disorder, which we will discuss next.

1.3.3 Anderson Insulators

The interaction of free, independent electrons with randomly distributed lattice defects can also lead to an insulating state. Several decades ago Anderson recognized [1.50] that disorder may lead to particle localization. Anderson localization is indeed most easily understood to result from the coherent (back-)scattering of electron waves in a random potential [1.51]. A true disorder-induced metal-to-insulator transition may appear in materials with a strongly perturbed lattice structure (structural disorder) or in materials with strong impurity scattering (substitutional or energetic disorder) [1.52]. Thus we may model the disorder either by a spatially random distribution of impurities of the same scattering strength (Edwards model [1.53]) or by a one-electron potential on a regular lattice for which the scattering strength randomly varies from site to site (Anderson model [1.50]). Since we have to sample many realizations of the disorder to obtain macroscopic physical quantities there should be no principle difference between the two forms of disorder, i.e., we assume that both models of disorder are in the same universality class in the vicinity of the metal–insulator transition.

The Anderson localization of electron waves is a quantum-mechanical coherence (or "memory") effect. The usual derivation of the classical Boltzmann transport theory for free, independent electrons in impure systems heavily relies on the Markov property: the result of the nth scattering process depends only on the conditions immediately before it, i.e., only on the result of the $(n-1)$th scattering process [1.54]. This assumption is not truly justified in quantum mechanics since the electron waves stay coherent under elastic scattering such that multiple scatterings and interference effects are possible and become important. A full description of this effect requires the calculation of the conductivity from a quantum-mechanical two-electron correlation function [1.4, Chap. 7]. The single-electron density of states tells us only whether there are states available close to the Fermi energy; it does not discriminate between localized and extended states. This information is contained in the two-electron correlation function.

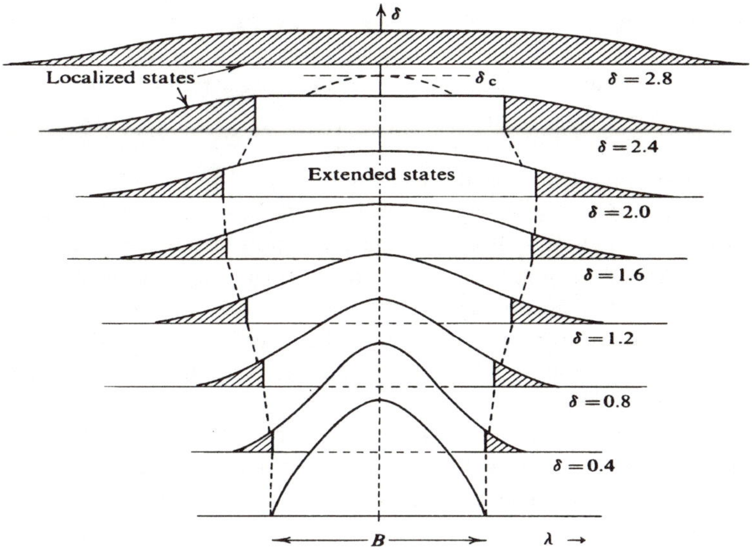

Fig. 1.6. Increasing the strength δ of the disorder potential in the Anderson model of disorder leads to a broadening of the density of states. The number of localized states increases until the mobility edges for electrons and holes coincide for δ_c. In the figure B denotes the bandwidth at zero disorder, and λ is the energy variable (from [1.52, Chap. 9.9]).

Localization effects are entirely dominant in one dimension: all but a finite number of one-electron states are strictly localized in a generic impure system [1.55], i.e., a wave function with maximum amplitude at site r_0 exponentially decays for large distances with an energy-dependent localization length $\xi(E)$,

$$|\psi(r, r_0; E)|^2 \sim \exp\left(-|r - r_0|/\xi(E)\right) \quad ; \quad |r - r_0| \gg \xi(E) \ . \tag{1.12}$$

Practically all states within a one-dimensional band are thus localized. This observation leads to experimental consequences only for samples whose linear size and inelastic scattering length due to electron–phonon interaction are large compared to the localization length $\xi(E)$. Otherwise, inelastic scattering processes dominate the transport properties. The theoretically postulated localization effect in one dimension [1.56] has been verified experimentally [1.57].

Banyai [1.58] and Mott [1.59] were the first to point out that in higher dimensions the disorder potential might not be strong enough to localize all states. In the presence of disorder the electrons in the lower-band tail of the resulting density of states are bound to deep impurity potentials and thus are localized. This argument naturally extends to the (anti-)bound (hole) states in the upper-band tail. Between two "mobility edges" [1.59] for electrons ($E_{\mathrm{m}}^{\mathrm{e}}$)

and holes ($E_{\rm m}^{\rm h}$) we find extended states, which contribute to the electron transport at $\omega \to 0$ if $E_{\rm F}$ lies in this region. Of course, these states are also influenced by the impurities even though their scattering strength is not large enough to render these states localized.

An idealized scenario for the position of the mobility edges as a function of the strength of the impurity potential δ is shown in Fig. 1.6. For a given electron concentration (fixed Fermi energy $E_{\rm F}$) we may induce a metal–Anderson insulator transition by a change of δ. A transition upon doping provides another route to the Anderson insulator. We may keep the impurity scattering strength fixed and tune the Fermi energy, which allows us to move it through a mobility edge. Finally, one may try to tune the effective bandwidth with the help of an external magnetic field [1.60] or an external electrical AC field (dynamic localization) [1.61].

The Anderson transition is evidently a quantum phase transition: the electrons' kinetic energy prefers a delocalized metallic state, and the local energetic disorder favors electron localization on those impurity sites that have the lowest energy levels. In fact, quantum interference effects inherently underlie the Anderson transition. The gap is defined as the energy separation between the Fermi energy and the mobility edge beyond which excitations can actually transport charge over macroscopic distances, $\Delta_{\rm QPT} = |E_{\rm m}^{\rm e,\,h} - E_{\rm F}|$.

The above considerations lead us to the following definition of an Anderson insulator.

For an Anderson insulator the disorder-induced localization of electrons leads to the occurrence of mobility edges that separate localized and delocalized states in a band. If the Fermi energy lies between a band edge and its corresponding mobility edge the single-electron states around the Fermi energy are localized.

An extensive literature exists on the theory and experiments for the metal–Anderson insulator transition; reviews are given in [1.62]–[1.68]. An example for the typical behavior of the conductivity at the Anderson transition is shown in Fig. 1.7. It shows that the phase transition is continuous in contrast to Mott's early ideas [1.69] (see also [1.1, Chap. 1.7]) of a jump discontinuity at the transition from a conducting to an insulating state at a "minimal conductivity" $\sigma_{\rm M}$ [1.70]. A continuous transition allows the application of concepts from the theory of critical phenomena, e.g., of scaling theories [1.71] and field-theoretic renormalization group methods [1.72] whose results are qualitatively reproduced in microscopic calculations (self-consistent Green function [1.73] and mode coupling theories [1.74]).

According to these approaches, extended one-electron states should exist only in dimensions larger than two (lower critical dimension $d_{\rm c} = 2$), which suggests the use of $\epsilon = d - d_{\rm c}$ as the expansion parameter for field-theoretic studies of the transition [1.72]. For all $d > 2$ the metal–Anderson insulator transition should occur. The metallic conductivity as a function of the carrier concentration vanishes at the transition according to a power law,

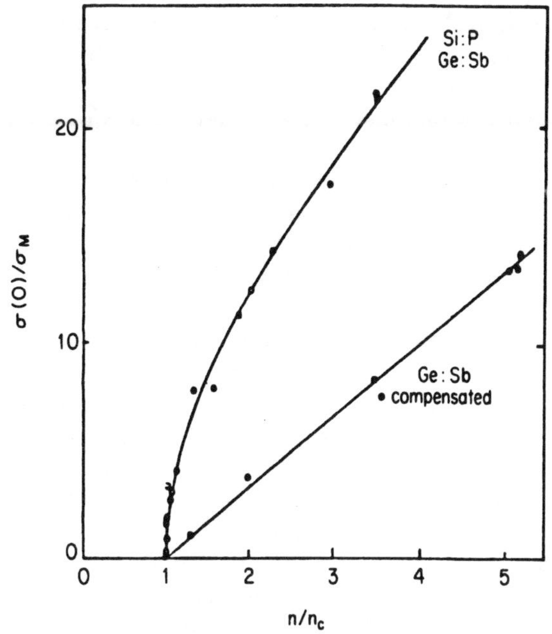

Fig. 1.7. Electrical conductivity of uncompensated Si:P and Ge:Sb, and compensated Ge:Sb, scaled to Mott's minimal conductivity σ_{M}, as a function of carrier concentration (from [1.62]).

$$\sigma^{\mathrm{DC}}/\sigma_{\mathrm{M}} = (1 - n/n_{\mathrm{c}})^{\nu} \; . \tag{1.13}$$

In three dimensions, in the absence of magnetic scattering, microscopic theories [1.73, 1.74] predict an exponent $\nu = 1$. This is also the result from non-linear sigma-model studies [1.72] to one-loop order. However, these theories are actually expansions around two dimensions, i.e., $\epsilon = d-2$ is supposed to be a small parameter. In fact, as found by Wegner [1.75], the result for the gap exponent ν to four-loop order in the ϵ expansion is given by

$$\nu = \frac{1}{\epsilon} - \frac{9}{4}\zeta(3)\epsilon^3 + \dots \quad , \tag{1.14}$$

where $\zeta(x)$ is the Riemann zeta function ($\zeta(3) \approx 1.202$). The large correction indicates that the ϵ expansion generates an asymptotic series. This might not be too surprising: on the one hand we expand around $d_{\mathrm{c}} = 2$ ($\epsilon = 0$), for which all states are localized and the Anderson transition is absent, and on the other hand we try to obtain information on the exponent of the Anderson transition for $\epsilon > 0$.

Equation (1.14) shows that three dimensions are out the reach of the ϵ expansion. In fact, if we naively used $\epsilon = 1$ in (1.14) we would obtain a negative result, and the exact Harris–Chayes criterion [1.76], $\nu \geq 2/d$, would already be violated for $\epsilon \gtrsim 0.586$. A comparison between (1.14) and the results form the numerical studies of the Anderson model on fractal geometries [1.77] indicates that the asymptotic expansion (1.14) may be used for $\epsilon \lesssim 0.3$. These numerical investigations find $\nu \approx 1.3$ in three dimensions, in slight disagreement with the experimental results; see below.

Normal metals do not display an Anderson metal-to-insulator transition since their bandwidth is large compared to the fluctuations of the lattice potential. Then the mobility edges are very close to the band edges (see Fig. 1.6) and the Fermi energy is far from the mobility edges. In addition, Coulomb effects are not decisive in normal metals (see Sect. 1.4.1). Nevertheless, all the extended states are also influenced by the (weak) disorder. Effects due to coherent backscattering are naturally most pronounced at low temperature. Not only are thermal excitations from localized to delocalized states strongly reduced but also the inelastic, coherence destroying scattering processes (electron–phonon scattering) are frozen out to a large degree. Thus the inelastic mean free path becomes much longer than the elastic scattering length at low temperature, enabling the coherent backscattering effects to be seen. The effects of disorder in metals ("weak localization") are experimentally observed and theoretically well understood. For example, the pronounced dependence of the conductivity on very weak external magnetic fields could be explained as the consequence of weak localization [1.51].

In contrast to weak localization the effects of strong disorder and, in particular, the Anderson transition itself appear to be strongly modified by the Coulomb interaction between the charge carriers and the charged impurities, and among themselves. For example, an additional (Coulomb) gap may open at the Fermi energy which reduces the (activated) transport in the insulating phase [1.78]–[1.80]. Its precise form is a matter of current debate; see [1.80] for a review. In many respects the insulating phase displays the features of a "Coulomb glass" [1.81, 1.82]. Thus, it appears natural that the Coulomb interaction also considerably influences the nature of the Anderson transition.

For doped uncompensated semiconductors [1.62], such as phosphorus-doped silicon (Si:P) or selenium-doped germanium (Ge:Se), the conduction takes place in the (half-filled) phosphorus impurity band. According to Fig. 1.7 these systems display an exponent of $\nu \approx 0.5$. In uncompensated semiconductors the donor atoms (P, Se) not only provide the conduction electrons but also serve as random scattering centers. The (long-ranged) Coulomb interaction between the positively charged donors and "their" electrons and/or the spin-exchange scattering between the electrons themselves obviously play an important role, at least in uncompensated materials.

The idea behind compensation is to weaken the Coulomb interaction by adding equal amounts of further donors (P or Sb, respectively) and acceptors (B). Although the number of carriers in the phosphorus band remains unchanged the number of impurity levels increases such that the carrier concentration and Coulomb correlations in the impurity band are reduced. In addition, the acceptors act as new scattering centers, which further favors disorder over interaction effects. Screening of the Coulomb attraction between a phosphorus ion and an electron by the additional donor and acceptor ions may also play a role in the subtle balance between disorder and Coulomb interactions. The net effect appears to be a reduction in the Coulomb interactions in the P (or Sb) impurity band such that an exponent $\nu = 1$ is recovered upon compensation [1.62]. It is difficult to reconcile the theoretical value for the Anderson model from numerical studies, $\nu_{\text{And. model}} \approx 1.3$, with the experimental value, $\nu_{\text{impurity}} = 0.5 \ldots 1$. Coulomb effects apparently play an important role in experiments, and only first steps have been made towards a theory of the "Anderson–Mott transition" [1.64, 1.65, 1.83].

There are, however, alternative ideas on how to solve the "conductivity exponent puzzle". The clue may also lie in the "proper" analysis of the data close to the transition. It is not that straightforward to determine the value of the exponent experimentally since the critical region cannot easily be specified [1.84]. The same data that lead to $\nu = 0.5$ in Fig. 1.7 can be used to extract $\nu \approx 1.3$ [1.84] but this analysis has been questioned [1.85]. So, the precise value of the critical exponent for the Anderson transition is a matter of current debate, not only from a theoretical but also from an experimental point of view.

In amorphous materials like hydrogenated amorphous silicon (a-Si:H) the carrier concentrations are comparably high but the structural disorder is equally strong in its effect such that these amorphous systems are understood as (dominantly) Anderson insulators [1.66]. At large carrier concentration, however, the influence of the Coulomb interaction is certainly important as well [1.79] and disorder and Coulomb effects may act together in the carrier localization in these materials. In addition, the influence of self-compensation in a-Si:H makes it difficult to observe the Anderson transition in amorphous materials. The interplay between phonons, Coulomb interactions, and disorder make the calculation of transport properties in these materials a fascinating but very difficult problem [1.66, 1.79, 1.80].

In the next two sections we focus on the case of pure electron–electron interactions as the driving mechanism behind a metal–insulator transition.

1.4 Mott Insulators: Basic Theoretical Concepts

In this section we introduce some basic theoretical concepts used to discuss the role of electron–electron interactions in the solid state. In metals its influence on low-energy properties is successfully incorporated in Landau's Fermi-liquid theory which, however, has to break down at the metal–insulator transition. To assess the strength of the electron–electron interaction relative to the electrons' kinetic energy we address the pair correlation function and the local magnetic moment. This quantity allows us to elucidate the principle difference between Slater insulators (self-consistently stabilized band-type insulators) and Mott insulators (correlated electron systems). We propose to classify further Mott insulators as Mott–Hubbard or Mott–Heisenberg insulators. As signature of electron correlations local moments exist in the metallic and in both insulating phases. In the Mott–Heisenberg insulator the pre-formed moments are additionally long-range ordered.

1.4.1 Electron–Electron Interaction in Metals

Thus far we have considered electrons that did not interact and thus moved independently of each other. Given the long-ranged and strong Coulomb interaction this seems to be a rather drastic and therefore unrealistic assumption. Nevertheless, many experiments can be explained qualitatively and even semi-quantitatively with the assumption of independent electrons, e.g., the de-Haas–van-Alphen effect and many other experiments on the Fermi surface in metals [1.5, Chap. 14].

The explanation for the seemingly marginal consequences of the electron–electron interaction in metals rests on two observations.

(i) In systems with itinerant electrons the Coulomb interaction between the charge carriers is very effectively screened. The screening length lies in the range of the inverse Fermi wave number k_{F}^{-1}, i.e., it is only several Ångströms long [1.4, Chap. 5].
(ii) In the presence of a Fermi surface the scattering rate between electrons near the Fermi surface with energy $E_{\mathrm{F}} + \omega$ vanishes proportional to ω^2 since the Pauli principle strongly reduces the number of scattering channels that are compatible with energy (and momentum) conservation [1.5, Chap. 17].

The second point is the basis for Landau's Fermi-liquid theory [1.86, 1.87]. *If* a Fermi surface exists for the interacting Fermi system as it did for the non-interacting case (Fermi gas), the excitations near the Fermi surface are long-living fermions ("Landau quasi-particles") in a one-to-one correspondence to the electron and hole excitations of the Fermi gas. Thus the energetically low-lying one-electron excitations of the Fermi gas and the Fermi liquid can be mapped completely onto each other. Since the Fermi energy in metals is almost always very much larger than typical thermal energies, only low-energy excitations are relevant for the thermodynamics and electrical transport properties. The residual interaction between Landau quasi-particles can effectively be described in terms of a few Landau parameters. Consequently, physical properties are only quantitatively modified in a Fermi liquid as compared to a Fermi gas, e.g., the effective mass of the charge carriers depends on the Landau parameters. However, there is no qualitative difference between a Fermi gas and a Fermi liquid, e.g., both describe metallic states.

Under which conditions a Fermi surface exists remains an open problem within the solely phenomenological Fermi-liquid theory. Microscopic calculations show that it does exist, if the perturbation theory in the electron–electron interaction converges (Luttinger theorem [1.88]). Naturally, perturbation theory itself does not allow for statements about the convergence of the series expansion, and we can argue only on physical grounds that the Fermi-liquid picture provides a self-consistent description of the influence of electron–electron interactions in metals: (i) the electrons are highly mobile in the crystal, thus (ii) their mutual interaction is well screened. Consequently,

(iii) we may apply Fermi-liquid theory for their description, which implies (iv) that the (residual, screened) electron–electron interaction does not qualitatively change the picture, which was the starting point of our analysis.

Clearly, this circle of argument has to break down at the metal–Mott insulator transition. Thus, the problem of the Mott transition is intimately related to the question of the stability of Fermi-liquid theory.

1.4.2 Exchange and Correlations

We have just seen that dynamic self-screening of itinerant electrons plays an important role in metals. This does not, however, mean that the interaction effects could generally be neglected, or merely subsumed into an effective single-electron description. In general, the repulsive electron–electron interaction tends to keep the electrons apart from each other. This localization tendency is in conflict with the electrons' kinetic energy that tries to spread out the electrons over the whole crystal. To assess the "true" strength of the electron–electron interactions as compared to their kinetic energy, we need to define a significant measure for their relative strength.

To keep a transparent notation we consider s electrons with spin $\sigma =\uparrow,\downarrow$ and electron density n (n is the ratio of the number of electrons to the number of ions, $n = N_{\mathrm{e}}/N_{\mathrm{I}}$). In second quantization [1.89, Chap. 1] we denote the Fermi creation and annihilation operators for σ electrons in position space as $\hat{\psi}^{+}_{\sigma}(\boldsymbol{r})$ and $\hat{\psi}_{\sigma}(\boldsymbol{r})$, respectively, and $\hat{n}_{\sigma}(\boldsymbol{r}) = \hat{\psi}^{+}_{\sigma}(\boldsymbol{r})\hat{\psi}_{\sigma}(\boldsymbol{r})$, $\hat{n}(\boldsymbol{r}) = \sum_{\sigma}\hat{n}_{\sigma}(\boldsymbol{r})$ are the corresponding particle density operators; see also Sect. 2.3.1. The average potential energy for the Coulomb interaction between these electrons is given by [1.4]

$$\langle\hat{V}\rangle = \sum_{\sigma,\sigma'}\int \mathrm{d}\boldsymbol{r}\,\mathrm{d}\boldsymbol{r}'\frac{e^2}{|\boldsymbol{r}-\boldsymbol{r}'|}\langle\hat{\psi}^{+}_{\sigma}(\boldsymbol{r})\hat{\psi}^{+}_{\sigma'}(\boldsymbol{r}')\hat{\psi}_{\sigma'}(\boldsymbol{r}')\hat{\psi}_{\sigma}(\boldsymbol{r})\rangle\,, \tag{1.15}$$

where $\langle\ldots\rangle$ denotes a thermal average over the quantum-mechanical states. From (1.15) we conclude that we need to know the probability distribution to find a σ' electron at $\boldsymbol{r}'$ when there already is a σ electron at $\boldsymbol{r}$ if we want to determine the strength of the electron–electron interaction.

This information is contained in the pair correlation function [1.4],

$$g_{\sigma,\sigma'}(\boldsymbol{r},\boldsymbol{r}') = \langle\hat{\psi}^{+}_{\sigma}(\boldsymbol{r})\hat{\psi}^{+}_{\sigma'}(\boldsymbol{r}')\hat{\psi}_{\sigma'}(\boldsymbol{r}')\hat{\psi}_{\sigma}(\boldsymbol{r})\rangle - \langle\hat{n}_{\sigma}(\boldsymbol{r})\rangle\langle\hat{n}_{\sigma'}(\boldsymbol{r}')\rangle\,. \tag{1.16}$$

This quantity differs from the expectation value we need for the evaluation of (1.15) by the subtraction of a product of two-fermion expectation values. The reason for this is two-fold: first, we expect the pair correlation function $g_{\sigma,\sigma'}(\boldsymbol{r},\boldsymbol{r}')$ to vanish at large electron separation, i.e., we have $g_{\sigma,\sigma'}(\boldsymbol{r},\boldsymbol{r}')\to 0$ for $|\boldsymbol{r}-\boldsymbol{r}'|\to\infty$ (provided no anomalous expectation values for Cooper pairs occur). Second, if the probability of finding a σ electron at $\boldsymbol{r}$ is independent of the σ' electron at $\boldsymbol{r}'$, the pair correlation function as defined in (1.16) will indeed vanish. Hence, $g_{\sigma,\sigma'}(\boldsymbol{r},\boldsymbol{r}')$ measures the correlations in the system.

There are two different contributions to the pair correlation function, namely the effect of the particle statistics (Pauli exclusion principle) for electrons of the same spin species, and "true" electron correlations induced by the electron–electron interaction. We discuss these two contributions separately.

Exchange. For fermions the Pauli principle applies and guarantees that two electrons of the same spin cannot occupy the same orbital. Hence, the pair correlation function for electrons with like spin is different from zero even if no electron–electron interaction was present. For example [1.90, Chap. 19], in the three-dimensional homogeneous Fermi-gas ground state we have

$$g_{\sigma,\sigma}(\boldsymbol{r},\boldsymbol{r}') = -\left[\left(\frac{3n}{2}\right)\frac{\sin x - x\cos x}{x^3}\right]^2 \qquad \text{(Fermi gas)} \tag{1.17a}$$

with $x = k_{\mathrm{F}}|\boldsymbol{r}-\boldsymbol{r}'|$. The Pauli principle itself thus reduces the probability to find a σ electron in the vicinity of another σ electron on the scale of the reciprocal Fermi wave number $1/k_{\mathrm{F}}$ ("Pauli hole" or "exchange hole"), an effect totally independent of the presence of electron–electron interactions and thus due solely to the Fermi statistics of the electrons. For electrons with different spin quantum numbers there is no exclusion principle and we instead find

$$g_{\sigma,-\sigma}(\boldsymbol{r},\boldsymbol{r}') = 0 \qquad \text{(Fermi gas)}, \tag{1.17b}$$

i.e., $\uparrow$ and $\downarrow$ spins are entirely uncorrelated in the Fermi gas.

To illustrate the consequences of the exchange hole we consider the Coulomb interaction as a perturbative effect on the Fermi gas. Electrons with like spin are already kept apart from each other by the Pauli principle, and this in turn reduces their mutual Coulomb interaction. To lowest order in perturbation theory the electron wave functions remain unchanged such that the electron–electron interaction among σ electrons is reduced compared to the interaction between σ and $-\sigma$ electrons. The corresponding (lowest-order) "energy gain" is called "exchange interaction" [1.5, Chap. 17]. In atoms the exchange interaction is responsible for the parallel orientation of the electron spins in open shells (Hund's first rule [1.5, Chap. 17]); in solid-state physics this energy contribution is invoked to explain ferromagnetism in transition metals (Stoner theory). Slater's views on the metal–insulator transition are also based on a mechanism driven essentially by the exchange interaction.

Correlations. For electrons with different spin quantum numbers the pair correlation function identically vanishes for all theories in which the eigenstates can be written as product states of $\uparrow$ and $\downarrow$ electrons, as is clear from (1.16). This was the second reason for subtracting there the product over the two-fermion terms. This not only applies to the case of non-interacting electrons but also to all mean-field theories in which the σ electrons move in the effective field of the $-\sigma$ electrons as in Stoner and Slater theory; see Sect. 3.1. Within such a description the electrons perform an uncorrelated motion.

The correlation function $g_{\sigma,-\sigma}(\boldsymbol{r},\boldsymbol{r}')$ between electrons of different spin accounts for the contributions beyond the exchange effect, i.e., it contains the true many-body effects. As we pointed out above, the Fermi statistics of the electrons gives rise to "exchange" between electrons of the same spin, which is signaled by a non-vanishing pair correlation function for equal spins. Unfortunately, this effect is sometimes misleadingly interpreted as the result of "correlations" instead of "exchange". We refer to systems with a non-vanishing pair correlation function between $\uparrow$ and $\downarrow$ electrons as "correlated electron systems". Naturally, the interaction also changes the correlation function between electrons of the same spin species. If the pair correlation functions for electrons of the same spin and electrons of opposite spin are comparable in size, the system is termed "strongly correlated". In general, a σ electron will not only be surrounded by an exchange hole but also by a "correlation hole" since the repulsive electron–electron interaction will also prevent electrons of different spin to come close to each other. In contrast to Slater's views Mott emphasizes the significance of correlations for the understanding of the metal–insulator transition.

1.4.3 Magnetic Moments

The pair correlation function contains the full information on the static electron spin and density structure factors. Unfortunately, it is not easily accessible either experimentally or theoretically. Hence we look for a less complete, i.e., a more local probe of our systems' properties. Thus we restrict ourselves to the terms $\boldsymbol{r}=\boldsymbol{r}'$ in the pair correlation function $g_{\sigma,\sigma'}(\boldsymbol{r},\boldsymbol{r}')$. To be specific we treat our spin-1/2 electrons as moving on a lattice with sites $\boldsymbol{R}$.

If we use natural units and the z axis as the quantization axis we define the (instantaneous) local magnetic moment in the z direction as

$$[M^z(\boldsymbol{R})]^2 = \langle \hat{S}^z(\boldsymbol{R})\hat{S}^z(\boldsymbol{R})\rangle \;, \tag{1.18}$$

where $\hat{S}^z(\boldsymbol{R}) = \left(\hat{n}_\uparrow(\boldsymbol{R}) - \hat{n}_\downarrow(\boldsymbol{R})\right)/2$. The other components of the local electron spin operator $\hat{\boldsymbol{S}}(\boldsymbol{R})$ are $\hat{S}^+(\boldsymbol{R}) = \hat{S}^x(\boldsymbol{R}) + \mathrm{i}\hat{S}^y(\boldsymbol{R}) = \hat{\psi}_\uparrow^+(\boldsymbol{R})\hat{\psi}_\downarrow(\boldsymbol{R})$ and $\hat{S}^-(\boldsymbol{R}) = \hat{S}^x(\boldsymbol{R}) - \mathrm{i}\hat{S}^y(\boldsymbol{R}) = \hat{\psi}_\downarrow^+(\boldsymbol{R})\hat{\psi}_\uparrow(\boldsymbol{R})$. The three components $M^{x,y,z}(\boldsymbol{R})$ are equivalent due to spin rotational invariance. If this symmetry is broken we assume that the magnetization vector is parallel to the z axis.

We propose to decompose the instantaneous magnetic moment into three positive definite parts,

$$[M^z(\boldsymbol{R})]^2 = [M^z(\boldsymbol{R})]^2_{\mathrm{FG}} + [M^z(\boldsymbol{R})]^2_{\mathrm{LRO}} + [M^z(\boldsymbol{R})]^2_{\mathrm{C}} \;, \tag{1.19a}$$

$$[M^z(\boldsymbol{R})]^2_{\mathrm{FG}} = \frac{1}{8}\langle \hat{n}(\boldsymbol{R})\rangle \left(2 - \langle \hat{n}(\boldsymbol{R})\rangle\right) \;, \tag{1.19b}$$

$$[M^z(\boldsymbol{R})]^2_{\mathrm{LRO}} = \frac{1}{2}\langle \hat{S}^z(\boldsymbol{R})\rangle^2 \;, \tag{1.19c}$$

$$[M^z(\boldsymbol{R})]^2_{\mathrm{C}} = -\frac{1}{2} g_{\uparrow,\downarrow}(\boldsymbol{R},\boldsymbol{R}) \;. \tag{1.19d}$$

The first term, the *Fermi-gas* contribution, is a consequence of the Pauli principle, and is independent of electron–electron interactions. This contribution can be ignored in our further discussions of local moments due to interactions, since it reflects the particle statistics only. The second term is non-zero only if the spin rotational symmetry is broken, e.g., as a result of ferromagnetism or antiferromagnetism. We denote this term as the contribution due to *long-range order*. The last term is the *correlation contribution* to the moment.

The *relative* local moment,

$$[M^z(\boldsymbol{R})]^2_{\rm rel} = [M^z(\boldsymbol{R})]^2 - [M^z(\boldsymbol{R})]^2_{\rm FG} = [M^z(\boldsymbol{R})]^2_{\rm LRO} + [M^z(\boldsymbol{R})]^2_{\rm C} \ , \quad (1.20)$$

arises inherently because of the electron–electron interaction. In effective single-electron approaches, e.g., Slater and Stoner theory, the correlation contribution to the relative moments is absent, and it is the importance of this term that is emphasized in Mott's view.

The local moments $|M^z(\boldsymbol{R})|$ as defined in (1.18) behave as free spins only when probed at short time scales or, equivalently, high energy scales. For example, the static spin susceptibility follows a Curie law with a Curie constant proportional to $[M^z(\boldsymbol{R})]^2$ only if the temperature is large compared to all other energy scales. The common experimental situation is different since we are often probing our system at low temperatures and over time scales that are very large compared to electronic time scales. For example, in a NMR experiment for the Knight shift in metals [1.5, Chap. 31] we probe the long-time behavior of the *time-dependent* moment–moment correlation function,

$$C^z_{\rm MM}(\boldsymbol{R},t) = \langle \hat{S}^z(\boldsymbol{R},t)\hat{S}^z(\boldsymbol{R})\rangle = \int_{-\infty}^{\infty} {\rm d}\omega\, {\rm e}^{-{\rm i}\omega t} C^z_{\rm MM}(\boldsymbol{R},\omega) \ , \quad (1.21)$$

where $\hat{A}(t) = \exp({\rm i}\hat{H}t)\hat{A}\exp(-{\rm i}\hat{H}t)$ defines a Heisenberg operator. The local moment of (1.18) is obtained in the limit of an instantaneous measurement,

$$[M^z(\boldsymbol{R})]^2 = C^z_{\rm MM}(\boldsymbol{R},t=0) = \int_{-\infty}^{\infty} {\rm d}\omega\, C^z_{\rm MM}(\boldsymbol{R},\omega) \ . \quad (1.22)$$

The correlation function $C^z_{\rm MM}(\boldsymbol{R},t)$ itself describes how a given local moment decays as a function of time due to its interaction with the medium. For example, in a Fermi gas a local moment formed by an $\uparrow$ electron decays on a time scale $1/E_{\rm F}$ since the electrons are itinerant. This time scale is very short compared to those of an NMR experiment such that in practical terms we effectively measure the *low-frequency* component, $C^z_{\rm MM}(\boldsymbol{R},\omega \to 0)$. For a Fermi gas, this quantity is much smaller than the local magnetic moment, as can be seen from the rather small Knight shift and Pauli susceptibility of metals. For a further discussion of the notion of local moments, see [1.91].

From this analysis we see that the relative local moments of (1.20) may not show up in low-energy properties in the metallic phase but they are nevertheless present at finite frequencies for all non-zero interaction strengths to fulfill the frequency sum-rule (1.22). A proper description of the metal–insulator transition where the low-energy (Fermi-liquid) picture breaks down must necessarily include the effects of local moments.

The notions of this subsection will be very helpful for the definition and distinction between Slater and Mott insulators. We introduce these theoretical concepts in the next two subsections.

1.4.4 Slater Insulator

Slater based his view of the insulating behavior due to electron–electron interactions on the features of band and Peierls insulators [1.92]. For a band insulator (see Sect. 1.3.1) the energy gap between the valence and conduction band is responsible for the lack of mobile charge carriers and thus for the vanishing DC conductivity. On the other hand the example of the Peierls insulators (see Sect. 1.3.2) shows that the band structure can be changed by the formation of a superlattice structure due to electron–lattice interactions. In close analogy to the Peierls mechanism interacting electrons may also generate a periodic modulation of their spatial charge and/or spin distribution – i.e., charge- and/or spin-density waves – which in this case is stabilized by the electron–electron interaction instead of the electron–lattice interaction.

The basic principle behind an antiferromagnetic Slater insulator can be explained most easily for s electrons on a bipartite lattice [1.92]. This is a lattice that could be separated into two interpenetrating sublattices (A and B) such that the nearest neighbor of any site are members of the opposite sublattice, e.g., simple cubic lattices in all dimensions. In addition, we assume for simplicity that the lattice is half filled such that there is one electron per site on average. Since $\uparrow$ and $\downarrow$ electrons mutually repel, the $\uparrow$ and $\downarrow$ electrons preferably arrange themselves on alternating lattice sites to minimize their interaction. Hence they form a spin-density wave whose wave vector $\boldsymbol{Q}$ is commensurate with the lattice. This spin-density wave is self-stabilizing since the $\uparrow$ electrons on the A sites experience an increased repulsion by the $\downarrow$ electrons on the B sites and thus preferably occupy the A sites. The same holds vice versa for the $\downarrow$ electrons on the B sites, and the self-consistent stabilization effect occurs. Since the electrons tend to avoid each other we have a gain in potential energy as compared to a translationally invariant state. This gain is balanced by a loss in kinetic energy because of the restricted electron motion.

In the corresponding band structure picture the lattice unit cell is doubled and the first Brillouin zone is cut in half. At the boundaries of the new (magnetic) Brillouin zone the bands split because of the $\boldsymbol{Q}$-periodic potential and the energy of the occupied levels is lowered. At half band-filling the band splitting causes a gap for charge excitations and the system is an insulator. An analogous situation is provided by an attractive local electron–electron interaction for which a commensurate charge-density wave is formed, which also leads to an insulating ground state.

Slater's theory can be formally obtained from a Hartree–Fock calculation for interacting electrons; see Sect. 3.1. Then we find that the (relative) magnetic moments form at the Hartree–Fock Néel temperature $T_{\mathrm{N}}^{\mathrm{HF}}$, which, for not too small interactions, is proportional to the interaction strength. Since correlations are absent in a Hartree–Fock description the moments order at the very same temperature they are formed. On a bipartite lattice and for hopping only between A and B sites a gap for charge excitations opens at the same temperature: a thermodynamic phase transition into an "insulator" occurs at the $T_{\mathrm{N}}^{\mathrm{HF}}$. Of course, as explained in Sect. 1.2, the conductivity does not instantly vanish at the transition but the temperature gradient of the resistivity becomes negative indicating the presence of a gap.

For general lattice types (e.g., "frustrated" hopping between sites of the same sublattice; non-bipartite lattices) the transition to the ordered state need not imply completely filled (sub-)bands. This leaves room for antiferromagnetic metallic states within Slater's (Hartree–Fock) theory.

We will argue below that this (self-consistent) band picture is too simple to describe real Mott–Heisenberg insulators. The insulating state is the consequence of electron exchange only in an over-simplified description. An understanding of real Mott insulators requires a theory for correlated electrons.

1.4.5 Mott Insulator

Mott himself describes a Mott insulator as a material that would be a metal if no moments were formed, and emphasizes that this notion "... depends on the existence of moments and not on whether or not they are ordered" [1.1, Chap. 4.8]. This is reflected, for example, in the fact that "for an antiferromagnetic insulator the most important property is that the magnitude of the Hubbard gap is not greatly affected when the temperature goes through the Néel point T_{N}" [1.1, Chap. 4.9].

Brandow [1.93, p. 662], however, equates Mott insulators and magnetic insulators. This definition appears to be too restrictive since Mott's decisive criterion is the existence of pre-formed local moments regardless of whether or not they are ordered. The moments appear since the electrons try to avoid each other to minimize their Coulomb repulsion (formation of a correlation hole); and the electrons may but need not localize on different sublattices as in Slater's view.

These consideration lead us to the following definition of a Mott insulator.

For a Mott insulator the electron–electron interaction leads to the occurrence of (relative) local moments. The gap in the excitation spectrum for charge excitations may arise either from the long-range order of the pre-formed moments (Mott–Heisenberg insulator) or by a quantum phase transition induced by charge and/or spin correlations (Mott–Hubbard insulator).

The notion of pre-formed moments even above the ordering temperature is very important not only for Mott insulators but for all systems with interacting itinerant electrons, for example in the description of itinerant ferromagnets [1.94].

1.4.6 Mott–Hubbard Insulator

Mott [1.93, 1.95] gave the following physically instructive example for a metal–insulator transition (see also [1.1, Chap. 4.3]) which is formalized in the Hubbard model (see Sect. 2.1) and also, more precisely, in the Harris–Lange model (see Sect. 6.2).

We consider the arrangement of hydrogen atoms with Bohr radius a_{B} on a d-dimensional cubic lattice with lattice constant a. The protons provide the L lattice sites for $N = L$ electrons (half filling). To set up the potential energy part of our Hamiltonian we make the (rather drastic) assumption that the effective electron–electron interaction will be local. The intra-atomic energy, ionization energy minus electron affinity, is called the "Hubbard" U,

$$U = E(\mathrm{H} \to \mathrm{H}^+) - \left|E(\mathrm{H} \to \mathrm{H}^-)\right| = E(\mathrm{H}^-) + E(\mathrm{H}^+) - 2E(\mathrm{H}) \,. \quad (1.23)$$

In this picture it is tacitly assumed that the electrons dominantly sit *on* and not *between* the lattice sites. In reality, hydrogen and halides form molecules of two atoms (H_2, F_2, Cl_2, I_2) with very high binding energies [$\mathcal{O}$(10eV)]. For hydrogen it has not become clear even at very high pressures whether metallic behavior has really been observed [1.96].

We further assume that electrons may hop between neighboring atoms. This "tight-binding" approximation for independent electrons [1.5, Chap. 10] (tunneling between nearest neighbors on a cubic lattice) results in a cosine dispersion relation

$$\epsilon(\boldsymbol{k}) = -2t \sum_{l=1}^{d} \cos(k_l a) \quad (1.24)$$

with a bandwidth $W = 2Zt$, where $Z = 2d$ is the number of nearest neighbors and $t > 0$ is the tunnel amplitude which depends on a/a_{B}. It has positive sign to guarantee the proper low-density limit (quadratic dispersion relation around $\boldsymbol{k} = \boldsymbol{0}$).

At large distances between the H atoms, $a \gg a_{\mathrm{B}}$, the overlap between the atomic wave functions is small, which implies $W \ll U$. In the ground state every lattice site will be singly occupied since we are at half band-filling. The situation is illustrated in Fig. 1.8. The energy $\mu^+(N) = E_0(N+1) - E_0(N)$ necessary to add another electron is given by $\mu^+(N = L) = U - W_1/2$ since charge excitations with energy U are mobile such that they form a band of width $W_1 = W_1(U/W)$; see Fig. 1.8a. This band is the "upper Hubbard band", which is generally *not* a one-electron band. Instead, it describes the spectrum of charge excitations for an extra electron added to the ground state

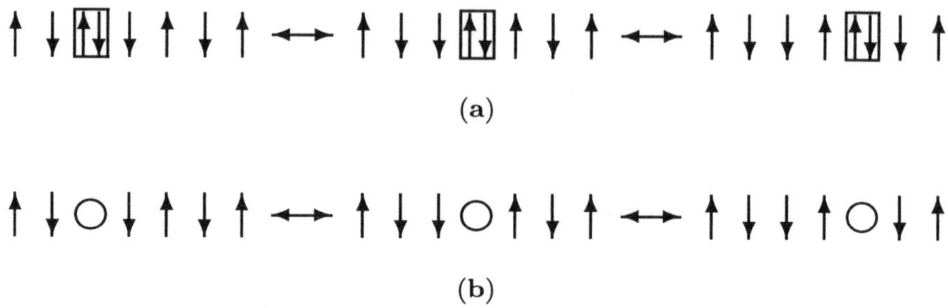

Fig. 1.8. Electrons (coherently) move (a) as double occupancies in the upper and (b) as holes in the lower Hubbard band when the number of double occupancies and holes is approximately conserved at strong coupling; the Hubbard bands are *charge* bands.

of the half-filled electron system. The energy required to remove an electron, $\mu^-(N) = E_0(N) - E_0(N-1)$, is given by $\mu^-(N=L) = +W_2(U/W)/2$. The corresponding spectrum for the removal of a charge from the half-filled ground state constitutes the "lower Hubbard band"; see Fig. 1.8b. At half band-filling and for $(W_1+W_2)/2 \ll U$ we expect that the chemical potential $\mu(N)$ is not continuous, but a gap for charge excitations occurs, $\Delta\mu(N=L) = (\mu^+ - \mu^-)(N=L) \approx U - (W_1+W_2)/2 > 0$. Since we find an energy gap for charge excitations at half band-filling, if $(W_1+W_2)/2 \approx W \ll U$, the system is an *insulator*. As we outlined in Sect. 1.1 a gap for a single charge excitation implies a vanishing DC conductivity if electron pairing is absent.

In a *Gedanken* experiment we may now reduce the distance between the hydrogen atoms, which increases the overlap in their atomic wave functions and corresponds to an enhancement of the screening and the electrons' tendency to delocalize. The combined bandwidth of the upper and lower Hubbard band W_1+W_2 increases accordingly: the two bands finally overlap for $a \approx a_B$, and the gap for charge excitation vanishes. As outlined in Sect. 1.4.1 the band theory of Sect. 1.3.1 should qualitatively apply for $W \gg U$, which predicts a single, half-filled band. Low-energy excitations hardly change the charge distribution in this situation and thus are gapless. For $W \gg U$ the system is a paramagnetic *metal.*

For intermediate values, around $U_c \approx W$, we expect a metal–insulator transition. This transition appears to be a metal–band insulator transition since the upper and lower Hubbard band begin to "overlap". However, these bands *do not represent* single-electron states; the transition is at heart a consequence of electron correlations as we shall see in the rest of the book, and not merely the result of a simple band crossing.

These considerations lead us to the following definition of a Mott–Hubbard insulator.

For a Mott–Hubbard insulator the electron–electron interaction leads to the formation of a gap in the spectrum for single charge excitations. The correlations force a quantum phase transition from a correlated metal to a paramagnetic Mott–Hubbard insulator, in which the local magnetic moments do not display long-range order.

As we pointed out before, we will observe drastic changes in the resistivity even at finite temperatures when we go through the quantum phase transition value for our external parameters (e.g., pressure) that control the relative strength of the electrons' kinetic and potential energy, U/W. For temperatures small compared to the single-electron gap, $k_{\mathrm{B}}T \ll \Delta\mu$, we may then speak of a "Mott–Hubbard insulator" in practical terms even at finite temperatures.

1.4.7 Mott–Heisenberg Insulator

In our above discussion of the Mott–Hubbard insulator we did not consider the possibility of an ordering of magnetic moments. Thus, the concept of the Mott–Hubbard insulator has to be supplemented by that of the Mott–Heisenberg insulator for which the insulating state displays long-range order.

What we totally ignored in the previous subsection was the exchange interaction between localized spins. Even for $U \gg W$ the tunneling of electrons to neighboring sites is allowed ("virtual" hopping processes). Formal perturbation theory in second order around the half-filled ground state with all electrons localized provides an antiferromagnetic coupling $J \sim t^2/U$ between the spin-1/2 electrons on neighboring lattice sites ("itinerant exchange") [1.97]. We further clarify the connection between models for itinerant electrons (Hubbard model) and models of magnetic insulators (Heisenberg model) in Sect. 2.5.

For large interactions, $U \gg W$, and half band-filling the energy scales for charge excitations ($\Delta\mu \approx U - W$) and spin excitations ($J \sim t^2/U$) are well separated. We are in the Mott–Hubbard insulating regime since the gap $\Delta\mu \approx U - W$ due to the quantum phase transition at $U_{\mathrm{c}} \approx W$ is finite. The spin energy scale, however, cannot be ignored for temperatures around or below that of the corresponding exchange interaction. The local magnetic moments in the Mott–Hubbard insulator may long-range order at the temperature $k_{\mathrm{B}}T_{\mathrm{N}} = \mathcal{O}(ZJ) \ll \Delta\mu$. The system goes over from the Mott–Hubbard insulator into the (antiferromagnetic) Mott–Heisenberg insulator in a thermodynamic phase transition at $T = T_{\mathrm{N}}$. According to Mott [1.1] the size of the local moments does not change much in the transition through the Néel temperature.

For small interactions, below the quantum phase transition at $U_{\mathrm{c}} \approx W$, the Mott–Hubbard gap is absent but correlations for all $U > 0$ lead to preformed magnetic moments. Again, these moments may long-range order at the Néel temperature, and we observe a thermodynamic phase transition from a paramagnetic (correlated) metallic state into the (antiferromagnetic) Mott–Heisenberg insulating state. Again, the magnitude of the local moments does not change (much) at the Néel temperature. We will further address this issue for $U \to 0$ in Sect. 3.1.

These considerations lead us to the following definition of the Mott–Heisenberg insulator.

The Mott–Heisenberg insulating state is the result of a thermodynamic phase transition in which the pre-formed local (magnetic) moments (antiferromagnetically) order below the critical (Néel) temperature.

Above the Néel temperature the Mott–Heisenberg insulator goes into a paramagnetic phase, which can either be a correlated metal or the Mott–Hubbard insulator.

We emphasize here that the term "Slater insulator" is often synonymously used for "Mott–Heisenberg insulator" in the sense of the above definition. In Slater's picture not only the long-range order but also the local magnetic moments are absent above the Néel temperature. In contrast, the moments are already present in the Mott–Heisenberg insulating state and remain in the paramagnetic metallic phase above the transition temperature. Since *pre-formed* moments, the important signature of electron correlations, order at the metal–insulator transition we introduced the concept of the "Mott–Heisenberg insulator" to provide a clear distinction between the ideas of Slater (self-consistent single-electron theory) and Mott (many-electron correlations).

1.5 Mott Insulators: Some Experimental Observations

In the last section of this chapter we list some experimentally observed types of Mott insulators, namely Wigner and Verwey insulators, multi-band Mott–Hubbard insulators, and charge-transfer insulators. At the end of this chapter we recall the necessity of full many-body theories for these and other correlated electron systems.

1.5.1 Wigner Insulator

More than 50 years ago Wigner [1.98] pointed out that free electrons in a homogeneously positively charged background (homogeneous electron gas or "jellium" model [1.4, Chap. 5]) may localize at low densities. In three dimensions the average kinetic energy per particle of completely delocalized electrons scales like $\langle\hat{T}\rangle_{d=3}/N_{\mathrm{e}} \sim n_{\mathrm{e}}^{2/3}$ for low electron densities $n_{\mathrm{e}} = N_{\mathrm{e}}/V$ and thus goes faster to zero than the average potential energy of an electron's Coulomb interaction with another electron at the average particle distance $n_{\mathrm{e}}^{-1/3}$, $\langle\hat{V}\rangle/N_{\mathrm{e}} \sim n_{\mathrm{e}}^{1/3}$. At low densities the Coulomb interaction

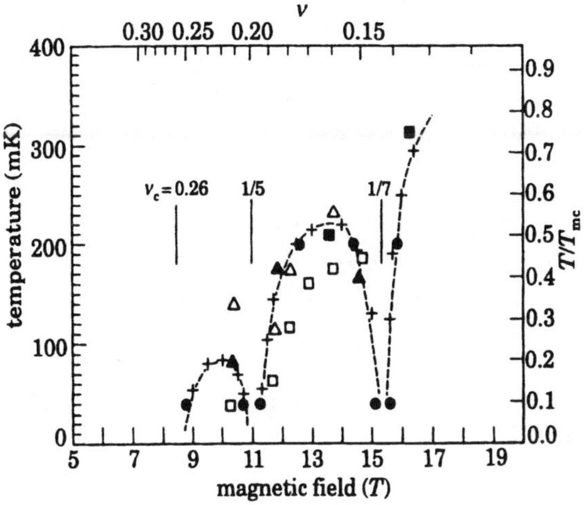

Fig. 1.9. Magnetic field versus temperature phase diagram for a two-dimensional electron gas in GaAs/AlGaAs heterostructures; the system is a correlated electron liquid above the dotted curves, below it is a Wigner lattice (from [1.105]).

should dominate the kinetic energy and the electrons should localize into the "Wigner lattice" at zero temperature. A variational estimate for the Wigner transition [1.99] gives a critical ratio $r_s = a_\mathrm{W}/a_\mathrm{B}$ of the Wigner lattice constant a_W to the Bohr radius a_B of $r_s(d=3) \approx 67$ in three dimensions and of $r_s(d=2) \approx 33$ in two dimensions. An exactly solvable one-dimensional model for Wigner crystallization can be found in [1.100]. Experimentally, a Wigner crystal is formed by electrons on a liquid Helium surface [1.101].

It is fairly obvious that the Wigner crystal is a charge-density wave state. Since it is pinned by impurities it is an insulator. The Wigner lattice is the state with minimum potential energy, and the kinetic energy results only in the electrons' zero-point motion around the lattice sites. The Wigner lattice breaks the continuous translational symmetry of the underlying Hamiltonian. Hence there will be a critical temperature for the melting of the Wigner lattice into a "Wigner liquid" which, since the electrons are mobile again, will be metallic. Thus we should observe a thermodynamic phase transition to occur from the Wigner insulator into a conducting state.

In order to demonstrate experimentally the existence of such a transition we have to investigate low-dimensional systems. Localization effects are stronger in lower dimensions since the average kinetic energy vanishes more rapidly as a function of density. For example, for a two-dimensional electron gas at low electron density we find $\langle\hat{T}\rangle_{d=2}/N_\mathrm{e} \sim n_\mathrm{e}$, where in three dimensions we had $\langle\hat{T}\rangle_{d=3}/N_\mathrm{e} \sim n_\mathrm{e}^{2/3}$. Hence, the Coulomb interaction is effectively stronger in low dimensions because of the restricted phase space. We may further reduce the effective dimension in a two-dimensional electron gas by applying a strong magnetic field, which restricts the motion of the electrons to the lowest Landau level such that we may hope to observe the Wigner insulator experimentally. A two-dimensional electron gas can be realized in heterostructures made out of gallium-arsenide and aluminum-gallium-arsenide (GaAs/AlGaAs).

The first evidence for a Wigner crystal in high magnetic fields was produced by Andrei et al. [1.102], although their results were regarded as insufficient in the beginning [1.103]. Now its existence has been confirmed by various methods, such as measurements of the conductivity [1.104] and of time-resolved photoluminescence [1.105]; for a recent review, see [1.106]. Near fractional filling factors the problem of its observation is the proximity of the Wigner lattice to a correlated quantum liquid, the fractional quantum Hall state. The whole phase diagram of the two-dimensional electron gas in a strong magnetic field is shown in Fig. 1.9, which unifies the results of various measurements on the two-dimensional electron gas in GaAs/AlGaAs heterostructures in a strong magnetic field in [1.104, 1.105]. Detailed variational and Monte-Carlo calculations for the two-dimensional quantum Hall gas give values for $r_s(d = 2)$ between 15 and 22, depending on the filling of the lowest Landau level [1.107], in good agreement with the experimentally observed critical electron densities [1.105].

1.5.2 Verwey Insulator

Similar to the Wigner insulator an electronic charge-density wave forms at the Verwey transition. The difference to the Wigner crystal lies in the much higher electron concentration such that the wave length of the charge-density wave is of the order of the lattice constant, and frequently commensurate with the underlying ion lattice.

A transition into an electronic charge-density wave was first observed by Verwey for magnetite (Fe_3O_4) at a temperature $T_V \approx 120\,K$ [1.109]. Magnetite has a spinel structure in which the oxygen ions occupy a face-centered cubic lattice with respect to which the spinel A and B sites are defined. The former are always occupied by Fe^{3+} ions. The spinel B sites are arranged in (011) sheets. A band structure calculation would predict a quarter-filled conduction band. Below the Verwey temperature one instead observes a commensurate charge-density wave whose wave vector is perpendicular to the (011) planes. These planes of spinel B sites are alternately dominantly occupied with Fe^{2+} and Fe^{3+} ions. The electrons are ferromagnetically ordered in the planes with an additional antiferromagnetic coupling (and ordering) between the iron ions on the A and B sites (ferrimagnetism [1.5, Chap. 33]). Early experimental results are displayed in Fig. 1.10. A review on more recent work on magnetite, in particular the doping dependence of the transport properties, i.e., the softening and the final destruction of the charge-density wave, can be found in [1.110]; see also [1.36, 1.37].

The Coulomb interaction between the electrons in magnetite drives the metal ferromagnetic at a Curie temperature of $T_C \approx 850\,K$, far above the Verwey temperature $T_V \approx 120\,K$. Since the electrons responsible for the electrical transport are (almost) fully ferromagnetically polarized at T_V we may describe the Verwey transition in magnetite in terms of spinless fermions [1.111]. Since the magnetic moments are aligned we need only to address the ordering

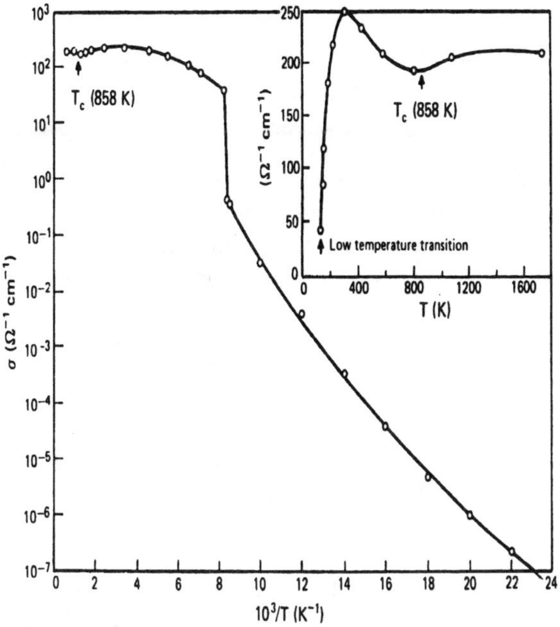

Fig. 1.10. Conductivity of magnetite (Fe_3O_4) as a function of the reciprocal temperature (after [1.108]).

of the charge degrees of freedom in (ferromagnetic) sheets due to the repulsive Coulomb interaction between the charges. We will study this model in some more detail in Sect. 5.4.5.

From a theoretical point of view the Wigner and Verwey insulators are very closely related concepts, only the wave length of the charge-density wave and the degree of electron localization differ. Whereas the underlying lattice structure plays little role for the Wigner insulator, it becomes important for the Verwey insulator where the charge-density wave is commensurate with the lattice structure.

1.5.3 Multi-Band Mott Insulator

Mott's considerations [1.1, Chap. 4.3], as sketched in Sect. 1.4, started from a single band of s electrons with variable distance between their orbitals. As explained in Sect. 1.4.1 the s electrons in most real materials under ambient conditions always form broad bands and their mutual Coulomb interaction is well screened. Hence, correlation effects in s bands are frequently almost negligible. Correlations are much more important for the d electrons of the transition metals (e.g., vanadium (V), iron (Fe), nickel (Ni), and copper (Cu)) because their bandwidth is substantially smaller than that of s bands in alkali metals under ambient conditions; see [1.112]–[1.114] for comprehensive reviews. However, the identification of the Hubbard parameter U is far more difficult than in Mott's (Gedanken) example, as we now discuss.

First of all, there are other interactions not included in a description based on a single Hubbard U since the d levels are degenerate: (i) Hund's rule couplings already present in the atom tend to maximize the total spin; (ii) crystal field effects tend to lift the (nominal) five-fold degeneracy of the atomic d levels. In general, these effects are not negligible.

Second, the local Hubbard U is considerably smaller than the estimate according to (1.23) would predict. In Sect. 1.4 we derived the Hubbard U from the difference between the ionization energy and the electron affinity. There, the calculation of the Hubbard U involves processes where we transfer an electron to infinity (ionization) or add another electron from infinity (electron affinity). In real d band materials, however, we may just deposit electrons on an energetically close s level, e.g.,

$$\begin{aligned} U &= E(3d^{n-1}4s^2) - E(3d^n 4s^1) - |E(3d^{n+1}4s^0) - E(3d^n 4s^1)| \\ &= E(3d^{n-1}4s^2) + E(3d^{n+1}4s^0) - 2E(3d^n 4s^1) \, . \end{aligned} \tag{1.25}$$

This mechanism considerably reduces the Hubbard U from its value for a single atom in the gas phase. In the solid state the creation of a local charge disproportion is thus partially screened by surrounding electrons.

An estimate of the screening effect for transition *metals* shows that $U \approx 2\,\mathrm{eV}$ is at most half the bandwidth W [1.112, Chap. 9.3]. Although the bandwidth W is much smaller in transition metals than in alkali metals the Coulomb interaction is too well screened to give rise to an insulating ground state. Although pure transition metals are metallic, their pronounced magnetic response already indicates that electron correlations play an important role in these metals.

To reduce the screening effect by electrons in other than the $3d$ bands we have to "chemically bind" the $4s$ electrons that are in wide bands in the transition metals. This can be achieved by oxidation; for recent reviews, see [1.36, 1.37]. In the band structure picture this can be understood as a hybridization of the $2p$ levels of the oxygen atoms with the $4s$ bands of the transition-metal atoms. The mainly oxygen-like bonding bands lie below the Fermi energy and are occupied, whereas the mostly $4s$-like antibonding bands are shifted above the Fermi energy and are unoccupied. Thus, the $4s$ bands are much less effective in screening the repulsion of the $3d$ electrons, and most of the transition-metal oxides are indeed insulators rather than metals. This qualitative picture has been verified experimentally for many transition-metal oxides such as manganese oxide (MnO) and nickel oxide (NiO), but calcium oxide (CaO) and titanium oxide (TiO) are still metallic.

For a more quantitative understanding we address the results of a band structure calculation [1.115], which are shown in Fig. 1.11. In the band picture all oxides would be metallic since the $3d$ bands would be partially filled, so that this prediction of band structure theory must not be trusted in detail. Nevertheless, the results are useful to estimate the importance of screening. For CaO and TiO the antibonding $4s$ bands (Γ_1 symmetry in the NaCl structure) and $3d$ bands (Γ_{12} and Γ_{25} symmetry) are predicted to overlap strongly and cross the Fermi energy. Thus, the band structure calculation indicates

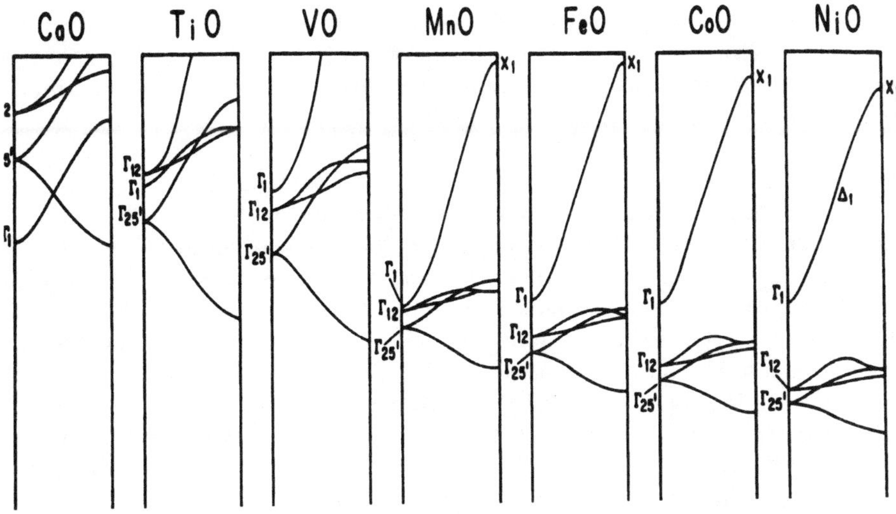

Fig. 1.11. Band structure of the antibonding $4s$ levels (Γ_1 symmetry in the cubic NaCl structure) and the $3d$ levels (e_g and t_{2g} with symmetry Γ_{12} and $\Gamma_{25'}$, respectively) in transition-metal mono-oxides in the [100] direction, calculated within a APW band-structure method with interaction corrections; the oxygen bands that are well below the Fermi energy are not shown (from [1.115]).

that the Hubbard interaction is screened well enough in CaO and TiO to keep the metallic behavior even in the presence of electron correlations. In contrast, for the transition-metal oxides from manganese (Mn) through copper (Cu) the $4s$ bands are split off the $3d$ bands. Thus, the Coulomb interaction in the (narrow) $3d$ bands becomes very important, and these transition-metal oxides are indeed found to be (Mott) insulators. The experimentally observed insulating ground state, which contrasts the predictions of band structure calculations, is a strong indication for the importance of correlation effects in these materials.

The bandwidth and the Coulomb interaction are each of the order of several electron volts. Since the experimentally changeable energy scales are of the order of a tenth of an electron volt, it is difficult to realize experimentally a metal–Mott insulator *transition* in transition-metal oxides. Hence we have to focus on oxides that already are at the brink of a metal–insulator transition. To find appropriate candidates we may again study the results of the band structure calculation as shown in Fig. 1.11. According to our reasoning above we would predict that vanadium oxides are at the boundary between metallic and insulating behavior since the $4s$ bands barely cross the $3d$ bands. This has been confirmed experimentally since both vanadium dioxide (VO_2) and vanadium sesquioxide (V_2O_3) undergo in practical terms a transition from a metallic to an insulating state at transition temperatures of $T \approx 340$K and $T \approx 150$K, respectively, which is accompanied by changes in volume and lattice structure. Hence, the oxides of vanadium, and its neighbors in the periodic table titanium and chromium, provide generic examples for materials with observable metal–Mott insulator transitions [1.1, Chap. 6].

Perhaps the most celebrated example for a material that displays a metal–Mott insulator transition is vanadium sesquioxide (V_2O_3), which crystalizes in a corundum structure with a V_2 pair per unit cell. As a result of crystal field splitting there are four $3d$ states per pair at the Fermi energy [1.116] that form two degenerate bands of e_g symmetry. The other two electrons of the V^{3+} pair form a quasi-molecular bond of a_g symmetry. Although this filled band is inert as far as transport is concerned, it can be polarized and contributes to the local magnetic moment. This view is confirmed by optical experiments [1.117]. Both band structure calculations [1.118] and a strong-coupling approach [1.119]–[1.121] to this material show that electrons in two partially filled bands participate in the metal–insulator transition. Thus we characterize the insulating state as a "multi-band Mott insulator".

By the application of pressure or by doping with titanium (Ti) we can slightly reduce the size of the lattice constant such that the bandwidth in V_2O_3 can be increased and the metallic phase is preferred. Analogously, we may add chromium (Cr) to increase the lattice constant, which reduces the bandwidth and stabilizes the insulating state. The apparent equivalence between doping and external pressure, as indicated in the phase diagram of (doped) V_2O_3 in Fig. 1.12, must be taken with care. Recent measurements on V_2O_3 [1.123] confirm that the doping with Ti or Cr not only changes the lattice constant, but also the carrier (and impurity) concentration in the bands since Ti and Cr are in different oxidation states compared to V. Further recent experiments on the photoemission spectrum and the optical conductivity of V_2O_3 are reported in [1.124] and [1.125], respectively.

The measurements that lead to the phase diagram of Fig. 1.12 confirm Mott's view on the stability of the local moment across the Néel temperature. The local magnetic moment of V_2O_3 has been determined experimentally [1.126] to be about 1.3 Bohr magnetons per vanadium in the antiferromagnetic state and thus to be about 10% larger than in the metallic state (1.2 Bohr magnetons). The magnetic moment of the metal is much larger than that of a free Fermi gas [1.126]. The metal is a strongly correlated electron system, and the antiferromagnetic phase is identified as Mott–Heisenberg insulator. The complex lattice structure and band degeneracy mean that the spin order is far from a simple Néel structure [1.127]. The most natural way to explain this spin structure is via the assumption that there is an additional orbital ordering in the occupation of the e_g orbitals [1.120]. A detailed description of V_2O_3 requires a theory of correlated electrons for degenerate bands.

Despite these complications the phase diagram of V_2O_3 in Fig. 1.12 is generic for Mott insulators inasmuch as it displays the features of both a

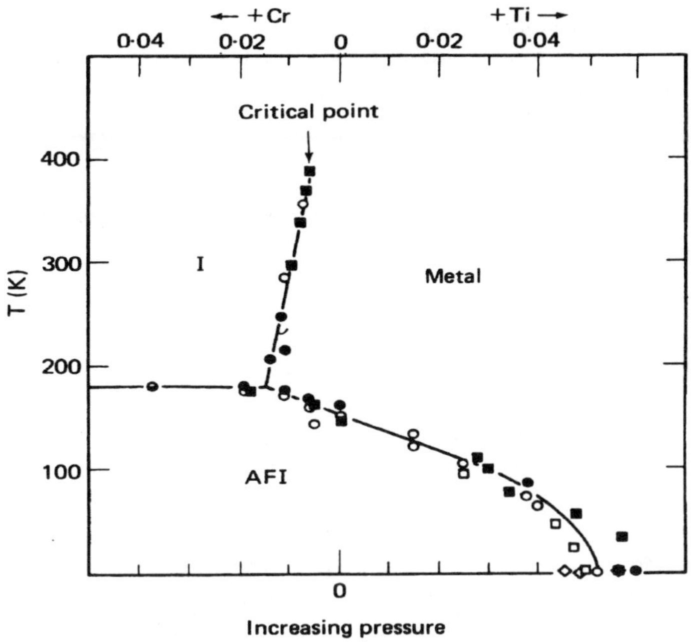

Fig. 1.12. Phase diagram of vanadium sesquioxide (V_2O_3) as a function of pressure and temperature. AFI: antiferromagnetic insulator. I: paramagnetic insulator. The open and closed symbols indicate the size of hysteresis effects upon approaching the transitions from above and below, respectively (after [1.122]).

quantum and a thermodynamic phase transition; see Sect. 1.2. The zero temperature quantum phase transition between the metal and the Mott–Hubbard insulator around a Cr concentration of 0.02 is hidden behind the ordered Mott–Heisenberg phase, which dominates at low temperatures. For temperatures $T \gtrsim 380\,\mathrm{K}$, well above the Néel temperature $T_N < 180\,\mathrm{K}$, there is a continuous crossover between the Mott–Hubbard insulator and the metal as outlined in the quantum phase transition scenario. For $T_N < T \lesssim T_c \approx 380\,\mathrm{K}$, however, there is a line of first-order transitions between metal and insulator that cannot be explained within the subsystem of interacting electrons alone. As observed the transition from the insulating to the metallic state is not isostructural but accompanied by a decrease in volume [1.19, 1.122]. Hence the lattice degrees of freedom in effect transmute the Mott–Hubbard quantum phase transition to a thermodynamic phase transition for $T_N < T \leq T_c$, where the lattice distortion plays the role of the order parameter [1.128]. This effect is not surprising in real materials. As we pointed out in Sect. 1.1 all other interactions of the electrons may and will modify the nature of the Mott transition. In the case of V_2O_3 the crossover region defined by the electronic quantum phase transition is dominated by the electron–lattice interaction.

The importance of electron correlations can also be seen in the physical properties of vanadium-underdoped vanadium sesquioxide, $V_{2-y}O_3$ [1.129]. As a function of defect concentration y and temperature this highly correlated electron system displays phase transitions from a metallic state into an antiferromagnetic insulator and into an antiferromagnetic metal. The complex physical properties of the transition-metal compounds can be understood only in terms of correlated electron theory for electrons in degenerate bands [1.119]–[1.121].

1.5.4 Charge-Transfer Insulator

The band structure scheme shown in Fig. 1.11 does not take into account the Coulomb splitting of the $3d$ bands into upper and lower Hubbard bands (Mott gap), which causes the insulating behavior of transition-metal oxides beyond VO. According to our arguments presented in the last subsection this splitting increases from V to Ni. In effect, the oxygen p bands that are not shown in the figure could energetically come to lie *between* the two Hubbard bands, an effect not incorporated in standard band structure calculations. In this case the energy for a charge excitation is no longer determined by the Hubbard U but instead by the "charge-transfer" energy Δ [1.114, 1.130]. The transport of an electron from a (strongly correlated) transition-metal d level to another transition-metal atom requires the Hubbard energy

$$U = E(d^{n+1}) + E(d^{n-1}) - 2E(d^n) \, , \tag{1.26}$$

whereas we have to afford the charge-transfer energy

$$\Delta = E(d^{n-1}) - E(d^n) + E(p^{n'+1}) - E(p^{n'}) \tag{1.27}$$

to shift an electron to an oxygen p level. If $U > \Delta$ holds, the insulator should no longer be denoted a (multi-band) Mott–Hubbard insulator, but rather a charge-transfer insulator since the energy gap for charge excitations is now determined by the charge-transfer energy Δ and no longer by the on-site interaction U.

To test these ideas one may analyze the Néel temperature of various transition-metal oxides. If t is the tunnelling matrix element between the transition metal and their neighboring oxygen atoms, one obtains an effective hopping $t_{\text{eff}} = t^2/\Delta$ between the transition-metal atoms, which generates an itinerant exchange $J = t_{\text{eff}}^2/U$. In addition, this exchange is supplemented by the "super-exchange" via the oxygen atoms [1.97], which is proportional to $J_S \sim t_{\text{eff}}^2/\Delta$. Both processes are fourth order in the tunnel matrix element t between the transition-metal and oxygen atoms. Since the Néel temperature of the transition-metal oxides is proportional to $J_{\text{total}} = J + J_S$ one can draw (semi-)quantitative conclusions on the contribution of the superexchange mechanism in various transition-metal oxides which in turn allows to estimate the relative size of U/Δ [1.114, 1.130]. It is found that the simple transition-metal oxides beyond iron oxide (FeO) are charge-transfer

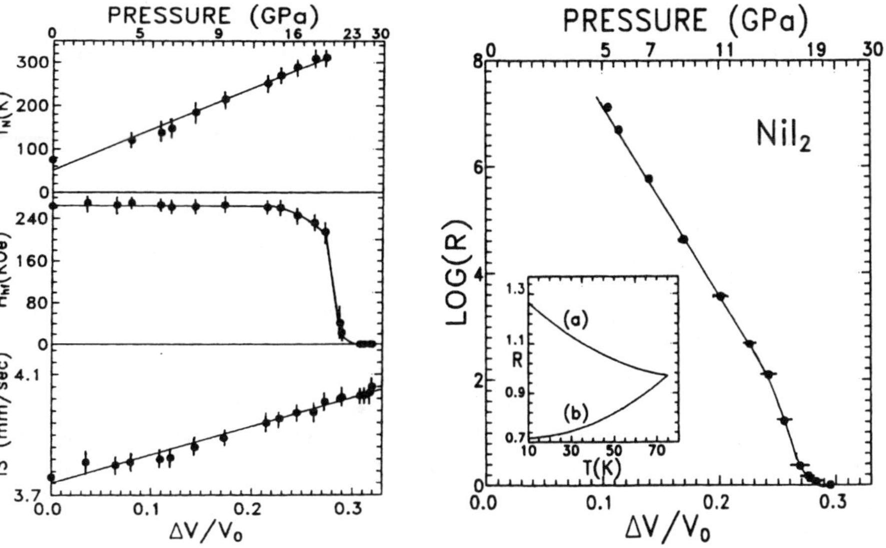

Fig. 1.13. *Left figure:* pressure dependence of the Néel temperature, the Mößbauer hyperfine field (local magnetic moment), and the isomer shift of NiI_2 at $T = 300$K. *Right figure:* pressure dependence of the resistance of NiI_2 at room temperature. The resistance saturates at a small value above the critical pressure of $p_c \approx 19$ GPa. *Inset:* resistance as a function of temperature (**a**) below p_c (insulating behavior), (**b**) above p_c (metallic behavior) (from [1.135]).

insulators. In the past it has been questioned that NiO is a Mott insulator [1.93, 1.131] but the size of the gap and the magnitude of the local moments [1.132, 1.133] call for a correlated electron description. An overview on the classification of the transition metals and their chalcogenides and halides can be found in [1.130], transition-metal pyrites are reviewed in [1.134].

An example for a metal–charge-transfer insulator transition is provided by Nickel iodine. Under pressure the antiferromagnetic insulator NiI_2 can be transformed into a paramagnetic metal [1.135], a quantum phase transition takes place. The corresponding experimental findings are displayed in Fig. 1.13. According to the interpretation of Pasternak et al. [1.135] the charge-transfer gap between the iodine p levels, and the upper Hubbard band of the nickel d levels closes at the transition. This results in a strong hybridization between the p and d levels such that the Hubbard U is reduced and the material becomes metallic and paramagnetic.

1.5.5 Whither Theory?

Band theory properly describes effects of the electron–ion interaction such as the formation of the lattice. Furthermore, it provides valuable indications for which materials electron–electron interactions become important such that a naive band structure will no longer be appropriate. Naturally, the question arises how to go beyond that approach since many materials of interest are correlated electron systems with pronounced local moments and pair correlations.

One way, followed by Slater and others, is to include electron–electron interactions on the level of effective single-electron theories. This approach to explaining insulators due to electron–electron interactions is popular for two reasons. First, it combines the two successful concepts of band and Peierls theory: the formation of an (antiferromagnetic) long-range order is invoked to explain the insulating behavior of the ground state. In fact, as we have repeatedly argued in this chapter, thermodynamic phase transitions may indeed be the reason for a metal–(Mott–Heisenberg) insulator transition. Second, Slater's theory can relatively easily be implemented in existing models for the *many-electron* problem since it reduces to a self-consistent *single-electron* theory.

The attractive conceptual simplicity of Slater's approach sharply contrasts the mathematical complexity of theories for Mott insulators. The experiments, however, clearly indicate that the metal–Mott insulator transition in its various forms can only be understood as correlation effect. Hence, non-perturbative many-body techniques have to be developed and tested against exactly solvable models and limits to make progress in this direction. In this book we will discuss modern models and methods for our present-day understanding of the Mott transition.

Not only the correlation-induced metal–Mott insulator transition requires a deeper understanding of interacting electron systems. The magnetism of transition metals and their oxides, the excitation spectrum of quasi one-dimensional electronic systems like polymers, the high-temperature superconductors, and the heavy fermion materials illustrate that strong correlation between electrons is a frequently encountered fundamental physical phenomenon for which a deeper understanding is clearly desirable. The "standard model" for theoretical studies of electron correlations is the Hubbard model, which we derive and scrutinize in the following chapters.

2. Hubbard Model

In the first section of this chapter we start from the general non-relativistic Hamilton operator of solid-state physics to derive the Hamilton operator for the purely electronic problem. There, the electrons are moving in the static, perfectly periodic lattice potential of the ions. This approximation is not crucial to the Mott transition since it is driven by electron–electron interactions.

In the second section we treat the electronic Hamiltonian within band structure theory. We begin with an independent electron approximation in which the electron–electron interaction is totally ignored. Although crude, this is a reasonable lowest-order approximation since the electron–ion interaction is responsible for the lattice formation. Next, we include the (classical) interaction of an electron with the surrounding charge distribution (Hartree theory). Finally, we discuss the local density approximation to density-functional theory in which effects of exchange and correlations are included. In principle, band structure theories are "ab initio" since they cope with the fully electronic problem. However, the interacting electron problem is always replaced by an effective single electron problem.

In the third section we derive the Hubbard model from the electronic Hamiltonian using Hubbard's two basic assumptions that only a single band has to be taken into account and that the interaction is purely local. Although the Hubbard model appears to be a caricature of the electronic problem it still contains the full complexity of the many-body problem and thus serves as a minimum model for interacting lattice electrons.

In the fourth section we emphasize the importance of the Hubbard model as key ingredient of most realistic models for real correlated electron (fermion) materials: polymers, transition metals and their oxides, fullerenes, high-temperature superconductors, heavy fermion materials, and liquid Helium. In all cases, however, at least one of Hubbard's basic assumptions is not fulfilled such that the Hubbard model itself presumably is too simple to describe a real physical system.

In the last section we discuss some qualitative properties of the Hubbard model. We specify its explicit and implicit parameters, classify common model variants, and present its global symmetries. We address its exactly solvable limits and show that the half-filled Hubbard model should indeed undergo a metal–Mott insulator transition as a function of the interaction strength. Finally, we collect some exact statements on magnetism in the Hubbard model and its derivatives.

2.1 Electronic Many-Particle Problem

In this first section we set up the general Hamilton operator of solid-state physics and gradually approximate it to derive the Hamiltonian for the purely electronic problem. Since the Mott transition is due to electron correlations we may indeed restrict ourselves to the problem of interacting electrons moving in a static, perfectly periodic lattice potential.

2.1.1 Hamilton Operator of Solid-State Physics

The only relevant elementary interaction in solid-state physics is the electromagnetic interaction between charged particles. For simplicity we start from a non-relativistic many-particle description of the solid state. This is a good approximation for materials with atomic numbers that are not too high and for which relativistic effects for the outer electrons may be disregarded. Some of these may later be taken into account by using relativistic correction terms as in the Pauli equation.

In a non-relativistic description the N_N nuclei of the constituents with atomic numbers Z_n and masses M_n at positions $\boldsymbol{R}_n$ ($n = 1, \ldots, N_\mathrm{N}$) have a kinetic energy T_N and a electrostatic interaction between different nuclei $V_{\mathrm{N-N}}$. Corresponding terms for the electrons (T_e, $V_{\mathrm{e-e}}$) at positions $\boldsymbol{r}_i$ ($i = 1, \ldots, N_\mathrm{e}$) and the interaction between electrons and nuclei $V_{\mathrm{e-N}}$ constitute the general nonrelativistic Hamilton operator of solid-state physics. In the absence of external fields and with $\boldsymbol{p} \equiv -i\partial/(\partial \boldsymbol{r})$, $\boldsymbol{P} \equiv -i\partial/(\partial \boldsymbol{R})$ we may write

$$
\begin{aligned}
H_{\mathrm{SS}} \;=\;& \underbrace{\sum_{n=1}^{N_\mathrm{N}} \frac{\boldsymbol{P}_n^2}{2M_n}}_{T_\mathrm{N}\text{: motion of nuclei}} \; \underbrace{+\frac{e^2}{2} \sum_{n\neq m=1}^{N_\mathrm{N}} \frac{Z_n Z_m}{|\boldsymbol{R}_n - \boldsymbol{R}_m|}}_{V_{\mathrm{N-N}}\text{: interaction between nuclei}} \\
& \underbrace{+\sum_{i=1}^{N_\mathrm{e}} \frac{\boldsymbol{p}_i^2}{2m}}_{T_\mathrm{e}\text{: motion of electrons}} \; \underbrace{+\frac{e^2}{2} \sum_{i\neq j=1}^{N_\mathrm{e}} \frac{1}{|\boldsymbol{r}_i - \boldsymbol{r}_j|}}_{V_{\mathrm{e-e}}\text{: interaction between electrons}} \\
& \underbrace{-e^2 \sum_{n=1}^{N_\mathrm{N}} \sum_{i=1}^{N_\mathrm{e}} \frac{Z_n}{|\boldsymbol{R}_n - \boldsymbol{r}_i|}}_{V_{\mathrm{e-N}}\text{: interaction between electrons and nuclei}} \; . \qquad (2.1)
\end{aligned}
$$

The Hamilton operator H_{SS} defines a quantum-mechanical many-particle problem.

The solution of the Schrödinger equation to the Hamilton operator H_{SS} is technically completely impossible such that we have to resort to approximations to make further progress. The maximum energy scale in the problem is the binding energy of electrons in inner shells, which is of the order of several 10^4 electron volts. However, we are interested in the behavior of the solid state at room temperatures or even at temperatures as low as several Kelvin or even milli-Kelvin, i.e., on an energy scale $\mathcal{O}(10^{-2} \ldots 10^{-4}\,\mathrm{eV})$ and below. If we were to examine directly the full Hamiltonian in (2.1) we would have to design an analytical approximation scheme that required a precision of 10^{-6} to 10^{-8}. Such an approach would then cover such different phenomena as the formation of atoms, their condensation into the solid state, various magnetic ordering phenomena, and superconductivity.

We are far remote from such a universal theory. Hence we have to concentrate on single aspects of the problem. Since the Mott transition is purely electronic in its origin we may almost entirely eliminate the influence of the nuclei on the electronic problem, as we now discuss.

2.1.2 Hamilton Operator of the Electronic Problem

In solid-state physics we are interested in the condensed state whose binding energy is about $\mathcal{O}(1 \ldots 10\,\mathrm{eV})$ per atom. For electrons whose binding energy to its nucleus is (much) larger than this energy scale we may use the results for single, isolated atoms to describe the formation of atoms or ions. Only the electron orbitals of the shells that are not completely filled will be substantially changed from their atomic structure during solidification. For condensed states it is thus sufficient to focus on the "valence electrons" with a binding energy of up to $\mathcal{O}(10\,\mathrm{eV})$ and study their interaction with themselves, with the ions, and the ion–ion interaction. We recall here that an individual atom alone poses a very complicated, and in fact unsolvable, many-particle problem. Hence, the reduction from the atomic energy scale of several ten keV down to the condensation energy scale of $\mathcal{O}(1 \ldots 10\,\mathrm{eV})$ implies a first step of approximation. However, core electrons do not play a role in the metal–Mott insulator transition, and we may further ignore them for our purposes.

The second step in the reduction of the energy scale, the formation of the crystal, is yet another, in its generality, unsolved problem. This does not come as a surprise in view of the innumerable variety of crystalline and amorphous structures in nature. In practice, we usually experimentally determine a crystal structure and then verify its stability, e.g., in band structure calculations. Often the overall chemical composition and knowledge about inorganic chemistry suffice for an educated guess on the ideal crystal structure. After the condensation we have reduced the energy scale for the global stability of the crystal to below about 1 eV per atom. As we will see below the electron volt also remains as the relevant energy scale for our studies of Mott insulators. In the context of the Mott transition the crystal structure is not unimportant but we may treat it as given for our analysis.

At temperatures well below the condensation energy scale it is further justified for Mott insulators to separate the motion of the electrons from the dynamics of the ions since the ion masses are much larger than the electron mass (adiabatic or Born–Oppenheimer approximation [2.1, Chap. 22]). As discussed in Sect. 1.4, the motion of the ions (phonons) does not play a role for the description of (pure) Mott insulators. Since we will not consider disorder effects we assume as a third step in our basic approximations that the immobile ions form an ideal lattice (Bravais lattice with a basis [2.1, Chaps. 4,7]). In this way we are left with a purely "electronic" problem of interacting electrons in a periodic potential,

$$H_{\text{electronic}} = T_{\text{e}} + V_{\text{e-I}} + V_{\text{e-e}} , \tag{2.2}$$

where $V_{\text{e-I}}$ is an (effective) perfectly periodic lattice potential which results from the replacement of nuclei by ions in the condensed state. This also implies that $H_{\text{electronic}}$ describes only "outer-shell" electrons which are not participating in the ion formation.

The electronic Hamiltonian in (2.2) describes the motion of electrons in the electrostatic field of the ions in the presence of electron–electron interactions. It is still an (unsolvable) quantum-mechanical many-body problem due to the electrons' mutual interaction. In principle, there are two distinct approaches for making further progress. On one hand we may design effective single-electron theories to derive a band structure. On the other hand we may further simplify the Hamiltonian for the electronic many-body problem down to model systems like the Hubbard model. These serve as standard models to develop and test new many-body techniques. We will further elaborate on these two approaches in the next two sections.

2.2 Band Structure Calculations

In band structure calculations the electron–electron interaction is approximated in such a way that the resulting problem becomes an effective single-electron problem. In this section we discuss three common schemes, namely (i) the independent electron approximation in which the electron–electron interaction is completely disregarded, (ii) the Hartree approximation in which only the interaction of an electron with the average electron density is taken into account, and (iii) the local density approximation to density-functional theory which provides the basis of modern band structure calculations. Finally, we briefly comment on possible improvements of the local density approximation. Density-function(al) theory and the local density approximation for lattice models such as the Hubbard model will be discussed in more detail in Sect. 3.2.

2.2.1 Independent Electrons

The seemingly crudest approximation we can make is to ignore totally the electron–electron interaction. This is not unreasonable since the electron–ion interaction and the delocalization of electrons are the main source of the lattice cohesion. Hence it is actually advisable to study independent electrons in the field of the ions first, $H_{\text{IE}} = T_{\text{e}} + V_{\text{e-I}}$. This approach also offers a practical advantage: at least in principle we can exactly solve the problem of independent electrons in a periodic potential to arbitrary accuracy since we are left with a single-electron problem.

The solution of the independent electron problem will provide us with a new set of electronic wave functions. Instead of working with plane waves in which T_{e} is diagonal we can then work with a basis which already includes the electron–ion interaction. This is important because the ions' non-retarded, barely screened Coulomb potential is rather singular both in its time and in its space dependence. In the new basis we can reanalyze the original problem $H_{\text{electronic}}$, which allows us to verify that the electron–ion interaction indeed dominates the electron–electron interaction. If this had not been so and the electron–electron interaction had been found to dominate the electron–ion interaction, the crystal would not have formed.

Hence, as a first approximation we completely drop the electron–electron interaction in the electronic problem and arrive at the Hamilton operator for independent electrons used in elementary band structure calculations,

$$H_{\text{IE}} = T_{\text{e}} + V_{\text{e-I}} = \sum_{i=1}^{N_{\text{e}}} H_{\text{IE}}(\boldsymbol{r}_i) \, . \tag{2.3}$$

We may numerically diagonalize H_{IE} for a given crystal structure with fixed ion positions $\boldsymbol{R}_n$ [2.1, Chap. 11] where $V_{\text{e-I}}$ represents an appropriate ionic potential. According to Bloch's theorem we may classify the single-electron wave functions (Bloch electrons) $\phi^{\text{IE}}_{\boldsymbol{k},b}(\boldsymbol{r})$ by their Bloch momentum $\boldsymbol{k}$ from the first Brillouin zone and their band index b. In our non-relativistic description the one-electron dispersion relation $\epsilon^{\text{IE}}(\boldsymbol{k}, b)$ is two-fold spin-degenerate ($\sigma = \uparrow, \downarrow$). Thus we may write the Schrödinger equation as

$$H_{\text{IE}}(\boldsymbol{r})\phi^{\text{IE}}_{\boldsymbol{k},b,\sigma}(\boldsymbol{r}) = \epsilon^{\text{IE}}(\boldsymbol{k}, b)\phi^{\text{IE}}_{\boldsymbol{k},b,\sigma}(\boldsymbol{r}) \, . \tag{2.4}$$

If electronic interaction effects were negligible, the band structure would be given by $\epsilon^{\text{IE}}(\boldsymbol{k}, b)$. In the independent electron approximation we have reduced our interacting electron problem to a Fermi-gas model. In the ground state all single-electron levels up to the Fermi energy E_{F} are occupied.

2.2.2 Hartree Theory

To go beyond this elementary band structure calculation we should first include in the interaction those terms that can still be treated in the framework of a single-electron theory. If the electrons were classical objects we would only have to include the interaction of an electron at $\boldsymbol{r}$ with the charge density $n(\boldsymbol{r}')$ formed by the other electrons. The corresponding "Hartree terms" describe this density–density interaction of a classical charge distribution. For example, crystal field effects are included in this approach.

In general, even the simplest quantum-mechanical many-body effect, the Fock term that reflects the exchange interaction, cannot be exactly treated in a lattice calculation. The inclusion of the non-local Fock term would result in substantial and often insurmountable numerical efforts. Furthermore, the exchange and the correlation contributions to the energy levels are of the same order of magnitude in (strongly) correlated electron systems. Since we do not want to prefer exchange over correlation effects we include solely the Hartree terms in our band structure calculation [2.2].

The Hartree band structure Hamilton operator thus becomes

$$H_0 = T_\mathrm{e} + V_{\mathrm{e-I}} + V_{\mathrm{e-e}}^{\mathrm{H}} \,, \tag{2.5a}$$

$$V_{\mathrm{e-e}}^{\mathrm{H}}(\boldsymbol{r}) = e^2 \int \mathrm{d}\boldsymbol{r}' \frac{n(\boldsymbol{r}')}{|\boldsymbol{r}-\boldsymbol{r}'|} \,. \tag{2.5b}$$

Here, $n(\boldsymbol{r}+\boldsymbol{R}) = n(\boldsymbol{r})$ is the local density, which has to be determined self-consistently ($\boldsymbol{R}$ is a lattice vector). We assume that the system is paramagnetic and does not form a charge- or spin-density wave since we do not want to investigate broken symmetry states at this point.

The one-electron wave functions $\phi_{\boldsymbol{k},b,\sigma}(\boldsymbol{r})$ obey the Schrödinger equation

$$H_0(\boldsymbol{r})\phi_{\boldsymbol{k},b,\sigma}(\boldsymbol{r}) = \epsilon(\boldsymbol{k},b)\phi_{\boldsymbol{k},b,\sigma}(\boldsymbol{r}) \,, \tag{2.6a}$$

which has to be fulfilled together with the self-consistency relation between the orbitals and the local density

$$n(\boldsymbol{r}) = \sum_{\epsilon(\boldsymbol{k},b)\le E_\mathrm{F},\sigma} |\phi_{\boldsymbol{k},b,\sigma}(\boldsymbol{r})|^2 \,. \tag{2.6b}$$

The latter equation is correct only for zero temperature. This is always a good approximation since the Fermi energy is usually large compared to typical experimental energy scales, e.g., the temperature.

2.2.3 Local Density Approximation

We have seen that it is possible to include partly the electron–electron interaction into a single-electron theory. The question arises whether an effective single-electron theory can be formulated which exactly reproduces the physical properties of the full many-body problem. In fact, Hohenberg and Kohn [2.3] showed that one can formulate a single-electron Hamiltonian whose ground-state energy, electron density, and Fermi energy agree with the corresponding quantities of the interacting electron problem $H_{\mathrm{electronic}}$; for a review, see [2.4]–[2.8]. In this density-functional theory the Hartree term in the interaction is supplied by the exchange-correlation potential $v_{\mathrm{xc}}(\boldsymbol{r})$, which can be written as a functional derivative of the density functional $E_{\mathrm{xc}}[n_{\mathrm{e}}]$ ($n(\boldsymbol{r})$ always denotes electron density n_{e} at $\boldsymbol{r}$).

Unfortunately, the exact density functional is hardly ever available since it requires detailed information on the exact ground-state energy as a function of the density. Hence, it is usually necessary to formulate an approximation to the exact density functional. The most popular is the "local density approximation" (LDA) [2.4]–[2.8]. Due to the Coulomb repulsion between electrons there should be a reduced probability for finding an electron next to another electron. In the framework of a linear response theory [2.9, Chap. 5] this probability density should be reduced up to an average screening length $R_{\mathrm{s}} \sim k_{\mathrm{F}}^{-1} \sim n_{\mathrm{e}}^{-1/3}$ in a metallic ground state to reduce the Coulomb interaction between the electrons, $\Delta V_{\mathrm{C}} \sim 1/R_{\mathrm{s}} \sim n_{\mathrm{e}}^{1/3}$. Explicitly, the density functional in the local density approximation becomes

$$E_{\mathrm{xc}}^{\mathrm{LDA}}[n_{\mathrm{e}}] = \int \mathrm{d}\boldsymbol{r}\, n(\boldsymbol{r}) e_{\mathrm{xc}}^{\mathrm{LDA}}\left(n(\boldsymbol{r})\right) , \tag{2.7a}$$

$$v_{\mathrm{xc}}^{\mathrm{LDA}}(\boldsymbol{r}) = \frac{\delta E_{\mathrm{xc}}^{\mathrm{LDA}}[n_{\mathrm{e}}]}{\delta n(\boldsymbol{r})} , \tag{2.7b}$$

where the exchange energy $e_{\mathrm{xc}}^{\mathrm{LDA}}$ depends only on the local electron density $n(\boldsymbol{r})$. According to the screening argument above, the dominant contribution to the exchange-correlation energy should be $e_{\mathrm{xc}}^{X\alpha}\left(n(\boldsymbol{r})\right) \sim \left(n(\boldsymbol{r})\right)^{1/3}$, as first proposed by Slater ("$X\alpha$ exchange") [2.10]. We see that the contribution $v_{\mathrm{xc}}^{\mathrm{LDA}}(\boldsymbol{r})$ simulates both the exchange hole between electrons of the same spin species and *also* the correlation hole between electrons of different spin species.

The form of $e_{\mathrm{xc}}^{\mathrm{LDA}}\left(n(\boldsymbol{r})\right)$ actually used in practical calculations slightly differs from the $X\alpha$ form of Slater [2.4, 2.7, 2.8]. In the case of a homogeneously smeared ionic background ("homogeneous electron gas" or "jellium model") one can actually calculate the ground-state energy as a function of the density within a high-density expansion. Other forms for the Coulomb screening in the homogeneous electron gas can be found in [2.9, Chap. 5]. The corresponding expressions then provide the basis for the LDA exchange energy $\epsilon_{\mathrm{xc}}^{\mathrm{LDA}}(n(\boldsymbol{r}))$. However, the dominant term is still of the $X\alpha$ form.

With these approximations for the form of the density functional the LDA Schrödinger equation becomes

$$\begin{aligned} H_{\mathrm{LDA}}(\boldsymbol{r})\phi_{\boldsymbol{k},b,\sigma}^{\mathrm{LDA}}(\boldsymbol{r}) &= \epsilon^{\mathrm{LDA}}(\boldsymbol{k},b)\phi_{\boldsymbol{k},b,\sigma}^{\mathrm{LDA}}(\boldsymbol{r}) \\ H_{\mathrm{LDA}}(\boldsymbol{r}) &= \Bigg[H_{\mathrm{IE}}(\boldsymbol{r}) + \underbrace{e^2 \int \mathrm{d}\boldsymbol{r}' \frac{n(\boldsymbol{r}')}{|\boldsymbol{r}-\boldsymbol{r}'|}}_{\text{Hartree term}} + \underbrace{\frac{\delta E_{\mathrm{xc}}^{\mathrm{LDA}}[n_{\mathrm{e}}]}{\delta n(\boldsymbol{r})}}_{v_{\mathrm{xc}}^{\mathrm{LDA}}(\boldsymbol{r})}\Bigg] . \end{aligned} \tag{2.8}$$

Again the local density and the exchange-correlation potential have to be determined self-consistently; see (2.6b) and (2.7).

At a first glance the assumptions behind the LDA seem to be rather drastic because in real systems the ion charges are fixed to the lattice points and do not provide a homogeneously charged background. To assess the quality and the limitations of the LDA we need to know the exact form of $E_{\mathrm{xc}}[n_{\mathrm{e}}]$. It can actually be derived from the exact (dynamic) pair correlation function, (1.16); see, e.g., [2.4, Chap. 3]. In fact, the pair correlation function varies strongly in the vicinity of an ion and is not smooth as in the jellium model. Hence, this local correlation function is poorly approximated by the LDA. However, only the integrated form of the pair correlation function $g_{\sigma,\sigma'}(\boldsymbol{r},\boldsymbol{r}')$ enters the calculation of the exchange energy $\epsilon_{\mathrm{xc}}(n(\boldsymbol{r}))$, which turns out to be a rather smooth function of $n(\boldsymbol{r})$. Hence, bulk quantities such as the binding energy are not too sensitive to the details of the LDA assumptions. This indicates that the LDA has a broader range of applications for ground-state properties than could be expected in the beginning [2.4, 2.7, 2.8].

Density-functional theory (and with it the LDA) only describes an equivalent one-electron substitute system with the same ground-state energy and electron density as its corresponding many-electron system. Although we may expect that ground-state properties are reasonably well described by local density approximation, we have to keep in mind that correlation and exchange are true many-body effects in an interacting electron system. Consequently, wave functions and excitation energies (band structures) may be used only if the prerequisites of Landau Fermi-liquid theory are fulfilled [2.4, 2.8, 2.11]; see Sect. 1.4. Typically, delocalized electron states are well reproduced by the LDA. For these we may start with a mean-field description of the electron–electron interaction; see the discussion in Sect. 1.4. Hence we expect the LDA to give a quantitative description of metals.

The comparison between calculated and measured binding energies and lattice constants of metals shows that an excellent agreement is obtained for alkali metals, with errors of a few percent. Within the LDA the errors in the binding energies for transition metals are of the order of about 1 eV per atom [2.5]–[2.8] which is about 10% to 20% of the experimental value. In this case we have to allow for a local electron density that also depends on the electron spin (local spin-density approximation, LSDA). This freedom in the description of the local electron density is important for the calculation of the band structure of all magnetic metals. The inclusion of magnetic effects in the framework of the LSDA yields a considerable reduction of the errors in the magnetic moments and in the bulk modulus but further *increases* the discrepancies in the binding energies. This behavior is attributed to deficiencies in the description of the inner atomic electron shells [2.6] and not to the LSDA itself.

2.2.4 Improvements to the Local Density Approximation

The LDA results for insulators are far less satisfactory in general. For example, the LDA predicts an energy gap for optical excitations of 0.3 eV for

NiO [2.12] whereas the experimental value is about 4.3 eV [2.13]. Insulators display "localized" electron states which are not well described by the LDA based on the homogeneous electron gas model. We shall comment further on this problem in Sect. 3.2.

As a first step to describe better the electrons' localized nature, we should correct for the electron self-interaction that is present in the LDA [2.14]. Various ideas to improve on the LDA in this respect are discussed in [2.8]. One possibility is to formulate the SIC-LSDA in which "self-interaction corrections" (SIC) to $E_{\mathrm{xc}}[n_{\mathrm{e}}]$ are formulated as [2.15]

$$\begin{aligned} E_{\mathrm{xc}}^{\mathrm{SIC-LSDA}}[n_{\mathrm{e}}] &= E_{\mathrm{xc}}^{\mathrm{LSDA}}[n_{\mathrm{e}}] - \sum_{\epsilon_i \leq E_{\mathrm{F}}} \left[\frac{e^2}{2} \int \mathrm{d}\boldsymbol{r}' \mathrm{d}\boldsymbol{r} \, \frac{n_i(\boldsymbol{r}) n_i(\boldsymbol{r}')}{|\boldsymbol{r} - \boldsymbol{r}'|} \right. \\ &\quad \left. + \int \mathrm{d}\boldsymbol{r} \, n_i(\boldsymbol{r}) e_{\mathrm{xc}}^{\mathrm{LSDA}} \left(n_i(\boldsymbol{r}), 0 \right) \right] . \end{aligned} \tag{2.9}$$

As a result of the correction terms the unphysical interaction of the electron with itself is subtracted from the Hartree term and the corresponding parts of the exchange-correlation term of the LSDA. The one-electron states may now localize such that we used a general one-electron index i instead of the band index $(\boldsymbol{k}, b, \sigma)$ for our notation. In the Schrödinger equation of the SIC-LSDA the local potential depends on the state i itself. The "proper" choice for the basis at the beginning of the self-consistency process thus becomes decisive for its final result since the SIC contributions are present for localized states only whereas they vanish for Bloch waves [2.15]. The SIC-LSDA gives a reasonable agreement with the measured band gaps and local moments in the transition-metal oxides. However, the SIC-LSDA tends to overestimate the local moments since the electrons truly localize in this description [2.15], as we shall further discuss in Sect. 3.2.

An alternative approach to "localized" electron states in insulators is the combination of the LDA and quantum-chemical cluster calculations. The purely electronic parameters for such a cluster calculation can be taken from the one-electron wave functions of a LDA band structure calculation [2.16]. The *intra-atomic* interaction parameters, however, are fitted to experiment, i.e., an additional local electron repulsion is explicitly incorporated to describe properly the insulating states in transition-metal oxides. This idea is further worked out in the "LDA + U" calculations proposed in [2.17].

The approaches discussed thus far include the Coulomb interaction between electrons only via some plausible ad hoc assumptions. The quality of these approximations on interacting electron systems can hardly be judged since the experimental situation is very complex, i.e., the measured quantities depend on many input parameters at the same time [2.4, 2.8]. In this complicated situation we prefer to study theoretical "standard models" with few changeable parameters to assess the quality of various approximation schemes. The Hubbard model serves as such a testing ground for many-body techniques; we will derive it in the next section.

2.3 Derivation of the Hubbard Model

In this section we derive the Hubbard model from the electronic Hamiltonian $H_{\text{electronic}}$, (2.2), after we have introduced the notation of second quantization, which is appropriate for interacting many-particle systems.

2.3.1 Second Quantization

For quantum many-particle problems it is convenient to work in the formalism of second quantization, in which the proper symmetry of the many-electron wave functions is included from the beginning in terms of commutation rules for the electron creation and annihilation operators [2.18, Chap. 1]. The creation and annihilation operators for an electron in the orbital $\phi_{\boldsymbol{k},b,\sigma}(\boldsymbol{r})$ are denoted by $\hat{c}^{+}_{\boldsymbol{k},b,\sigma}$ and $\hat{c}_{\boldsymbol{k},b,\sigma}$, respectively. They obey anti-commutation relations for fermions ($[\hat{A},\hat{B}]_{+} = \hat{A}\hat{B}+\hat{B}\hat{A}$):

$$\left[\hat{c}^{+}_{\boldsymbol{k},b,\sigma},\hat{c}^{+}_{\boldsymbol{k}',b',\sigma'}\right]_{+} = 0\,, \tag{2.10a}$$

$$\left[\hat{c}_{\boldsymbol{k},b,\sigma},\hat{c}_{\boldsymbol{k}',b',\sigma'}\right]_{+} = 0\,, \tag{2.10b}$$

$$\left[\hat{c}^{+}_{\boldsymbol{k},b,\sigma},\hat{c}_{\boldsymbol{k}',b',\sigma'}\right]_{+} = \delta_{\boldsymbol{k},\boldsymbol{k}'}\delta_{b,b'}\delta_{\sigma,\sigma'}\,. \tag{2.10c}$$

For electrons in narrow bands it is useful to work with the creation operators for Wannier electrons. They are related to the creation operators for Bloch electrons $\hat{c}^{+}_{\boldsymbol{k},b,\sigma}$ by Fourier transformation with respect to the crystal momentum $\boldsymbol{k}$ from the (first) Brillouin zone ($L = N_{\text{N}}$ is the number of lattice sites):

$$\hat{c}^{+}_{\boldsymbol{R},b,\sigma} = \sqrt{\frac{1}{L}}\sum_{\boldsymbol{k}} \exp\left(-\mathrm{i}\boldsymbol{k}\boldsymbol{R}\right)\hat{c}^{+}_{\boldsymbol{k},b,\sigma}\,. \tag{2.11}$$

$\hat{c}^{+}_{\boldsymbol{R},b,\sigma}$ creates an electron with the Wannier wave function

$$\Phi_{\boldsymbol{R},b,\sigma}(\boldsymbol{r}) = \sqrt{\frac{1}{L}}\sum_{\boldsymbol{k}} \exp\left(-\mathrm{i}\boldsymbol{k}\boldsymbol{R}\right)\phi_{\boldsymbol{k},b,\sigma}(\boldsymbol{r})\,. \tag{2.12}$$

For narrow bands [2.1, Chap. 10] the Wannier orbitals are almost identical with atomic orbitals. Hence the Wannier orbitals properly describe well-localized states, $\Phi_{\boldsymbol{R},b,\sigma}(\boldsymbol{r}) \approx 0$, if $|\boldsymbol{R}-\boldsymbol{r}| \gg a$ holds (a is the lattice constant). Despite possible hybridizations between atomic orbitals in the solid state the assignment of an atomic-level notation to bands is frequently justified, i.e., we may classify a band according to its corresponding atomic level, e.g., as "4s band".

The field operator for a σ electron ($\sigma = \uparrow,\downarrow$) at position $\boldsymbol{r}$,

$$\hat{\psi}^{+}_{\sigma}(\boldsymbol{r}) = \sum_{\boldsymbol{k},b}\phi^{*}_{\boldsymbol{k},b,\sigma}(\boldsymbol{r})\hat{c}^{+}_{\boldsymbol{k},b,\sigma} = \sum_{\boldsymbol{R},b}\Phi^{*}_{\boldsymbol{R},b,\sigma}(\boldsymbol{r})\hat{c}^{+}_{\boldsymbol{R},b,\sigma}\,, \tag{2.13}$$

creates a σ electron at the point $\boldsymbol{r}$ in real space. With these definitions the electronic Hamilton operator $\hat{H}_0$ of the Hartree band structure calculation in second quantization becomes [2.18, Chap. 1]

$$\begin{aligned}
\hat{H}_0 &= \sum_{\sigma} \int \mathrm{d}\boldsymbol{r}\, \hat{\psi}_{\sigma}^{+}(\boldsymbol{r}) H_0(\boldsymbol{r}) \hat{\psi}_{\sigma}(\boldsymbol{r}) \\
&= \sum_{\sigma} \sum_{\substack{k_1,b_1\\k_2,b_2}} \hat{c}^{+}_{\boldsymbol{k}_1,b_1,\sigma} \hat{c}_{\boldsymbol{k}_2,b_2,\sigma} \int \mathrm{d}\boldsymbol{r}\, \phi^{*}_{\boldsymbol{k}_1,b_1,\sigma}(\boldsymbol{r}) H_0(\boldsymbol{r}) \phi_{\boldsymbol{k}_2,b_2,\sigma}(\boldsymbol{r}) \qquad (2.14)\\
&= \sum_{\boldsymbol{k},b,\sigma} \epsilon(\boldsymbol{k},b) \hat{n}_{\boldsymbol{k},b,\sigma} ,
\end{aligned}$$

where we used the Schrödinger equation (2.6a) and the orthonormality of the wave functions $\phi_{\boldsymbol{k},b,\sigma}(\boldsymbol{r})$ in the last step. $\hat{n}_{\boldsymbol{k},b,\sigma} = \hat{c}^{+}_{\boldsymbol{k},b,\sigma} \hat{c}_{\boldsymbol{k},b,\sigma}$ is the occupation number operator for electrons in the state $(\boldsymbol{k}, b, \sigma)$.

With the matrix elements for the electron transfer between two lattice sites

$$t(\boldsymbol{R}-\boldsymbol{R}',b) = \frac{1}{L} \sum_{\boldsymbol{k}} \exp\left(\mathrm{i}\boldsymbol{k}\left(\boldsymbol{R}-\mathbf{R}'\right)\right) \epsilon(\boldsymbol{k},b) \qquad (2.15)$$

we may write $\hat{H}_0$ as

$$\hat{H}_0 = \sum_{\boldsymbol{R}_1,\boldsymbol{R}_2,b,\sigma} t(\boldsymbol{R}_1-\boldsymbol{R}_2,b) \hat{c}^{+}_{\boldsymbol{R}_1,b,\sigma} \hat{c}_{\boldsymbol{R}_2,b,\sigma} . \qquad (2.16)$$

With the help of the matrix elements for the Coulomb interaction

$$\begin{aligned}
&V\Big((\boldsymbol{R}_1,b_1,\sigma);(\boldsymbol{R}_2,b_2,\sigma')\Big|(\boldsymbol{R}_3,b_3,\sigma');(\boldsymbol{R}_4,b_4,\sigma)\Big) \qquad (2.17)\\
&\quad = e^2 \int \mathrm{d}\boldsymbol{r}\mathrm{d}\boldsymbol{r}' \frac{\Phi^{*}_{\boldsymbol{R}_1,b_1,\sigma}(\boldsymbol{r}) \Phi^{*}_{\boldsymbol{R}_2,b_2,\sigma'}(\boldsymbol{r}') \Phi_{\boldsymbol{R}_3,b_3,\sigma'}(\boldsymbol{r}') \Phi_{\boldsymbol{R}_4,b_4,\sigma}(\boldsymbol{r})}{|\boldsymbol{r}-\boldsymbol{r}'|}
\end{aligned}$$

we may write the electron–electron interaction in second quantization:

$$\begin{aligned}
\hat{V} =& \hat{V}_{\mathrm{e-e}} - \hat{V}^{\mathrm{II}}_{\mathrm{e-e}} \\
=& \frac{1}{2} \sum_{\sigma,\sigma'} \sum_{\substack{R_1,R_2,R_3,R_4\\b_1,b_2,b_3,b_4}} V\left((\boldsymbol{R}_1,b_1,\sigma);(\boldsymbol{R}_2,b_2,\sigma')\middle|(\boldsymbol{R}_3,b_3,\sigma');(\boldsymbol{R}_4,b_4,\sigma)\right) \\
& \times \Big[\hat{c}^{+}_{\boldsymbol{R}_1,b_1,\sigma} \hat{c}^{+}_{\boldsymbol{R}_2,b_2,\sigma'} \hat{c}_{\boldsymbol{R}_3,b_3,\sigma'} \hat{c}_{\boldsymbol{R}_4,b_4,\sigma} \\
& - \langle \hat{c}^{+}_{\boldsymbol{R}_1,b_1,\sigma} \hat{c}_{\boldsymbol{R}_4,b_4,\sigma} \rangle \hat{c}^{+}_{\boldsymbol{R}_2,b_2,\sigma'} \hat{c}_{\boldsymbol{R}_3,b_3,\sigma'} \qquad (2.18)\\
& - \hat{c}^{+}_{\boldsymbol{R}_1,b_1,\sigma} \hat{c}_{\boldsymbol{R}_4,b_4,\sigma} \langle \hat{c}^{+}_{\boldsymbol{R}_2,b_2,\sigma'} \hat{c}_{\boldsymbol{R}_3,b_3,\sigma'} \rangle \\
& + \langle \hat{c}^{+}_{\boldsymbol{R}_1,b_1,\sigma} \hat{c}_{\boldsymbol{R}_4,b_4,\sigma} \rangle \langle \hat{c}^{+}_{\boldsymbol{R}_2,b_2,\sigma'} \hat{c}_{\boldsymbol{R}_3,b_3,\sigma'} \rangle \Big] ,
\end{aligned}$$

which we had to correct for the Hartree terms already included in $\hat{H}_0$. Note that there are no Hartree corrections for $(\boldsymbol{R}_1, b_1, \sigma) = (\boldsymbol{R}_2, b_2, \sigma')$ and $(\boldsymbol{R}_3, b_3, \sigma') = (\boldsymbol{R}_4, b_4, \sigma)$. Although the kinetic energy operator is diagonal in the band index we must, at least in principle, include all inter-band interactions in the potential energy operator.

2.3.2 Hubbard Model

The derivation below follows [2.19]–[2.21]. An early review on the Hubbard model is provided by Cyrot [2.22]; see Long [2.23] for a more recent overview.

The position of the bands in terms of energy relative to each other is determined by $t(\mathbf{0}, b) = (1/L)\sum_{\boldsymbol{k}} \epsilon(\boldsymbol{k}, b)$. This quantity contains not only effects of the electron–ion interactions but also the influence of crystal fields of the ligands via the Hartree contributions. According to the Hartree band structure calculations we should find only few partly filled bands at the Fermi energy, $t(\mathbf{0}, b) \approx E_{\mathrm{F}}$. Bands far above or below the Fermi energy will be only slightly modified by the electron–electron interaction $\hat{V}$ since for them $V\big((\boldsymbol{R}_1, b_1, \sigma); (\boldsymbol{R}_2, b_2, \sigma')\big|(\boldsymbol{R}_3, b_3, \sigma'); (\boldsymbol{R}_4, b_4, \sigma)\big)/\big(E_{\mathrm{F}} - t(\mathbf{0}, b)\big)$ should be a small parameter. This assumption relies on the fact that the electron–ion interaction dominates the electron–electron interaction; see Sect. 2.2. In the ideal case there is only *one* band at E_{F} and all other bands are energetically far from the Fermi energy. This is Hubbard's first assumption [2.20].

Hubbard's second assumption is that only the maximal term in $\hat{V}$ is taken into account. This term stems from the intra-atomic Coulomb interaction. If we have to take into account only a single band at the Fermi energy the intra-atomic contribution ("Hubbard U") reads

$$\begin{aligned} U &= V\big((\boldsymbol{R}, \sigma); (\boldsymbol{R}, \sigma')\big|(\boldsymbol{R}, \sigma'); (\boldsymbol{R}, \sigma)\big) \\ &= e^2 \int \mathrm{d}\boldsymbol{r}\mathrm{d}\boldsymbol{r}' \, \frac{|\Phi_{\boldsymbol{R},\sigma}(\boldsymbol{r})|^2 \, |\Phi_{\boldsymbol{R},\sigma'}(\boldsymbol{r}')|^2}{|\boldsymbol{r} - \boldsymbol{r}'|} \,, \end{aligned} \tag{2.19}$$

which is the same on all lattice sites. The Hartree orbitals of Sect. 2.2.2, which are the basis for our considerations, are calculated for a translationally and spin-rotationally invariant system. Thus we may use $\langle \hat{c}^{+}_{\boldsymbol{R},\sigma} \hat{c}_{\boldsymbol{R},\sigma} \rangle = n/2$ as further simplification.

If we neglect all Coulomb matrix elements in $\hat{V}$ except for the on-site term, according to Hubbard's second assumption we obtain the Hamilton operator for the Hubbard model,

$$\hat{H} = \sum_{\boldsymbol{R}_1, \boldsymbol{R}_2, \sigma} t(\boldsymbol{R}_1 - \boldsymbol{R}_2) \hat{c}^{+}_{\boldsymbol{R}_1,\sigma} \hat{c}_{\boldsymbol{R}_2,\sigma} + U \sum_{\boldsymbol{R}} (\hat{n}_{\boldsymbol{R},\uparrow} - n/2)\,(\hat{n}_{\boldsymbol{R},\downarrow} - n/2) \ . \tag{2.20a}$$

Often the zero energy is set to the "center of gravity" of the band in the absence of electron–electron interactions, i.e., $t(\mathbf{0}) - Un = 0$, and the constant energy term $Un^2/4$ is dropped from the considerations. This frequently-used alternative formulation of the Hubbard Hamiltonian reads

$$\begin{aligned}\hat{H} &= \sum_{\boldsymbol{R}_1 \neq \boldsymbol{R}_2, \sigma} t(\boldsymbol{R}_1 - \boldsymbol{R}_2) \hat{c}^+_{\boldsymbol{R}_1,\sigma} \hat{c}_{\boldsymbol{R}_2,\sigma} + U \sum_{\boldsymbol{R}} \hat{n}_{\boldsymbol{R},\uparrow} \hat{n}_{\boldsymbol{R},\downarrow} \\ &= \hat{T} + U\hat{D} ,\end{aligned} \tag{2.20b}$$

where we introduced the operator for the number of double occupancies $\hat{D} = \sum_{\boldsymbol{R}} \hat{n}_{\boldsymbol{R},\uparrow} \hat{n}_{\boldsymbol{R},\downarrow}$.

The Hubbard model in its above form, (2.20b), was independently introduced and used by Gutzwiller [2.24] and Kanamori [2.25] at about the same time to study the possibility for ferromagnetism in transition metals. We briefly mention the results of Kanamori's T-matrix method in Sect. 2.5.5, and more extensively discuss Gutzwiller's variational approach in Sect. 3.4.

Even before the Hubbard model was "officially" introduced by Hubbard, Gutzwiller, and Kanamori the importance of the local interaction terms had been recognized. For example, it is implicitly contained in Mott's discussion on the origin of the gap in Nickel-oxide (NiO) [2.26]. In fact, the controversy between "itinerant" versus "local" electron properties has a long tradition in the discussion for the appropriate description of solid-state magnetism [2.27]. Also the explanation of the itinerant exchange and super-exchange mechanism in magnetic insulators by Anderson [2.19] is based on the Hubbard Hamiltonian that was explicitly formulated only years later.

2.4 Hubbard Model and Real Materials

In this section we briefly show that the approximations that lead to the Hubbard model are too drastic to allow for the description of real materials. We indicate those minimal extensions necessary for a qualitatively correct description of polymers, transition metals and their oxides, fullerenes (C_{60}), high-temperature superconductors, heavy fermion substances, and liquid Helium. The importance of the Hubbard model itself lies in the fact that it is the conceptually simplest and thus the standard model of correlated electron systems.

2.4.1 Polymers

Models with a variable local interaction U and long-range Coulomb interactions were introduced and used a long time ago in physical chemistry by Pariser, Parr, Pople, and others for the description of the electron–electron interaction in polymers (PPP model) [2.28]. However, in the original PPP model the kinetic energy term of the Hubbard model is absent.

For polymers Hubbard's first assumption is justified. As an example we consider the simplest conceivable polymer, trans-polyacetylene $(CH)_x$ [2.29].

Three of the four carbon $2s^1p^3$ orbitals are sp^2 hybridized and form σ bonds to nearest-neighbor carbon atoms and a hydrogen atom. The remaining p^z electrons form molecular π orbitals along the chain, i.e., a one-dimensional half-filled band. Polymers thus are important examples for (one-dimensional) correlated electron systems. Since an (almost ideal) one-dimensional system suffers a Peierls distortion in the ground state the Hubbard model must be extended to the Peierls–Hubbard model which allows for the coupling between electrons and lattice distortions [2.30]. Electron correlations are found to strengthen the Peierls distortion of the ground state [2.31, 2.32] and strongly influence the excitation spectrum of polymers.

A more serious defect of the Hubbard model is the neglect of all long-range Coulomb interactions such that Hubbard's second assumption is not justified in polymers. Excitons as observed in these materials [2.29] are an indication that long-range Coulomb interactions are crucial to describe polymers properly. Unfortunately, it is not so easy to estimate the size of the terms that have been neglected in the Hubbard model. The choice of the one-electron Hamiltonian is decisive for an estimate of the size of the correction terms. For example, the inclusion of the Hartree contributions in $\hat{H}_0$ reduces the values for the nearest-neighbor interaction as compared to the value obtained from standard tight-binding orbitals; for a discussion, see [2.33] and [2.34].

Hubbard [2.20] used weakly overlapping hydrogen wave functions to obtain the following estimates for the interaction matrix elements ($\boldsymbol{\tau}$ is a nearest-neighbor vector):

$$U = V\left((\boldsymbol{R},\sigma);(\boldsymbol{R},\sigma')\big|(\boldsymbol{R},\sigma');(\boldsymbol{R},\sigma)\right) \approx 20\text{eV}\ , \tag{2.21a}$$

$$V = V\left((\boldsymbol{R},\sigma);(\boldsymbol{R}+\boldsymbol{\tau},\sigma')\big|(\boldsymbol{R}+\boldsymbol{\tau},\sigma');(\boldsymbol{R},\sigma)\right) \approx 6\text{eV}\ , \tag{2.21b}$$

$$X = V\left((\boldsymbol{R},\sigma);(\boldsymbol{R},\sigma')\big|(\boldsymbol{R}+\boldsymbol{\tau},\sigma');(\boldsymbol{R},\sigma)\right) \approx 0.5\text{eV}\ , \tag{2.21c}$$

$$\begin{aligned} W &= V\left((\boldsymbol{R},\sigma);(\boldsymbol{R},\sigma')\big|(\boldsymbol{R}+\boldsymbol{\tau},\sigma');(\boldsymbol{R}+\boldsymbol{\tau},\sigma)\right) \\ &= V\left((\boldsymbol{R},\sigma);(\boldsymbol{R}+\boldsymbol{\tau},\sigma')\big|(\boldsymbol{R},\sigma');(\boldsymbol{R}+\boldsymbol{\tau},\sigma)\right) \approx 0.025\text{eV}\ , \end{aligned} \tag{2.21d}$$

all other matrix elements being exponentially small. For small bandwidths it is thus seen that the Hubbard interaction dominates, but we also find that the nearest-neighbor interaction is not negligibly small, especially because we still have to take the number of nearest neighbors into account. Even if the Hubbard X and W terms are small, the contribution of the V terms is important [2.33, 2.34]. If the bandwidth is large compared to the Coulomb interaction, we may safely neglect the Hubbard X and W terms but we have to include the long-range parts of the Coulomb repulsion [2.28, 2.32, 2.35].

Any realistic model for polymers has to include a term for a local and at least a nearest-neighbor interaction. This also applies for other quasi one-dimensional conductors such as the charge-transfer salts or metal–halide compounds; a review is given in [2.36].

2.4.2 Transition Metals and Their Oxides

In the transition metals with their narrow d bands the Wannier orbitals are well localized and the hopping and Coulomb matrix elements decay fast with distance. Hence, Hubbard's second assumption is well satisfied. Since other, e.g., s bands are close in energy and the system thus has a large polarizability, the intra-atomic Coulomb interaction between the d electrons is strongly reduced. As a consequence of this effect of strong screening by other band electrons the Hubbard U_d is reduced to a value of $U_d \approx 2$–$3\,\mathrm{eV}$ and thus is only half as big as the bandwidth [2.37, 2.38]. Electron correlations due to the Hubbard interaction do not play a decisive role in these metals.

In transition metals, however, Hubbard's first assumption of a single band is not justified. Hence we have to consider the many-band problem of (2.14). In real materials with considerable bandwidths those interaction matrix elements are small that are non-diagonal in position space, like the Hubbard W and X terms. In addition, interaction terms which are non-diagonal in the band index are also considerably smaller than their diagonal counterparts. To a good approximation we may thus consider the band index b as a good quantum number. This allows us to restrict ourselves to the two cases $b_1 = b_4$, $b_2 = b_3$ (direct term $U_{b,b'}^{\sigma,\sigma'}$) and $b_1 = b_3$, $b_2 = b_4$ (exchange term $J_{b,b'}^{\sigma,\sigma'}$, $b \neq b'$). The maximal terms in $\hat{V}$ still result from the the intra-atomic Coulomb interaction

$$U_{b,b'}^{\sigma,\sigma'} = e^2 \int \mathrm{d}\boldsymbol{r}\mathrm{d}\boldsymbol{r}' \frac{|\Phi_{\boldsymbol{R},b,\sigma}(\boldsymbol{r})|^2 \, |\Phi_{\boldsymbol{R},b',\sigma'}(\boldsymbol{r}')|^2}{|\boldsymbol{r}-\boldsymbol{r}'|} \,, \tag{2.22a}$$

$$J_{b,b'}^{\sigma,\sigma'} = e^2 \int \mathrm{d}\boldsymbol{r}\mathrm{d}\boldsymbol{r}' \frac{\Phi^*_{\boldsymbol{R},b,\sigma}(\boldsymbol{r})\Phi^*_{\boldsymbol{R},b',\sigma'}(\boldsymbol{r}')\Phi_{\boldsymbol{R},b,\sigma'}(\boldsymbol{r}')\Phi_{\boldsymbol{R},b',\sigma}(\boldsymbol{r})}{|\boldsymbol{r}-\boldsymbol{r}'|} \,. \tag{2.22b}$$

The terms $J_{b,b'}^{\sigma,\sigma'} > 0$ lift the spin degeneracy at the atom and maximize the atomic spin (Hund's first rule) [2.19]. In the crystal the Hund's rule coupling introduces strong correlations between the electrons in different bands and ultimately leads to a (ferro-)magnetic metallic ground state. A brief review on theories for magnetic metals can be found in [2.39, 2.40].

Hund's rule coupling also plays an important role in transition-metal oxides; see [2.41, 2.42] for a recent review. Vanadium-sesquioxide (V_2O_3, Sect. 1.4), the most prominent example for a metal–Mott–Hubbard insulator transition, displays a local moment both in its metallic and in its insulating states. The peculiar spin correlations in the paramagnetic phases and the antiferromagnetic insulator indicate that a realistic description of the magnetic (and conducting) properties of V_2O_3 requires a two-band model with a ferromagnetic Hund's rule coupling $J_{b,b'}^{\sigma,\sigma'}$ [2.43]–[2.46].

In general, we need to include several bands in a realistic description of transition-metal oxides, sulfides, and halides like Nickel-iodine (NiI_2), and their corresponding metal–Mott insulator transitions. All these materials are charge-transfer insulators; see Sect. 1.4. For this class of materials we should formulate a theory based on a multi-band rather than a one-band Hubbard model description.

2.4.3 Fullerenes

Fullerenes are a new class of materials based on carbon. The prototype is C_{60}, a soccer-ball shaped molecule made out of twenty carbon hexagons and twelve pentagons. As in the case of polymers the backbone of the molecule is stabilized by the σ bonds between the carbon atoms and there is an unbound p^z electron for every carbon [2.47]. This π electron cloud is polarizable and allows the van-der-Waals interaction to condense the C_{60} balls into a stable lattice (fcc structure at low temperatures).

In addition to their potential technological applications the fullerenes display fascinating electronic properties; in particular, the alkali-doped compounds like K_3C_{60} become superconducting at unusually high temperatures ($T_c \gtrsim 30\,\mathrm{K}$). From early on [2.48] it was suspected that electron correlations on the C_{60} balls are important for an understanding of the material. This point of view was supported by an analysis of spectroscopic data on solid C_{60} [2.49]. In an elementary tight-binding (Hückel) band structure theory the 60 π electrons per ball result in a closed-shell situation with a gap between the highest, five-fold orbitally degenerate occupied (h_u) and the lowest, three-fold degenerate unoccupied levels (t_{1u}). In [2.49] it is argued that the observed optical gap is considerably larger than the gap predicted from a non-interacting electron approach. In fact, the intra-ball Hubbard repulsion ($U \approx 1.6\,\mathrm{eV}$) was estimated to be four times as large as the t_{1u} bandwidth ($W \approx 0.4\,\mathrm{eV}$) as a result of inter-ball tunneling. Hence, solid C_{60} could be a strongly correlated material.

To set up an appropriate Hamilton operator for solid C_{60} we follow [2.50] and focus on the doped materials (e.g., K_3C_{60}) for notational simplicity. Various conceptually similar Hamiltonians have been proposed in [2.48]–[2.51]. A single C_{60} molecule is the host for up to six electrons in the t_{1u} band at bare energy $\epsilon_{t_{1u}}$ with a pairwise Hubbard interaction U:

$$\begin{aligned}\hat{H} &= \sum_{\boldsymbol{R},\alpha,\sigma} \epsilon_{t_{1u}} \hat{n}_{\boldsymbol{R},\alpha,\sigma} + \sum_{\boldsymbol{R},\boldsymbol{T},\alpha,\beta,\sigma} t_{\alpha,\beta}\, \hat{c}^+_{\boldsymbol{R}+\boldsymbol{T},\alpha,\sigma} \hat{c}_{\boldsymbol{R},\beta,\sigma} \\ &\quad + U \sum_{(\alpha,\sigma)<(\beta,\sigma')} \hat{n}_{\boldsymbol{R},\alpha,\sigma} \hat{n}_{\boldsymbol{R},\beta,\sigma'} \,. \end{aligned} \tag{2.23}$$

In principle, long-range Coulomb terms should also be taken into account.

The essential difference from other multi-band models is the lack of the Hund's rule terms of (2.22b). Since the diameter of the balls is much larger than an atomic size, Pauli principle (exchange) effects on the balls are much weaker than in atoms. Nevertheless, a multi-band description is still mandatory, and a simple one-band Hubbard model approach is not sufficient.

It is still a matter of ongoing debate whether a strong correlation view is indeed necessary. Elaborate band structure calculations predict an optical gap of the correct size [2.52], and NMR Knight shift measurements in the superconducting state could consistently be explained within BCS theory [2.53]. It remains to be seen which viewpoint will prevail for the fullerenes.

2.4.4 High-Temperature Superconductors

High-temperature superconductors with their characteristic copper-oxide layers also belong to the class of correlated electron systems. Like in lanthanum copper-oxide ($LaCuO_4$) the undoped layers are antiferromagnetic insulators with a high Néel temperature, a typical indication for strong electron correlations [2.11, 2.54]. Therefore, we consider it rather likely that electron correlations will play an important role in the microscopic mechanism of high-temperature superconductivity.

The copper-oxide planes in the high-temperature superconductors cannot be properly described by the Hubbard model. For these materials the charge-transfer energy between the copper and oxygen levels is too small to allow for a realistic description by a one-band model. In this case it seems advisable to start from a three-band model of one copper and two oxygen levels per unit cell [2.54] ("Emery model" or "Varma–Schmitt-Rink–Abrahams model"):

$$\begin{aligned}
\hat{H} \;=\; & U_d \sum_{\boldsymbol{R}} \hat{d}^{+}_{\boldsymbol{R},\uparrow}\hat{d}_{\boldsymbol{R},\uparrow}\hat{d}^{+}_{\boldsymbol{R},\downarrow}\hat{d}_{\boldsymbol{R},\downarrow} + \epsilon_d \sum_{\boldsymbol{R},\sigma} \hat{d}^{+}_{\boldsymbol{R},\sigma}\hat{d}_{\boldsymbol{R},\sigma} \\
& + U_p \sum_{\boldsymbol{R}',\alpha} \hat{p}^{+}_{\boldsymbol{R}',\uparrow,\alpha}\hat{p}_{\boldsymbol{R}',\uparrow,\alpha}\hat{p}^{+}_{\boldsymbol{R}',\downarrow,\alpha}\hat{p}_{\boldsymbol{R}',\downarrow,\alpha} + \epsilon_p \sum_{\boldsymbol{R}',\sigma,\alpha} \hat{p}^{+}_{\boldsymbol{R}',\sigma,\alpha}\hat{p}_{\boldsymbol{R}',\sigma,\alpha} \\
& + V \sum_{\langle \boldsymbol{R},\boldsymbol{R}'\rangle,\sigma,\sigma',\alpha} \hat{d}^{+}_{\boldsymbol{R},\sigma}\hat{d}_{\boldsymbol{R},\sigma}\hat{p}^{+}_{\boldsymbol{R}',\sigma',\alpha}\hat{p}_{\boldsymbol{R}',\sigma',\alpha} \qquad (2.24) \\
& - t_{pd} \sum_{\langle \boldsymbol{R},\boldsymbol{R}'\rangle,\sigma,\alpha} \left(\hat{d}^{+}_{\boldsymbol{R},\sigma}\hat{p}_{\boldsymbol{R}',\sigma,\alpha} + \hat{p}^{+}_{\boldsymbol{R}',\sigma,\alpha}\hat{d}_{\boldsymbol{R},\sigma} \right) \\
& - t_{pp} \sum_{\langle \boldsymbol{R}',\boldsymbol{R}''\rangle,\sigma,\alpha} \hat{p}^{+}_{\boldsymbol{R}',\sigma,\alpha}\hat{p}_{\boldsymbol{R}'',\sigma,\alpha} \, .
\end{aligned}$$

In this model we have not only a Hubbard U_d between the copper d electrons at energy ϵ_d on the lattice sites $\boldsymbol{R}$ of the square lattice but also a Hubbard U_p for the oxygen p electrons with energy levels ϵ_p on the lattice sites $\boldsymbol{R}'$ ($\boldsymbol{R}''$) on the lines between the square latticed points $\boldsymbol{R}$ in the $\alpha = x, y$ directions. In addition, we allow for a nearest-neighbor interaction V between the copper and oxygen electrons. The nearest-neighbor copper-oxygen and the oxygen-oxygen orbitals, indicated by the notation $\langle \ldots \rangle$, may hybridize with each other via the matrix elements (t_{pd}, t_{pp}) which allow for electron transport.

Band structure calculations indicate that the Hubbard interaction on the d sites with $U_d \approx 10\,\mathrm{eV}$ dominates over the interaction between the p electrons, $U_p \approx 4\,\text{eV}$, and that U_d is at least of the order of or even larger than the bandwidth [2.55]. Extensive Monte-Carlo calculations to the three-band Hubbard model essentially confirm the trends in the band structure parameters where U_p turns out to be a rather insensitive parameter [2.56]. These model calculations confirm that a three-band model is indeed the minimum model if we want to reproduce the results of the spectroscopic measurements [2.57]. At least on the energy scale of $\mathcal{O}(1\,\text{eV})$ a mapping to an effective one-band model is not possible. It is conceivable, though, that the low-energy physics of the superconductor itself could be well approximated by some effective one-band theory [2.58, 2.59]. In general, however, a multi-band description is required.

2.4.5 Heavy Fermion Materials

Further typical correlated electron systems are rare earth compounds such as $CeAl_3$ and actinide compounds such as UPt_3. Their metallic low-temperature state displays a linear specific heat coefficient that is a factor of $10^2 \dots 10^3$ larger than in ordinary metals. This corresponds to a very high electronic density of states at the Fermi level or a very large effective mass ("heavy fermions"). Other physical properties further confirm the existence of strong correlations in these materials; for a review, see [2.60, 2.61].

The physics of these materials is again dominated by the local repulsion of electrons, this time the f electrons of the partly filled $4f$ or $5f$ shell are strongly correlated. By a small but relevant hybridization of the f electrons with the conduction band electrons the f electron levels are effectively shifted to the Fermi energy. As a result of their dominantly localized nature this contribution of the f electrons gives rise to an enormous effective mass of the charge carriers at the Fermi energy. This cooperative effect appears only at small temperatures. At high temperatures the f electrons behave like localized, uncoupled electron spins in partly filled atomic shells.

The corresponding model that describes this situation is the "periodic Anderson model" or "Varma–Yafet model" [2.62]:

$$\begin{aligned}\hat{H}_{\text{PAM}} &= \sum_{\boldsymbol{k},\alpha} \epsilon(\boldsymbol{k}) \hat{c}^{+}_{\boldsymbol{k},\alpha} \hat{c}_{\boldsymbol{k},\alpha} + \epsilon_f \sum_{\boldsymbol{R},\alpha} \hat{n}^f_{\boldsymbol{R},\alpha} + \frac{U}{2} \sum_{\boldsymbol{R}} \sum_{\alpha \neq \beta} \hat{n}^f_{\boldsymbol{R},\alpha} \hat{n}^f_{\boldsymbol{R},\beta} \\ &\quad - \sqrt{\frac{1}{N_f L}} \sum_{\boldsymbol{k},\boldsymbol{R},\alpha} V_{\boldsymbol{k}} \left[\exp\left(-\mathrm{i}\boldsymbol{k}\boldsymbol{R}\right) \hat{c}^{+}_{\boldsymbol{k},\alpha} \hat{f}_{\boldsymbol{R},\alpha} + \text{h.c.} \right] . \end{aligned} \tag{2.25}$$

Here, $\hat{c}_{\boldsymbol{k},\alpha}$ denote the conduction band electrons with dispersion $\epsilon(\boldsymbol{k})$ and $\hat{f}_{\boldsymbol{R},\alpha}$ describe the f electrons whose local interaction is modeled by a Hubbard term. As a simplification we assumed that the N_f-fold orbital degeneracy ($\alpha, \beta = 1, \dots, N_f$) will not be lifted by the hybridization $V_{\boldsymbol{k}}$.

The model given in (2.25) is the generalization of the (Fano–)Anderson model which describes a single impurity with a f electron in a host metal [2.63] and is also used for the description of "valence fluctuations". Similar to the Hubbard model both the periodic and the single-impurity Anderson models reduce to two other frequently studied models in the limit of strong interactions ($U \to \infty$) [2.64]. These models are the Kondo and the Kondo lattice model [2.65], which describe the interaction of magnetic spins with conduction electrons. Further details on these correlated electron systems may be found in [2.61]. Again, the Hubbard interaction is seen to be an important ingredient in a model Hamilton operator for a real system. However, a two-band model is minimally required for a realistic description of heavy fermion materials.

2.4.6 Liquid Helium

Owing to its two protons and one neutron in its nucleus, ^{3}He ("Helium-3") is a fermion. Furthermore, the electron charge distribution is spherical symmetric and hardly polarizable due to the closed 1s shell. Consequently, there is only a weak van-der-Waals attraction between the atoms. In addition, ^{3}He is a rather light element such that the zero-point motion is large. For these reasons ^{3}He solidifies only under a pressure of about 34 bar at temperatures below 1 K. ^{3}He thus is best characterized as a quantum liquid that displays additional, most interesting suprafluid properties at temperatures below 1 mK; for a review, see [2.66].

Normal fluid Helium-3 is among the few physical systems that almost entirely fulfill the requirements of the Hubbard model. The particles experience a strong Coulomb repulsion at small distances because of the shell electrons, which results in an exchange and correlation hole in the pair correlation function of the size of the atomic diameter. At low temperatures the average particle distance is only slightly larger than the diameter of a single atom. If we *interpret* the average particle distance as the radius of a Wigner–Seitz cell we obtain a lattice gas model for this correlated Fermi system [2.67] in which the fermions are restricted to a single band (Hubbard's first assumption). If we further ignore the effects of the attractive interaction beyond the formation of the "lattice", only the intra-cell repulsion, i.e., a purely local interaction remains such that Hubbard's second assumption also holds. Indeed it could be shown [2.67] that the Hubbard model can successfully be applied to liquid ^{3}He as an "almost localized Fermi liquid".

Analogously there are studies of the Hubbard model for *bosons*, which might be applied to liquid, normal fluid ^{4}He. Although we cannot argue that the Hubbard model really describes normal fluid or suprafluid ^{4}He we may use it as a simple reference model for the comparison between various numerical techniques [2.68] as well as for basic considerations on the transition between the suprafluid and the (Mott) insulating state; see [2.69].

In the case of liquid Helium the problem with a Hubbard model description is the lack of an underlying fixed lattice structure. We may hope, however, that this assumption might not be too drastic for global ground-state properties like the compressibility or magnetic susceptibility.

2.4.7 Relevance of the Hubbard Model

As outlined above the Hubbard interaction is an important ingredient in many models for real materials. We cannot expect, though, that the Hubbard model itself properly describes an experimental situation. On the other hand the model principally contains all necessary terms to describe the basic properties of correlated electron systems. The kinetic energy operator reflects the electrons' itinerant features (metallic conductivity) and the interaction operator forces the electrons' correlated motion (correlated metal) or even their "localization" (insulator). Despite some of its unrealistic simplifications the model should allow us to understand qualitatively the basic properties of correlated Fermi systems, such as the metal–Mott insulator transition.

In this qualitative sense the Hubbard model serves as "the" standard model of correlated electron systems, and has the same conceptional importance for interacting electrons as the Ising model for classical statistical mechanics. The low number of explicit and implicit parameters provides the ideal condition for a thorough test of the power and quality of analytical and numerical methods. In this book we choose the issue of the metal–Mott insulator transition to critically examine the usefulness of the model and the applicability of approximate methods.

2.5 Model Properties

In this section we first tabulate the explicit and implicit variables of the Hubbard model. Next we classify various common variants of it. Then we discuss its global symmetries and some of its basic properties, in particular its potential to describe the Mott transition. At the end of this section we collect some exact statements on magnetism in the Hubbard model and its variants.

2.5.1 Model Parameters

To facilitate our discussion we repeat the Hamilton operator for the Hubbard model that we derived in the third section:

$$\hat{H} = \sum_{\boldsymbol{R}_1 \neq \boldsymbol{R}_2, \sigma} t(\boldsymbol{R}_1 - \boldsymbol{R}_2) \hat{c}^{+}_{\boldsymbol{R}_1,\sigma} \hat{c}_{\boldsymbol{R}_2,\sigma} + U \sum_{\boldsymbol{R}} \hat{n}_{\boldsymbol{R},\uparrow} \hat{n}_{\boldsymbol{R},\downarrow} = \hat{T} + U\hat{D} \,. \quad (2.26)$$

In the following we use the bandwidth W of the one-electron band with dispersion relation $\epsilon(\boldsymbol{k})$, (2.15), as our energy scale.

Explicit model parameters obviously are

1. the interaction strength U/W;
2. the dimension d ($d = 1, 2, 3, \ldots$) and the lattice structure;
3. the hopping matrix elements $t(\boldsymbol{R}_1 - \boldsymbol{R}_2)$ and the corresponding one-electron dispersion relation $\epsilon(\boldsymbol{k})$. On the derivation of the Hubbard model we employed well-localized Wannier orbitals such that we should have

$$t(\boldsymbol{R}_1 - \boldsymbol{R}_2) = \begin{cases} -t & , \quad \text{if } \boldsymbol{R}_1 - \boldsymbol{R}_2 = \boldsymbol{\tau} \\ 0 & , \quad \text{otherwise} \end{cases} \tag{2.27}$$

   for the standard Hubbard model, where $\boldsymbol{\tau}$ denotes all vectors to the nearest neighbors on the lattice.

In addition we have implicit parameters in the problem, namely:

1. the density of σ electrons ($\sigma =\uparrow, \downarrow$), $n_\sigma = N_\sigma / L$, where $L = N_{\mathrm{N}}$ denotes the number of sites (nuclei or ions) and $N \equiv N_{\mathrm{e}} = N_\uparrow + N_\downarrow$ is the total number of electrons;
2. the temperature $k_{\mathrm{B}}T/W$.

The convention $t > 0$ in (2.27) guarantees that the dispersion relation displays a minimum at the Γ point for almost all lattice types. In addition, we will be interested in repulsive interactions and thus assume $U \geq 0$, if not specified otherwise.

The model defined in this manner will be called "the" or the "standard" Hubbard model in the following.

2.5.2 Model Variants

Frequently the operators in the Hubbard model are modified and/or new interactions are added. These changes are made to describe more realistic situations or, in most cases, to find exactly solvable models. A model will still be assigned to the class of Hubbard models in this book if it is a *one-band model* whose interaction term includes a dominant local part, i.e., the Hubbard interaction $U\hat{D}$. This does not imply that the physics described by these models has to be identical or only similar to the standard Hubbard model even though this is often hoped for. *Multi-band models*, which are used to study the high-temperature superconductors (Emery model [2.54]) or heavy fermion materials (periodic Anderson model [2.61]), will not be subsumed under the class of Hubbard-type models even though the Hubbard interaction is a central part of these models.

The variations on the Hubbard model can be cast into three categories.

1. "Modified" Hubbard models with non-zero hopping matrix elements beyond nearest-neighbor distances. Here, the electron transfer is no longer determined by (2.27). Examples are the one-dimensional $1/r$ and the $1/\sinh r$ Hubbard model which are discussed in Sect. 4.2 and 4.3, respectively.

2. "Simplified" Hubbard models in which one electron species is localized, e.g., $t_\downarrow \equiv 0$. The main representative of this category is the Falicov–Kimball model [2.70] which is briefly discussed in Sect. 5.4.
3. "Extended" Hubbard models in which further interaction terms have been added to the standard Hubbard model. Results for the Hubbard model with additional nearest-neighbor interactions and bond-charge interactions can be found in Sects. 6.1 and 6.2.

These three basic categories cover a multitude of model variants whose construction allows for exact statements or even exact solutions. Examples are models where the number of double occupancies is conserved, see Sect. 6.2, and models for which the kinetic and potential energy operators commute, see Sect. 6.3.

2.5.3 Symmetries

For the one-dimensional Hubbard model infinitely many conserved quantities exist, which proves its integrability [2.71]. Even 25 years after the Bethe-ansatz solution new symmetries could still be found [2.72]. We discuss one-dimensional (modified) Hubbard models in more detail in Chap. 4.

In higher dimensions the model is not integrable and only a few global symmetries are known to exist for the Hubbard model in dimensions $d > 1$. To carry out our symmetry considerations it is convenient in some cases to supply the Hamilton operator with an additional term which sets the chemical potential to zero at half band-filling. We thus prefer to study the Hubbard model in the form

$$\begin{aligned} \hat{H}' &= \hat{H} - U\hat{N}/2 + UL/4 \qquad (2.28) \\ &= \hat{T} + U\sum_{\boldsymbol{R}}\left(\hat{n}_{\boldsymbol{R},\uparrow} - \frac{1}{2}\right)\left(\hat{n}_{\boldsymbol{R},\downarrow} - \frac{1}{2}\right) , \end{aligned}$$

instead of $\hat{H}$ in (2.26).

Spin-Rotational Invariance. $\hat{H}'$ is invariant against global spin rotations whose generators form the $SU(2)$ spin algebra,

$$\hat{S}^z = \sum_{\boldsymbol{R}} \left[\hat{n}_{\boldsymbol{R},\uparrow} - \hat{n}_{\boldsymbol{R},\downarrow}\right]/2 , \qquad (2.29a)$$

$$\hat{S}^+ = \sum_{\boldsymbol{R}} \hat{c}^+_{\boldsymbol{R},\uparrow}\hat{c}_{\boldsymbol{R},\downarrow} , \qquad (2.29b)$$

$$\hat{S}^- = \sum_{\boldsymbol{R}} \hat{c}^+_{\boldsymbol{R},\downarrow}\hat{c}_{\boldsymbol{R},\uparrow} = \left(\hat{S}^+\right)^+ , \qquad (2.29c)$$

where $\hat{S}^+ = \hat{S}^x + \mathrm{i}\hat{S}^y$ and $\hat{S}^- = \hat{S}^x - \mathrm{i}\hat{S}^y$ hold. The magnitude of the total spin $\hat{\boldsymbol{S}}$ and its z component $\hat{S}^z$ are good quantum numbers.

Charge-Rotational Invariance. On bipartite lattices with symmetric hopping matrix elements $t(\boldsymbol{R}_1, \boldsymbol{R}_2) = t(\boldsymbol{R}_2, \boldsymbol{R}_1)$ between only A and B lattice sites (non-frustrated hopping) the Hamilton operator $\hat{H}'$ displays another global $SU(2)$ symmetry. The generators of the charge $SU(2)$ are defined as

$$\hat{C}^z = \sum_{\boldsymbol{R}} \left[\hat{n}_{\boldsymbol{R},\uparrow} + \hat{n}_{\boldsymbol{R},\downarrow} - 1\right]/2 , \tag{2.30a}$$

$$\hat{C}^+ = \sum_{\boldsymbol{R}} (-1)^{\boldsymbol{R}} \hat{c}^+_{\boldsymbol{R},\uparrow} \hat{c}^+_{\boldsymbol{R},\downarrow} , \tag{2.30b}$$

$$\hat{C}^- = \sum_{\boldsymbol{R}} (-1)^{\boldsymbol{R}} \hat{c}_{\boldsymbol{R},\downarrow} \hat{c}_{\boldsymbol{R},\uparrow} = \left(\hat{C}^+\right)^+ , \tag{2.30c}$$

where we have set $(-1)^{\boldsymbol{R}} = +1$ on the A lattice sites and $(-1)^{\boldsymbol{R}} = -1$ on the B lattice sites. These operators also commute with $\hat{H}'$ and the operators for the spin $SU(2)$ algebra. $\hat{C}^+$ and $\hat{C}^-$ are raising and lowering operators for $\hat{H}$ with eigenvalues U since they create and annihilate a double occupancy, respectively. Altogether the Hubbard model $\hat{H}'$ is $SO(4) = SU(2) \times SU(2)/Z_2$ invariant since the eigenvalues of $\hat{C}^z$ and $\hat{S}^z$ are both integer or half odd-integer at the same time [2.73]. Eßler, Korepin, and Schoutens [2.74] only recently showed that the generators of the charge $SU(2)$ can be used to determine a complete set of 4^L eigenstates for the one-dimensional Hubbard model. Another application of the charge symmetry will be discussed later; see Sect. 6.2.

Electron–Hole Transformations. Very helpful for the study of the Hubbard model $\hat{H}$ are the electron–hole transformations $\mathcal{T}_1$ for all types of lattices and $\mathcal{T}_2$, $\mathcal{T}_{3,\sigma}$ for bipartite lattices. $\mathcal{T}_1$ is defined as

$$\mathcal{T}_1 : \quad \hat{c}^+_{\boldsymbol{R},\sigma} \mapsto \hat{c}_{\boldsymbol{R},\sigma} \quad ; \quad \hat{c}_{\boldsymbol{R},\sigma} \mapsto \hat{c}^+_{\boldsymbol{R},\sigma} \qquad \sigma = \uparrow, \downarrow \tag{2.31}$$

and transforms all electrons into holes and vice versa. When we apply the transformation to the Hubbard Hamilton operator in subspaces with fixed electron numbers we can deduce [2.75, 2.76] that

$$\hat{H}(N_\uparrow, N_\downarrow, U, t) \equiv \hat{H}(L - N_\uparrow, L - N_\downarrow, U, -t) - U(L - N_\uparrow - N_\downarrow) . \tag{2.32}$$

On bipartite lattices we may freely choose the sign of the hopping amplitude t. Under the transformation $\mathcal{T}_2$

$$\mathcal{T}_2 : \quad \hat{c}^+_{\boldsymbol{R},\sigma} \mapsto (-1)^{\boldsymbol{R}} \hat{c}^+_{\boldsymbol{R},\sigma} \quad ; \quad \hat{c}_{\boldsymbol{R},\sigma} \mapsto (-1)^{\boldsymbol{R}} \hat{c}_{\boldsymbol{R},\sigma} \qquad \sigma = \uparrow, \downarrow \tag{2.33}$$

the kinetic energy operator changes its sign if the electrons tunnel between only A and B lattice sites ("non frustrated hopping") which is indeed the case for the standard Hubbard model. The potential energy operator stays invariant. On bipartite lattices with non-frustrated hopping we thus find [2.75, 2.76]

$$\hat{H}(N_\uparrow, N_\downarrow, U, t) \equiv \hat{H}(N_\uparrow, N_\downarrow, U, -t) \tag{2.34}$$

such that the sign of t may be chosen at will.

Another simplification results from the electron–hole transformation that acts only on one spin species,

$$\mathcal{T}_{3,\sigma}: \quad \hat{c}^{+}_{\mathbf{R},\sigma} \mapsto (-1)^{\mathbf{R}} \hat{c}_{\mathbf{R},\sigma} \quad ; \quad \hat{c}_{\mathbf{R},\sigma} \mapsto (-1)^{\mathbf{R}} \hat{c}^{+}_{\mathbf{R},\sigma} \,. \tag{2.35}$$

The kinetic energy operator stays invariant and the operator for the potential energy changes its sign. Including the constants we find

$$\begin{aligned} \hat{H}(N_\uparrow, N_\downarrow, U) &\equiv \hat{H}(L - N_\uparrow, N_\downarrow, -U) + UN_\downarrow \\ &\equiv \hat{H}(N_\uparrow, L - N_\downarrow, -U) + UN_\uparrow \,. \end{aligned} \tag{2.36}$$

These symmetries allow us to restrict ourselves to the subspace of less than half band-filling ($N_\uparrow + N_\downarrow \leq L$) and a non-negative z component of the spin ($S^z = [N_\uparrow - N_\downarrow]/2 \geq 0$) for the Hubbard model on bipartite lattices [2.76].

As an application we can easily show that the band is half-filled at all temperatures for the chemical potential $\mu(N = L) = U/2$. We have

$$\begin{aligned} \langle \hat{N} \rangle(\mu, T) &= \frac{1}{\beta} \frac{\partial}{\partial \mu} \mathrm{Tr}\left[\exp\left(-\beta(\hat{H} - \mu \hat{N}) \right) \right] \\ &= \frac{1}{\beta} \frac{\partial}{\partial \mu} \mathrm{Tr}\left[\exp\left(-\beta\left(\hat{H} + U(L - \hat{N}) - \mu(2L - \hat{N}) \right) \right) \right] \\ &= 2L + \frac{1}{\beta} \frac{\partial}{\partial \mu} \mathrm{Tr}\left[\exp\left(-\beta\left(\hat{H} - (U - \mu)\hat{N} \right) \right) \right] \\ &= 2L - \langle \hat{N} \rangle(U - \mu, T) \,, \end{aligned} \tag{2.37}$$

where we used both transformations $\mathcal{T}_{3,\sigma}$ in the second step and the definition in the last step of the derivation. Obviously, $\mu = U/2$ forces $\langle \hat{N} \rangle(U/2, T) = L$ at all temperatures. An insulator is characterized by a finite range of μ values around $\mu = U/2$, for which $\langle \hat{N} \rangle(\mu, T = 0) = L$ holds (gap criterion). At finite temperatures $\langle \hat{N} \rangle(\mu, T) = L$ is fulfilled only for $\mu = U/2$ since the jump in the chemical potential is smeared out by thermal excitations.

2.5.4 Basic Properties of the Hubbard Model

The Hubbard model can be treated exactly in some limiting cases, which we will now discuss.

Fermi-Gas Limit. First of all, a complete solution is possible for vanishing interactions, $U = 0$. The remaining kinetic energy operator, $\hat{H}(U = 0) = \hat{T}$, is diagonal in momentum space and the model describes a free Fermi gas whose properties are well understood; see [2.1, Chap. 2]. In particular, the free Fermi gas is an (ideal) metal.

Atomic Limit. The model is also easily solved in the atomic limit, $t = 0$, since $\hat{H}(t = 0) = U\hat{D}$ is now diagonal in position space [2.20]. A lattice site can be occupied with two (double occupancy), one (single occupancy with spin σ), and zero electrons (hole). For $N \leq L$ only single occupancies and holes are present in the ground state whereas for $N > L$ only double and single occupancies occur. Excited states can be classified according to the number of double occupancies and the calculation of the free energy density in the grand canonical ensemble in the presence of an external magnetic field $\mathcal{H}_0$ is readily performed with the result ($\beta = 1/(k_B T)$, $g\mu_B \equiv 1$)

$$f_{\text{atom}}(U, T, \mu, \mathcal{H}_0) = -T \ln \left(1 + \mathrm{e}^{\beta(\mu-\mathcal{H}_0)} + \mathrm{e}^{\beta(\mu+\mathcal{H}_0)} + \mathrm{e}^{\beta(2\mu-U)}\right) . \tag{2.38a}$$

From that one can verify that the average density of doubly occupied sites, $\overline{d} = \langle\hat{D}\rangle/L$, obeys

$$\mathrm{e}^{-\beta U} = \frac{\overline{d}(1-n-\overline{d})}{(n_\uparrow - \overline{d})(n_\downarrow - \overline{d})} . \tag{2.38b}$$

This is a law of mass action between doubly occupied sites and holes on one hand and singly occupied sites on the other hand, which is controlled by the ratio of the interaction strength to the temperature. The ground state and all excited states are highly degenerate with respect to the spin and charge degrees of freedom since neither the position of a hole or double occupancy nor the position of a singly occupied site in the lattice has any influence on the spectrum. Since the lattice sites do not communicate with each other the system is an insulator.

Infinite-U Limit. More interesting than the atomic limit is the limit of strong interactions, $U \gg W$. We stress that the atomic limit and the limit of strong interactions are different since in the former case there is no coupling between the lattice sites whereas in the limit $U \to \infty$ a finite coupling between the lattice sites is kept. Electron hole symmetry means we may concentrate on less than or equal half band-filling. In lowest-order perturbation theory around the limit of infinitely strong interactions ($U = \infty$) the dynamics is restricted to the subspace of the lowest possible number of double occupancies, i.e., to $D = 0$ for $N \leq L$. The operator for projection onto this subspace is the Gutzwiller projector (see also Sect. 3.4)

$$\hat{P}_{D=0} = \prod_{\boldsymbol{R}} [1 - \hat{n}_{\boldsymbol{R},\uparrow}\hat{n}_{\boldsymbol{R},\downarrow}] . \tag{2.39}$$

To lowest order in perturbation theory [2.77, Chap. 16] in the parameter W/U and for $n \leq 1$ we obtain in the subspace of no double occupancies

$$\begin{aligned} \hat{H}^{(1)} &= \hat{P}_{D=0}\hat{T}\hat{P}_{D=0} \\ &= \sum_{\boldsymbol{R}_1 \neq \boldsymbol{R}_2,\sigma} t(\boldsymbol{R}_1 - \boldsymbol{R}_2)(1 - \hat{n}_{\boldsymbol{R}_1,-\sigma})\hat{c}^+_{\boldsymbol{R}_1,\sigma}\hat{c}_{\boldsymbol{R}_2,\sigma}(1 - \hat{n}_{\boldsymbol{R}_2,-\sigma}) . \end{aligned} \tag{2.40}$$

The model (2.40) is termed the "$\hat{T}$ model", the "infinite-U Hubbard model", or the "Nagaoka problem". For notational simplicity the projection operators in (2.40) are often formally omitted.

The exact solution of this model is impossible in general [2.22]. There are two exceptions: (i) a chain in one dimension [2.78], and (ii) a single hole in arbitrary dimensions, $N = L - 1$ [2.79]. In case (i) the electrons cannot hop over each other and the system is, as in the atomic limit, 2^N-fold spin-degenerate. Thus the spectrum is equivalent to a fully ferromagnetically polarized system since the spin of an electron plays no role. In addition to the spin degeneracy we thus obtain the free energy density of a free Fermi gas of holes at density $n_\mathrm{h} = 1 - n$. In case (ii) the system is fully ferromagnetically polarized, i.e., the state with total spin $S = N/2$ ("Nagaoka state") is a – possibly degenerate – ground state of the system. Although a single hole bears no resemblance to the thermodynamic limit the Nagaoka state seems to be surprisingly stable even for finite hole densities [2.80]. This example already indicates that the hole motion described by the $\hat{T}$ model depends on the position and the polarization of the electron spins in a rather subtle manner.

For less than half band-filling, $N < L$, the model (2.40) should describe an ideal metal since the holes are mobile charge carriers. However, a formal proof does not exist in $1 < d < \infty$ dimensions [2.81] since the motion of the holes strongly influences the spin background. Nevertheless, a metallic ground state is very plausible for all dimensions because delocalized holes minimize the energy of the $\hat{T}$ model. Obviously, the case of half band-filling is special since holes are completely absent, which implies $\hat{P}_{D=0}\hat{T}\hat{P}_{D=0} \equiv 0$. In this system the chemical potential is discontinuous at $N = L$ because the annihilation of an electron requires a finite energy $\mu^-(N = L) = W_2/2$ whereas the creation of an electron requires the energy $\mu^+(N = L) = U - W_1/2 \to \infty$. For $n_\uparrow + n_\downarrow = n = 1$ and $U = \infty$ we thus have an *insulating ground state.*

Large-U Limit. The 2^L-fold degeneracy of the ground state for $N = L$ is lifted in second-order (degenerate) perturbation theory. For all electron densities $n < 1$ we find [2.77]

$$\hat{H}^{(2)} = \hat{H}^{(1)} - \frac{1}{U}\hat{P}_{D=0}\hat{T}\hat{P}_{D=1}\hat{T}\hat{P}_{D=0} \tag{2.41}$$

where $\hat{P}_{D=1}$ projects onto the subspace with one double occupancy. For the construction of higher-order terms in the perturbation expansion we have to apply a more systematic approach [2.82, 2.83]. In the subspace of no double occupancies we may equally write

$$\begin{aligned}\hat{H}^{(2)} &= \hat{T} + \sum_{\boldsymbol{R}_1 \neq \boldsymbol{R}_2} \frac{2\,|t(\boldsymbol{R}_1 - \boldsymbol{R}_2)|^2}{U}\left[\hat{\boldsymbol{S}}_{\boldsymbol{R}_1}\cdot\hat{\boldsymbol{S}}_{\boldsymbol{R}_2} - \frac{1}{4}\hat{n}_{\boldsymbol{R}_1}\hat{n}_{\boldsymbol{R}_2}\right] \\ &\quad - \sum_{\boldsymbol{R}_1\neq\boldsymbol{R}_2\neq\boldsymbol{R}_3\neq\boldsymbol{R}_1} \frac{t(\boldsymbol{R}_1 - \boldsymbol{R}_2)t(\boldsymbol{R}_2 - \boldsymbol{R}_3)}{U} \\ &\quad \times \sum_{\sigma,\sigma'} \lambda_{\sigma,\sigma'}\hat{c}^{+}_{\boldsymbol{R}_1,\sigma}\hat{c}^{+}_{\boldsymbol{R}_2,-\sigma}\hat{c}_{\boldsymbol{R}_2,-\sigma'}\hat{c}_{\boldsymbol{R}_3,\sigma'}\,,\end{aligned} \tag{2.42}$$

where $\hat{\boldsymbol{S}}_{\boldsymbol{R}}$ are Heisenberg spin operators for spin $S = 1/2$ (compare Sect. 1.4), and we have set $\lambda_{\sigma,\sigma} = +1$, $\lambda_{\sigma,-\sigma} = -1$. If we restrict ourselves to the first and second terms in (2.42) we obtain the "tJ model" which is frequently used for the description of the copper-oxide planes in the high-temperature superconductors [2.59]. Actually, this model was introduced by van Vleck a long time ago ("minimum polarity model") in order to describe itinerant ferromagnetism [2.84].

Neither the first nor the third terms in (2.42) appear at half band-filling since they do not lead to a final state without a double occupancy. For half band-filling we thus find the effective Hamiltonian for strong correlations to be given by ($n_\uparrow + n_\downarrow = n = 1$)

$$\hat{H}^{(2)}(n=1) \equiv \hat{H}_{\text{Heis}} = \sum_{\boldsymbol{R}_1 \neq \boldsymbol{R}_2} \frac{2\left|t(\boldsymbol{R}_1 - \boldsymbol{R}_2)\right|^2}{U} \left[\hat{\boldsymbol{S}}_{\boldsymbol{R}_1} \cdot \hat{\boldsymbol{S}}_{\boldsymbol{R}_2} - \frac{1}{4}\right] . \tag{2.43}$$

The Hubbard model at half band-filling and large interactions becomes equivalent to the antiferromagnetic Heisenberg model with the exchange coupling [2.19]:

$$J(\boldsymbol{R}_1 - \boldsymbol{R}_2) = \frac{2\left|t(\boldsymbol{R}_1 - \boldsymbol{R}_2)\right|^2}{U} > 0 . \tag{2.44}$$

The Heisenberg model (2.43) is the "standard model" for the description of magnetic insulators [2.1, Chap. 32]. Hence the Hubbard model at half band-filling and strong interactions describes an *antiferromagnetic insulator* in all dimensions.

Metal–Mott Insulator Transition. We have convinced ourselves that the Hubbard model at half band-filling describes a metal for $U = 0$ and an antiferromagnetic insulator for $U \gg W$. Thus we are confident that the model is indeed capable of describing a Mott transition at a critical interaction strength U_{c}. Of course, both the value of U_{c} and the nature of the transition remain undetermined by these general considerations. For arbitrary dimension the value for U_{c} can be found only for rather special cases, e.g., at (unphysically) strong magnetic fields [2.85].

Mott's example in Sect. 1.4 described only the charge degrees of freedom whereas the Hubbard model also contains an antiferromagnetic exchange interaction. Hence, the predictions on the nature of the Mott transition based on Mott's example must not be taken too literally for the Hubbard model. As we will see in Sect. 4.1, in one dimension the transition does not occur at $U_{\text{c}} \approx W$, but the metallic phase is limited to $U = 0$ ($U_{\text{c}} = 0^+$). Strictly speaking the one-dimensional Hubbard model does not describe a Mott transition because the metallic phase is not present in a finite range of interactions. We present further details on the exact solution of the standard and some modified Hubbard models in Chap. 4. An explicit one-dimensional realization of Mott's example in the strict sense is discussed in Sect. 6.2.

2.5.5 Magnetism

Although very few exact statements exist on the metal–Mott insulator transition in the Hubbard model there is more information available on its magnetic phase diagram. From early on one of the main objectives in Hubbard model studies was the description of itinerant magnetism [2.24, 2.25]; see also [2.39, 2.40]. Recent reviews for exact results on the magnetic phases of the Hubbard model are provided by Lieb [2.86] and Tasaki [2.87].

For some models of classical statistical mechanics, such as the Ising spin model, the phase diagram can be qualitatively determined from symmetry considerations [2.88]. Quantum spin models such as the antiferromagnetic Heisenberg model of (2.43) already pose a considerably more difficult task. For example, on bipartite lattices the ground state for the antiferromagnetic Ising model with nearest-neighbor interaction is a simple Néel state in all dimensions such that the Ising model displays antiferromagnetic long-range order even in one dimension. For the corresponding spin-1/2 Heisenberg chain there is no long-range antiferromagnetic order. It could be proven only that the ground state displays long-range order in two dimensions for $S \geq 1$ [2.89] but spin $S = 1/2$ remains an open problem at the moment in $d = 2$ dimensions. Néel-type order has been proven to exist in $d \geq 3$ for all $S \geq 1/2$.

In the Hubbard model our quantum-mechanical objects are not local spins but mobile electrons such that we have to expect that a qualitative but exact analysis of the ground-state phase diagram as a function of the density and interaction strength is even more difficult than for the Heisenberg model. Nevertheless some exact statements can be found on magnetism in the Hubbard model; the most important results are listed below [2.86, 2.87]. As usual, any type of magnetic order at finite temperatures is excluded in one and two dimensions (Mermin–Wagner theorem [2.90]). Hence we focus on ground-state properties for which this "no-go theorem" does not apply.

Absence of Ferromagnetism. It turns out that ferromagnetism is the exception rather than a generic feature of one-band models.

1. Lieb–Mattis theorem: the (possibly degenerate) ground state of a one-dimensional system with nearest-neighbor hopping and arbitrary interactions is a total spin singlet ($S = 0$) or a spin doublet ($S = 1/2$) for even or odd electron number [2.91, 2.92]. This theorem shows that the functional form of the hopping rather than that of the interaction is decisive for ferromagnetism in one dimension.
2. Kanamori's result: in the limit of small electron densities, $n \to 0$, the Hubbard model is paramagnetic for all U [2.25, 2.92] if the perturbation theory in the electron density n converges. Kanamori's result from a T-matrix calculation is questionable for dimensions $d \leq 2$ since the

restricted phase space in low dimensions prohibits convergence in one dimension[1] and renders it doubtful in two dimensions.

3. Lieb's ferrimagnetism: on bipartite lattices (e.g., the copper-oxide planes) with L_A A and L_B B lattice sites and at half band-filling with $N = L = L_A + L_B$ electrons the ground state of the repulsive Hubbard model has the total spin $S = |L_B - L_A|/2$ [2.93] whereas the Hubbard model with attractive interaction displays superconducting correlations [2.94]. These results generally exclude ferromagnetism in the ground state at half band-filling for all extended Hubbard models with real hopping amplitudes $t(\boldsymbol{R}_1, \boldsymbol{R}_2) = t(\boldsymbol{R}_2, \boldsymbol{R}_1)$ between A and B lattice sites. At best ferrimagnetism [2.1, Chap. 33] is realized for $L_B \neq L_A$. For $L_A = L_B$ the total spin is zero but the theorem indicates the strong antiferromagnetic correlations between the two sublattices.

Models with Ferromagnetic Ground States. The restrictions of the above theorems can be overcome in various ways as the following examples indicate.

1. Nagaoka theorem: the ground state for a single hole in the half-filled band for $U = \infty$ is fully ferromagnetically polarized, $S = S^{\max}$ (Nagaoka state) [2.79, 2.87, 2.95]. As already mentioned the Nagaoka state is surprisingly stable at finite hole densities [2.80]. For example, a mesoscopic number of holes $N_{\mathrm{h}} \lesssim L^{1/(2d)}$ in d dimensions is not sufficient to destabilize it in the thermodynamic limit [2.96]. Nevertheless, no proof exists that a finite density of holes in the $\hat{T}$ model still leads to the Nagaoka state. It is conceivable that a partly ferromagnetically polarized state exists in the Hubbard model for a small but finite hole density.
2. Mielke's flat-band ferromagnetism: a ferromagnetic ground state does exist in the Hubbard model for some electron density $n > n_{\mathrm{c}}$ for special lattices such as the Kagomé lattice [2.97]. In these "flat-band models" the dispersion relation is such that ferromagnetic states are degenerate with the Fermi sea at $U = 0$, and they become ground states for arbitrarily small $U > 0$.
3. Müller-Hartmann's split-band ferromagnetism in one dimension: a modified Hubbard model was recently constructed by Müller-Hartmann whose low-density ground state should be ferromagnetic [2.98]. Here, the dispersion of the model has two degenerate minima such that the model effectively reduces to a two-band (or two-chain) model.
4. Tasaki's split-band ferromagnetism: a set of one-dimensional models with additional next-nearest-neighbor hopping was constructed by Tasaki [2.87]. These models also belong to the class of "split-band models", i.e., as in Müller-Hartmann's example the bare dispersion relation is split into an upper and a lower band. At quarter filling and strong enough

[1] Perturbation theory to finite order in U gives a Fermi liquid with a quasi-particle peak in the spectral function. In the exact solution the system is a Luttinger liquid with singularities at the two different velocities for spin and charge excitations; see Sect. 4.4.1.

interactions these models display ferromagnetism since the interaction between ↑ and ↓ spins results in energetically costly electron transfers between the two bands; they are absent for a ferromagnetic ground state which then is a ferromagnetic *insulator*.
Tasaki's ferromagnetism is apparently stable away from the quarter-filled case [2.99]. Hubbard models with additional next-nearest-neighbor hopping [2.100] are presently the best candidates for stable itinerant ferromagnetism in one dimension; for a recent review, see [2.101].
5. Models for ferromagnetic insulators: Hubbard-type models in arbitrary dimensions with a fully ferromagnetically polarized, insulating ground state can be constructed by adding more terms to the electron hopping and interaction [2.102]. The results may not necessarily be taken as proof that ferromagnetism exists in realistic one-band Hubbard models; instead they mark extreme parameter regions for extended Hubbard models; see also Sect. 6.1.3.

The exact results cited above show that there are tendencies to both ferromagnetic and antiferromagnetic order in the Hubbard model. In addition to these standard magnetic structures [2.1, Chap. 33] special forms of magnetic order are conceivable, in particular for two dimensions, e.g., spiral topologies [2.103, 2.104]. Correspondingly, the magnetic phase diagram of the Hubbard model as a function of the interaction strength and the electron density could include a large variety of magnetic phases. Hartree–Fock mean-field calculations and other approximate treatments of the Hubbard model typically produce such a rich phase diagram. We will discuss the most important approximation schemes on the Hubbard model in the next chapter.

3. Approximate Methods

Despite its conceptional simplicity the Hubbard model provides a complicated interacting many-electron problem which, in general, cannot be solved exactly. Therefore, approximate methods are employed which are presumed to give a qualitatively correct picture of the physics contained in the model. Here we are interested in the metal–insulator transition and other phenomena beyond perturbation theory. Hence, we have to design feasible non-perturbative approaches that also allow for the description of possible phase transitions.

Given the results of such an approach we have to face a new problem: each approximate method for the Hubbard model introduces a second level of approximation to reality. When we judge the quality of an (often uncontrolled) approximate method we have to distinguish between predictions about the physics of *the Hubbard model* on one hand, and speculations about the properties of *real materials* on the other hand.

To circumvent this new problem in this book we will investigate only whether a chosen approximate treatment yields reasonable results for the Hubbard model. Hence, we will not compare those results to experimental data, as is frequently done to justify an approximation. For the same reason we will not consider model calculations on metallic hydrogen or lithium, which are conceivable experimental realizations of the bare Hubbard model.

We begin this chapter with an examination of effective single-electron theories. In the first section we discuss the Hartree–Fock theory, the simplest single-electron theory, which already gives Slater's theory of the metal–insulator transition, and Stoner's theory of ferromagnetism. In the second section we summarize the results of density-functional theory and the local density approximation for the Hubbard model.

In the following three sections we examine approximate many-particle techniques for the Hubbard model. The third section is concerned with the results of the Green-function decoupling method, a many-particle method that was used by Hubbard in his basic papers. In the fourth section we investigate the properties of the Hubbard model with the help of Gutzwiller-correlated variational wave functions. The fifth section is dedicated to the Slave Boson approach. All these many-particle approaches predict metal–Mott–Hubbard insulator transitions at half band-filling at a finite interaction strength, in agreement with Mott's basic concepts.

3.1 Hartree–Fock Theory

In this section we first introduce the method and some statements on the Hubbard model that are exact within Hartree–Fock theory. The next two subsections are concerned with Slater's theory of the antiferromagnetic insulator and Stoner's theory of the ferromagnet. The resulting phase diagram is displayed in the fourth subsection. Finally, we discuss the advantages and problems as well as possible improvements to Hartree–Fock theory.

3.1.1 Method and Exact Statements

In Hartree–Fock theory the four-fermion term of the (Hubbard) interaction is split into two-fermion terms by replacing a pair of a creation and an annihilation operator by its expectation value. For the Hubbard model with repulsive interaction we do not have to consider (anomalous) expectation values for two creation or two annihilation operators. For the double occupancy operator in the Hubbard interaction one then obtains

$$\hat{D} = \sum_{\boldsymbol{R}} \hat{n}_{\boldsymbol{R},\uparrow}\hat{n}_{\boldsymbol{R},\downarrow} \quad \mapsto \quad \hat{D}^{\mathrm{HF}} = \hat{D}^{\mathrm{H}} + \hat{D}^{\mathrm{F}} , \tag{3.1a}$$

$$\hat{D}^{\mathrm{H}} = \sum_{\boldsymbol{R}} \left[\hat{n}_{\boldsymbol{R},\uparrow}\langle\hat{n}_{\boldsymbol{R},\downarrow}\rangle + \langle\hat{n}_{\boldsymbol{R},\uparrow}\rangle\hat{n}_{\boldsymbol{R},\downarrow} - \langle\hat{n}_{\boldsymbol{R},\uparrow}\rangle\langle\hat{n}_{\boldsymbol{R},\downarrow}\rangle\right] , \tag{3.1b}$$

$$\hat{D}^{\mathrm{F}} = -\sum_{\boldsymbol{R}} \left[\hat{S}^{+}_{\boldsymbol{R}}\langle\hat{S}^{-}_{\boldsymbol{R}}\rangle + \hat{S}^{-}_{\boldsymbol{R}}\langle\hat{S}^{+}_{\boldsymbol{R}}\rangle - \langle\hat{S}^{+}_{\boldsymbol{R}}\rangle\langle\hat{S}^{-}_{\boldsymbol{R}}\rangle\right] , \tag{3.1c}$$

where $\langle\ldots\rangle$ denotes the thermodynamic average over the eigenstates of the single-electron Hartree–Fock Hamilton operator

$$\hat{H}^{\mathrm{HF}} = \hat{T} + U\hat{D}^{\mathrm{HF}} . \tag{3.2}$$

Hence the approximation results in a single-electron problem which, however, has to be solved self-consistently.

The classical or Hartree contribution $\hat{D}^{\mathrm{H}}$ is given by expectation values of density operators on the same lattice site. The Fock contribution $\hat{D}^{\mathrm{F}}$ typically contains non-diagonal expectation values in the space and/or spin indices. In both terms we subtracted products of two averages to ensure that $\langle\hat{D}\rangle = \langle\hat{D}^{\mathrm{HF}}\rangle$ holds within Hartree–Fock theory.

In general, the determination of the minimal free energy in the space of all self-consistent solutions for all electron densities n and interaction strengths U is a mathematically unsolvable problem. Nevertheless there are some rigorous results on the Hubbard model with nearest-neighbor hopping [3.1] that are exact within Hartree–Fock theory defined by (3.1) and (3.2):

1. for $U = \infty$ ($\hat{T}$ model, (2.40)) the ground state is fully ferromagnetically polarized for all $N \leq L - 1$;

2. for $n = N/L = 1$ (half-filled band) and on bipartite lattices with non-frustrated hopping the relation

$$\langle \hat{n}_{\boldsymbol{R},\sigma} \rangle = \frac{1}{2} \left(1 + m(-1)^{\boldsymbol{R}} \lambda_{\uparrow,\sigma}\right) , \tag{3.3}$$

with $\lambda_{\sigma,\sigma} = +1$ and $\lambda_{\sigma,-\sigma} = -1$, holds. Here, $m = |\langle \hat{n}_{\boldsymbol{R},\uparrow} \rangle - \langle \hat{n}_{\boldsymbol{R},\downarrow} \rangle|$ denotes the sublattice magnetization. $(-1)^{\boldsymbol{R}} = \pm 1$, depending on whether $\boldsymbol{R}$ belongs to the A or B sublattice.

For a half-filled band this implies that we may choose the axis of quantization such that the Fock contributions according to (3.1c) vanish due to the fact that the spin in the z direction is conserved. Hence, the Hartree and Hartree–Fock equations become identical and describe the Fermi gas for $m = 0$ or a state with long-range antiferromagnetic Néel-type order for $m > 0$ (spin-density wave). For exact statements for negative U, and other hopping matrix elements $t(\boldsymbol{R}_1, \boldsymbol{R}_2)$, see [3.1].

In all except the two cases listed above no exact statements are available for the Hubbard model with nearest-neighbor hopping and repulsive interaction. To keep the Hartree–Fock problem tractable one has to make *further assumptions* for the expectation values in (3.1). Typically one assumes that the translational and/or spin symmetry are broken in a simple manner.

3.1.2 Slater's Theory of the Antiferromagnet

For a half-filled band the best Hartree–Fock state for the Hubbard model with nearest-neighbor hopping on bipartite lattices has a finite sublattice magnetization m and thus antiferromagnetic long-range Néel-type order [3.1]; see (3.3). The number of $\uparrow$ and $\downarrow$ electrons is equal, $n_\uparrow = n_\downarrow = n/2 = 1/2$. For this state the self-consistency problem can actually be solved for all band fillings [3.2, 3.3]. We present the calculations in Appendix A.1.

On a d-dimensional simple cubic lattice the theory describes two bands in the magnetic Brillouin zone in which the vectors $\boldsymbol{k}$ and $\boldsymbol{k} + \boldsymbol{Q}$ from the first Brillouin zone are identical, and $\boldsymbol{Q} = \boldsymbol{G}/2 = (\pi, \ldots, \pi)/a$ is half a reciprocal lattice vector. The dispersion relation of the two bands is given by

$$E^{\pm}(\boldsymbol{k}) = \frac{1}{2} \left[\epsilon(\boldsymbol{k}) + \epsilon(\boldsymbol{k} + \boldsymbol{Q}) \pm \sqrt{[\epsilon(\boldsymbol{k}) - \epsilon(\boldsymbol{k} + \boldsymbol{Q})]^2 + (Um)^2} \right] , \tag{3.4}$$

where m is the sublattice magnetization, which has to be determined self-consistently; see Appendix A.1. These two bands are usually interpreted as the upper and lower Hubbard bands.

The Hubbard model for bipartite lattices has a peculiar property if the hopping is non-frustrated, i.e., restricted to hopping between lattice sites on different sublattices. Then, the dispersion relation obeys the so-called "perfect nesting" property,

$$\epsilon(\boldsymbol{k}) = -\epsilon(\boldsymbol{k} + \boldsymbol{Q}) \tag{3.5}$$

for all $\boldsymbol{k}$ from the magnetic Brillouin zone. Equation (1.24) gives an example for the simple cubic lattice. As a consequence, the Fermi energy for half-filling is zero and the Fermi surface defined by the wave vectors $\epsilon(\boldsymbol{k}_{\mathrm{F}}) = E_{\mathrm{F}} = 0$ is identical to the surface of the magnetic Brillouin zone. Hence the Fermi surface perfectly fits into the magnetic Brillouin zone. It is seen from (3.4) that a band gap $\Delta\mu^{\mathrm{HF}} = Um$ opens for *all* $U > 0$ since $m(U > 0) > 0$ holds.

Equation (A.1.11) of Appendix A.1 shows that the band gap in Hartree–Fock theory is exponentially small for small U/t,

$$\Delta\mu^{\mathrm{HF}}(U \ll t, d=1) \quad \sim \quad \exp\left(-\frac{2\pi t}{U}\right) , \tag{3.6a}$$

$$\Delta\mu^{\mathrm{HF}}(U \ll t, d=2) \quad \sim \quad \exp\left(-\sqrt{\frac{E_{\mathrm{c}}}{U}}\right) , \tag{3.6b}$$

$$\Delta\mu^{\mathrm{HF}}(U \ll t, d=3) \quad \sim \quad \exp\left(-\frac{1}{UD_{\sigma,0}(0)}\right) . \tag{3.6c}$$

The prefactors are independent of U and $E_{\mathrm{c}} = \mathcal{O}(t)$ is a cut-off energy; see also [3.4]. $D_{\sigma,0}(0)$ denotes the density of states of one spin species at the Fermi energy in the paramagnetic state. For systems with "perfect nesting" we obtain a (thermodynamic) phase transition into an insulating state with antiferromagnetic long-range order at $U_{\mathrm{c}}^{\mathrm{HF}}(n=1) = 0^{+}$. For non-bipartite lattices, e.g., for a face-centered cubic lattice, one obtains $U_{\mathrm{c}}^{\mathrm{HF}}(n=1) > 0$ instead. In the latter case, however, the validity of (3.3) is only an assumption. The optimum Hartree–Fock ground state for a non-bipartite lattice like the face-centered cubic lattice has yet to be determined.

For large interactions we find

$$\Delta\mu^{\mathrm{HF}}(U \gg t) = U - \mathcal{O}\left(t^2/U\right) , \tag{3.7}$$

independent of dimension and lattice structure. In Hartree–Fock theory the electrons are fixed onto their respective sublattice sites for large interaction strengths, and there is *no* bandwidth contribution to the gap that should be present according to Mott's general ideas. We expect $\Delta\mu \approx U - (W_1 + W_2)/2$ where $W_{1,2}$ are the bandwidths of the upper and lower Hubbard bands for the charge excitations; see Sect. 1.4.6. Charge excitations are *not* mobile for large U/t within the Hartree–Fock description, $W_1^{\mathrm{HF}} = W_2^{\mathrm{HF}} = 0$, and only the exchange interaction correction to the gap of order $J \sim t^2/U$ lowers the gap from its atomic limit value, $\Delta\mu^{\mathrm{atom}} = U$.

For less than half band-filling the Slater theory describes an antiferromagnetic metal for all $U < \infty$. For $U = \infty$ the bandwidth of both bands $E^{\pm}(\boldsymbol{k})$ goes to zero. Hence we have $U_{\mathrm{c}}^{\mathrm{Slater}}(n < 1) = \infty$. In this limit, however, the ground-state energy of the antiferromagnetic Hartree–Fock solution is larger than the ground-state energy of the fully ferromagnetically polarized state; see Sect. 3.1.1. Consequently, a metal–insulator transition does not take place within the exact Hartree–Fock theory for $n < 1$ and large U/t. Instead, there will be a transition from an antiferromagnetic into a ferromagnetic metal at a finite interaction strength; see Sect. 3.1.4.

A discussion of the thermodynamics can be found, e.g., in [3.3, 3.5]. Here, the important point is that the Hartree–Fock theory describes a thermodynamic phase transition from the antiferromagnetic into a paramagnetic state at the critical (Néel) temperature T_{N}. Even for small but finite interactions the Hartree–Fock Néel temperature is proportional to U, $k_{\mathrm{B}}T_{\mathrm{N}}^{\mathrm{HF}} \sim U$. This transition temperature is unrealistically high since U is a large energy scale, of the order of electron volts. This wrong prediction had to be expected since correlations are absent in the Hartree–Fock approach. Within this theory local moments form and order at the same temperature $T_{\mathrm{N}}^{\mathrm{HF}}$. In reality, however, the Hartree–Fock value $T_{\mathrm{N}}^{\mathrm{HF}}$ serves as an estimate for the temperature below which (relative) moments are present. These (correlated) moments order at the true Néel temperature T_{N} (much) below $T_{\mathrm{N}}^{\mathrm{HF}}$. For example, for $U > W$, the ordering temperature is of the order of the exchange interaction, $T_{\mathrm{N}} \sim t^2/U$, and much smaller than $T_{\mathrm{N}}^{\mathrm{HF}} \sim U$, the temperature above which moments are thermally destroyed.

3.1.3 Stoner's Theory of the Ferromagnet

Another kind of symmetry breaking in the Hartree–Fock equations corresponds to Stoner's model for ferromagnetism [3.6]. In this case the spin symmetry is broken but the system remains translationally invariant. Stoner's assumption for the expectation values reads

$$\langle \hat{n}_{\boldsymbol{R},\uparrow} \rangle = n_{\uparrow} \geq n_{\downarrow} = \langle \hat{n}_{\boldsymbol{R},\downarrow} \rangle \, . \tag{3.8}$$

As already mentioned in Sect. 3.1.1 this assumption becomes exact within the Hartree–Fock approximation for $n < 1$ in the limit $U \to \infty$.

In Stoner's theory the single-electron bands are simply shifted by a spin-dependent but momentum-independent constant,

$$\epsilon_{\sigma}(\boldsymbol{k}) = \epsilon(\boldsymbol{k}) + \frac{U n_{-\sigma}}{2} \, . \tag{3.9}$$

The electron density n and the total magnetization m at temperature $T = 0$ are calculated from the self-consistency equations as

$$n_{\uparrow} = \int_{-W/2}^{E_{\mathrm{F}}(n,m)-U(n-m)/2} \mathrm{d}\epsilon D_{\sigma,0}(\epsilon) \, , \tag{3.10a}$$

$$n_{\downarrow} = \int_{-W/2}^{E_{\mathrm{F}}(n,m)-U(n+m)/2} \mathrm{d}\epsilon D_{\sigma,0}(\epsilon) \, , \tag{3.10b}$$

$$n = n_{\uparrow} + n_{\downarrow} \, , \tag{3.10c}$$

$$m = n_{\uparrow} - n_{\downarrow} = R(m) \, . \tag{3.10d}$$

Equation (3.10c) defines the Fermi energy $E_\mathrm{F}(n,m)$ and (3.10d) allows us to determine the magnetization self-consistently. $D_{\sigma,0}(\epsilon)$ and E_F^0 denote the density of states of one spin species and the Fermi energy for $U=0$, respectively.

If we are interested in the critical interaction strength for the transition from a paramagnetic into a ferromagnetic state, we just have to expand (3.10d) for small magnetization. The function on the right-hand side of the last equation, $R(m)$, is antisymmetric in m since $E_\mathrm{F}(n,m)$ is symmetric. Furthermore, the limiting value of $R(m)$ is n when we let m go to n. Hence, a solution with $m>0$ will exists if the inclination of the curve $R(m)$ as a function of m near $m=0$ is larger than unity, the inclination of the left-hand side of (3.10d). The expansion of $R(m)$ in lowest order in m gives the criterion $UD_{\sigma,0}\big(E_\mathrm{F}(n,m=0)-Un/2\big)\geq 1$. We use (3.10c) and see that $E_\mathrm{F}(n,m=0)-Un/2=E_\mathrm{F}^0$ holds. Hence we find the "Stoner criterion for ferromagnetism"

$$UD_{\sigma,0}(E_\mathrm{F}^0)\geq 1\,. \tag{3.11}$$

At $U_\mathrm{c}^\mathrm{Stoner}=[D_{\sigma,0}(E_\mathrm{F}^0)]^{-1}$ there should be a transition from a paramagnetic metal into a ferromagnetic metal. For the Hubbard model Stoner's theory does not predict a metal–insulator transition for $n<1$ for any interaction strength. For half-filling it describes a transition to a standard band insulator at $U=\infty$.

The ground-state energy is linear in U/t below the Stoner transition,

$$(E_0^\mathrm{Stoner}/L)(U\leq[D_{\sigma,0}(E_\mathrm{F}^0)]^{-1})=E_\mathrm{Fermi\ sea}+\frac{Un^2}{4}\,, \tag{3.12}$$

because the interaction term contains only the Hartree contribution $Un^2/4$ of the paramagnetic Fermi gas. There are no correlation contributions, which is typical for a Hartree–Fock calculation.

3.1.4 Simplified Hartree–Fock Phase Diagram

To determine a Hartree–Fock phase diagram for the Hubbard model it is not sufficient to investigate independently phases with broken symmetries. Instead we have to analyze the stability of these phases relative to each other. This can be achieved by a comparison of their respective ground-state or free energies. Alternatively, we may calculate the magnetic susceptibility as a function of momentum and frequency. Divergences in this quantity as a function of the model parameters n and U/t indicate a transition to another phase. The latter method offers the advantage that the analysis may be restricted to a given phase for which the calculations can actually be carried out. However, only the *local* stability of that phase is thereby tested. A state with a *globally* lower energy may thus be missed.

Shortly after the Hubbard model was introduced Penn [3.7] carried out a detailed Hartree–Fock calculation for a three-dimensional simple cubic lattice

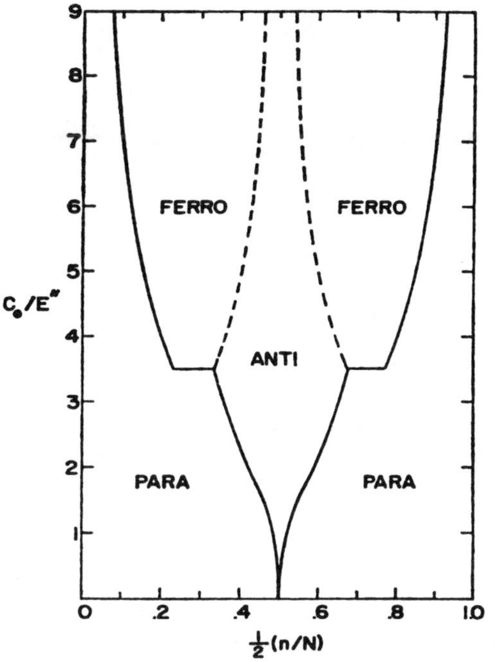

Fig. 3.1. Simplified phase diagram of the Hubbard model for a three-dimensional simple cubic lattice in Hartree–Fock theory with paramagnetic, ferromagnetic, and antiferromagnetic phases. *Lines* (*dashed lines*) indicate second-order (first-order) phase transitions. X axis: half of the electron density, $n/2$. Y axis: interaction strength U, normalized to the hopping matrix element t (from [3.7]).

involving ferromagnetic, antiferromagnetic, ferrimagnetic, and spiral states at zero temperature. A simplified magnetic ground-state phase diagram without ferrimagnetic and spiral phases is shown in Fig. 3.1. There is a qualitative agreement between the phase diagram in Fig. 3.1 and the exact statements about the Hartree–Fock phase diagram listed in Sect. 3.1.1. Antiferromagnetism is stable near half band-filling, and there are ferromagnetic regions beyond the Stoner value $U_{\mathrm{c}}^{\mathrm{Stoner}} = [D_{\sigma,0}(E_{\mathrm{F}}^{0})]^{-1}$ which become larger as U/t increases. Finally, ferromagnetism is the only stable phase within Hartree–Fock theory for all $n < 1$ for $U/t = \infty$.

Usually the ground-state energy is a differentiable function of U/t at transitions between the paramagnetic and a magnetically ordered phase. This is no longer the case for transitions between phases with topologically different magnetic structures. In analogy to the standard theory of phases transitions as a function of temperature, the transitions in Fig. 3.1 between paramagnet and ferromagnet or antiferromagnet are termed "second order" and the transition between ferromagnetic and antiferromagnetic order are of "first order". The latter transition can be identified only by a comparison of the ground-state energies of the respective phases. As discussed above, the analysis of the magnetic susceptibilities does not signal an instability in this case [3.7].

As well as the states considered so far we may include more complicated spin structures such as ferrimagnetism or spiral states in our analysis. These additional phases, however, do not substantially change the Hartree–Fock phase diagram [3.7]. In addition, there are no rigorous statements on the real existence of these phases within Hartree–Fock theory. Hence we did not include them in Fig. 3.1. Analytical [3.8] and numerical [3.9] studies show that Hartree–Fock ground states with a rich magnetic structure like spin polarons and magnetic domain walls are conceivable, especially in two dimensions. We may thus conclude that not even the Hartree–Fock phase diagram of the Hubbard model has been fully resolved.

3.1.5 Advantages and Problems

Hartree–Fock theory offers several advantages.

(i) It obeys a variational principle. The free energy in Hartree–Fock theory is always an upper bound for the exact free energy [3.10, 3.11].
(ii) It can be derived from a generating functional in the sense of Baym and Kadanoff [3.12]. Hence, it is a "conserving approximation" which automatically exactly fulfills all sum rules. Consequently, the Hartree–Fock theory cannot yield unphysical results that violate any conservation laws.
(iii) It is perturbatively controlled [3.13]. It can be formulated in the language of Feynman diagrams for a perturbation expansion in U/t or n [3.14, Chap. 4] which shows that the theory is exact up to and including the order U/t, and n^2, respectively.

The last advantage of the Hartree–Fock theory is also its biggest problem: all correlations are missing. The theory does not allow perturbatively controlled statements about the physical properties of the Hubbard model for appreciable U/t or densities n since higher-order corrections in U/t or n have not been taken into account. In particular, we find that

(i) for small interaction strengths there are no contributions in order U^2/t to the ground-state energy, for neither the antiferromagnetic nor the ferromagnetic state;
(ii) for large interaction strengths the Hartree–Fock ground state is ferromagnetic for *all* $0 < n < 1$. However, Kanamori [3.15] used a perturbation expansion in the density n to show that the ground state of the Hubbard model in dimensions $d \geq 3$ is paramagnetic for small electron densities for arbitrarily large U/t;
(iii) for $n = 1$ and $U \to \infty$ the Hartree–Fock ground state for the Hubbard model continuously transforms into the Néel state. This "classical" spin state lacks all quantum-mechanical spin fluctuations which are typical for the antiferromagnetic Heisenberg model. In addition, the upper and lower Hubbard bands have zero bandwidth for large interactions at half-filling. A true localization of the $\uparrow$ and $\downarrow$ electrons on their corresponding sublattices is unphysical.

Hartree–Fock theory overestimates the tendencies to symmetry breaking in the Hubbard model since correlations between the electrons are totally ignored. Consequently, only a symmetry breaking can reduce the interaction energy, either by reducing the number of available pairs of $\uparrow$ and $\downarrow$ electrons like in Stoner's theory, or by fixing the electrons onto different sublattices like in Slater's theory. This typically results in a severe overestimation of the Curie and Néel temperature by several orders of magnitude.

Hartree–Fock theory generally overestimates the importance of the magnetic degrees of freedom for the Hubbard model. Within Slater's and Stoner's theories the spin degrees of freedom always dominate the charge degrees of freedom. This can most clearly be seen for large U/t and half-filling where the reduction of the energetically unfavorable charge fluctuations (double occupancies) is the result of only the broken spin symmetry in Hartree–Fock theory. This scenario for the Hubbard model is unlikely for large interaction strengths since the energy scale for charge fluctuations U is much larger than the energy scale for spin fluctuations $J \sim t^2/U$. On the contrary, we expect to observe the signatures of a metal–Mott–Hubbard insulator transition driven by the charge degrees of freedom.

For small interaction strengths and/or low enough temperatures the spin degrees of freedom will give rise to a thermodynamic phase transition into a Mott–Heisenberg insulator with long-range order of the magnetic moments. These moments persist above the Néel temperature T_{N} in the correlated metal or the Mott–Hubbard insulator, and are thermally destroyed at the (much higher) Hartree–Fock Néel temperature $T_{\mathrm{N}}^{\mathrm{HF}} \sim U$.

Hartree-Fock theory indicates the temperatures below which we need to include local moments into our description of the Hubbard model. With the help of this simple theory we can guess possible ordered phases and draw a qualitative (magnetic) phase diagram of the Hubbard model. Hence we can conjecture which thermodynamic phase transitions could cause a metal–insulator transition. However, Hartree–Fock theory cannot provide a deeper understanding of magnetic moments and the metal–Mott insulator transition, which is actually dominated by electron–electron correlations.

3.1.6 Improvements

Any improvement in Hartree–Fock theory must aim at the inclusion of correlation effects. In the following we will focus on half band-filling on bipartite lattices with non-frustrated hopping where the optimum Hartree–Fock state is known to be the antiferromagnetic spin-density wave.

Weak Interactions. For small interaction strengths we may use perturbation theory to second order in U/t to derive systematically corrections to the Hartree–Fock results [3.5, 3.16].

Both Slater's and Stoner's theories give the correct result for the ground-state energy to lowest order in U/t but the antiferromagnetic state is smaller in energy by exponential terms; see (3.12) and (A.1.14). This raises the question whether Hartree–Fock theory can formally become exact for $U \to 0$.

The definite answer to this question is *no* [3.5, 3.17] because the energy difference between the antiferromagnetic Hartree–Fock state and the paramagnetic Fermi gas is proportional to the square of the energy gap and is thus of order $(\Delta\mu^{\mathrm{HF}})^2 \sim U^2$. Hence it is inconsistent to consider only the Hartree contributions to the ground-state energy. We have to include the diagram in second order for a perturbatively consistent theory.

The second-order diagram leads to a finite correlation function between different spins of order U. Consequently, the local magnetic moments due to correlations are of order U as well and are much bigger than the (exponentially small) contributions due to the long-range order. Local magnetic moments are present in the Hubbard model for *all* $U > 0$, and they are preserved above the Néel temperature, even though their long-range order vanishes.

In effect, the second-order term not only yields correlation contributions to the ground-state energy but also renormalizes all Hartree parameters like the sublattice magnetization m and the size of the energy gap. In the framework of a $1/d$ expansion [3.5] there is a uniform renormalization by a factor $q < 1$ ($q \approx 0.3$ in $d = 3$); for the negative-U Hubbard model, see [3.18]. Evidence for a renormalization of the Hartree parameters is also found in numerical studies of the infinite-dimensional Hubbard model [3.19]. Apparently, in $d > 2$ the sublattice magnetization stays finite. Statements on one and two dimensions seem forbidding since the series expansion in $1/d$ might not converge there, and the results may qualitatively change for low dimensions. In fact, there is no magnetic long-range order in the one-dimensional Hubbard model, and the presence of magnetic order in the two-dimensional Hubbard model at zero temperature is still an unsettled issue.

Strong Interactions. In the limit of large interaction strengths it is natural to set up a perturbation expansion in t/U. Unfortunately, at half band-filling the degeneracy of the ground state is only lifted in second-order degenerate perturbation theory, and the lowest-order problem corresponds to the Heisenberg model; see Sect. 2.5. In Hartree–Fock theory the antiferromagnetic spin density for the Hubbard model smoothly evolves into the Néel state as Hartree–Fock ground state of the Heisenberg model. Since the Néel temperature for the Heisenberg model is known to be of the order of the exchange coupling $J \sim t^2/U$, one can extend the Heisenberg model approach to the Hubbard model to derive a good estimate for the Néel temperature at *all* interaction strengths. This idea is carried out in [3.20].

The Néel state is the true ground state of the Heisenberg model in the limit of infinite dimensions, or infinite length of the spin vector $|\hat{\boldsymbol{S}}(\boldsymbol{R})| = \sqrt{S(S+1)}$. Therefore, Hartree–Fock theory provides a solid starting point for a $1/d$ expansion in the limit of strong interactions. For example, the ground-state energy and the phase diagram of the extended Hubbard model at half band-filling for large (and small) interaction strengths can be derived along these lines [3.5].

Alternatively, we may include spin waves into our Hartree–Fock theory for the Heisenberg model which represents the lowest-order corrections in $1/S$. For the Hubbard model this implies that we need to self-consistently include all ring diagrams [3.14, Chap. 4]. A complete self-consistent calculation for the Hubbard model can be performed only with considerable numerical effort [3.21]. Although the ring diagrams give the major correlation corrections for the free electron gas at high densities [3.14, Chap. 4] this condition is not fulfilled for the Hubbard model since the average electron distance is always at least of the order of a lattice constant. Hence, the particular inclusion of solely ring diagrams is justified only for small interaction strengths (second-order perturbation theory in U/t) and spin-wave theory for strong interaction at half band-filling.

Conserving Approximations. We have seen that the inclusion of certain classes of diagrams (e.g., ring diagrams) may not give satisfactory, i.e., controlled results for the parameter region of most interest to us, namely intermediate interaction strengths $U \approx W$.

Hartree–Fock theory is the lowest-order conserving approximation in the sense of Baym and Kadanoff [3.12]. Thus, it would be much more natural to evaluate the next order *conserving approximation*. Unfortunately, this implies the self-consistent evaluation of the so-called parquet diagrams, which include ring and ladder diagrams as subclasses. The corresponding numerical calculations pose an almost unsurmountable task such that there are no conclusive results available at the moment. Any simplification of the equations necessarily makes them non-conserving and, more severely, may violate fundamental requirements like Fermi statistics or time reversal invariance which is expressed in the "crossing symmetry" of the diagrams against exchange of fermion lines. A brief account of conceptional and numerical problems in the treatment of the parquet equations is given in [3.22]. We shall not comment further on these approaches in this book.

3.2 Local Density Approximation

In this section we derive the lattice version of density-functional theory and its corresponding local density approximation. Both approaches can be carried through exactly for the one-dimensional Hubbard model, which allows us to assess their potential for the description of the Mott transition. Furthermore, we study the lattice generalization of Slater's $X\alpha$ approximation to the exact exchange energy functional. We test the corresponding $X\alpha$ local density approximation and the $X\alpha$ local spin-density approximation with and without self-interaction corrections against the exact results for the one-dimensional Hubbard model. We find that the ground-state energy density and the Mott–Hubbard gap in the model are not very well reproduced by the various local density approximations.

3.2.1 Site Occupation Functional Theory

Density-functional theory and the local density approximation were originally designed for the continuum. Consequently, we first have to formulate properly these approaches *on the lattice* before we can assess their applicability for the Hubbard model. There are several (non-equivalent) ways to derive the lattice density-functional theory and the lattice local density approximation. A systematic way is the site occupation functional theory [3.23], which canonically extends Levy's ideas of constrained search [3.24], as we now discuss.

As is well known from elementary quantum mechanics [3.25, Chap. 18] we can always find the ground-state energy as the minimum of the energy functional

$$E[\Psi] = \langle\Psi|\hat{H}|\Psi\rangle \tag{3.13a}$$

in the Hilbert space of all normalized states $|\Psi\rangle$,

$$E_0 = \underset{|\Psi\rangle}{\mathrm{Min}}\left\{E[\Psi]\right\} . \tag{3.13b}$$

For lattice electrons the Hamiltonian can be written as

$$\hat{H} = \hat{H}_0 + \sum_{\boldsymbol{R}} V(\boldsymbol{R})\hat{n}_{\boldsymbol{R}} , \tag{3.14}$$

where Hamiltonian $\hat{H}_0$ describes the electrons' kinetic energy and their mutual interaction. The "external" electron–ion lattice potential couples to the electron densities only. Consequently, it is possible in this case to split the Ritz minimization procedure into two steps.

In the first step we determine the minimum of the energy functional in the subspace of normalized states $|\Psi\{n(\boldsymbol{R})\}\rangle$ with the *same* local densities $n(\boldsymbol{R}) = \langle\Psi\{n(\boldsymbol{R})\}|\hat{n}_{\boldsymbol{R}}|\Psi\{n(\boldsymbol{R})\}\rangle$. The curly brackets indicate the function(al) dependence of the wave functions on the densities. Within the subspace of fixed local densities we determine the minimum energy of $\hat{H}_0$,

$$F[n(\boldsymbol{R})] = \underset{|\Psi\{n(\boldsymbol{R})\}\rangle}{\mathrm{Min}}\left\{\langle\Psi\{n(\boldsymbol{R})\}|\hat{H}_0|\Psi\{n(\boldsymbol{R})\}\rangle\right\} . \tag{3.15a}$$

The density function(al) $F[n(\boldsymbol{R})]$ depends on the properties of $\hat{H}_0$, not of $\hat{H}$.

In the second step we vary the densities $n(\boldsymbol{R})$ to optimize $\hat{H}$. We use a Lagrange multiplier λ to include the constraint of a fixed total number of electrons,

$$N = Ln = \sum_{\boldsymbol{R}} n(\boldsymbol{R}) . \tag{3.15b}$$

The constraint implies that only $L-1$ of the densities can be chosen freely. Thus, we solve the L equations

$$\frac{\delta F[n(\boldsymbol{R})]}{\delta n(\boldsymbol{R})} - \lambda + V(\boldsymbol{R}) = 0 \tag{3.15c}$$

to obtain the ground-state densities $n_0(\boldsymbol{R})$ compatible with (3.15b). Naturally, λ is nothing but the (exact) chemical potential.

The exact ground-state energy becomes

$$E_0 = F\left[n_0(\boldsymbol{R})\right] + \sum_{\boldsymbol{R}} V(\boldsymbol{R}) n_0(\boldsymbol{R}) \,. \tag{3.15d}$$

The derivation shows that this approach allows only statements about the ground-state energy and the local electron densities. Unfortunately, the variational problem (3.15) that determines the ground-state energy is by no means simpler than the solution of the Schrödinger equation itself.

The essential idea behind the Hohenberg–Kohn–Sham theory [3.26] is the following. The many-electron problem still contained in $\hat{H}_0$ is replaced by a one-electron Hamiltonian $\widetilde{H}_0$, and the actual external potential $V(\boldsymbol{R})$ is replaced by an effective one-electron potential $V^{\text{eff}}(\boldsymbol{R})$, i.e., we analyze

$$\widetilde{H} = \widetilde{H}_0 + \sum_{\boldsymbol{R}} V^{\text{eff}}(\boldsymbol{R}) \hat{n}_{\boldsymbol{R}} \tag{3.16a}$$

and demand that $\hat{H}$ in (3.14) and $\widetilde{H}$ in (3.16a) give the same ground-state energy, $\widetilde{E}_0 = E_0$, and densities, $\widetilde{n}_0(\boldsymbol{R}) = n_0(\boldsymbol{R})$. Equation (3.16a) is known as the assumption of "V representability". If $\widetilde{H}_0$ is a two-electron Hamiltonian, a one-body potential $V^{\text{eff}}(\boldsymbol{R})$ is not sufficient; see, e.g., [3.27].

For the new representation $\widetilde{H}$ of the Hamiltonian $\hat{H}$ we find the corresponding functional $\widetilde{F}\left[n(\boldsymbol{R})\right]$, which is determined by $\widetilde{H}_0$. To guarantee that the local densities are indeed the same we have to fulfill

$$\frac{\delta \widetilde{F}\left[n(\boldsymbol{R})\right]}{\delta n(\boldsymbol{R})} - \lambda^{\text{eff}} + V^{\text{eff}}(\boldsymbol{R}) = 0 \tag{3.16b}$$

for the densities $\widetilde{n}_0(\boldsymbol{R}) \equiv n_0(\boldsymbol{R})$, which we determined by (3.15b) and (3.15c). Hence, the effective external potential is obtained from

$$V^{\text{eff}}(\boldsymbol{R}) = V(\boldsymbol{R}) - \lambda + \lambda^{\text{eff}} + \left.\frac{\delta (F - \widetilde{F})\left[n(\boldsymbol{R})\right]}{\delta n(\boldsymbol{R})}\right|_{\{n(\boldsymbol{R})\} = \{n_0(\boldsymbol{R})\}} \,. \tag{3.16c}$$

The advantages of this alternative approach become apparent if $\widetilde{H}$ can be treated (numerically) exactly and if a good approximation for the *difference* $(F - \widetilde{F})[n(\boldsymbol{R})]$ can be found.

Since the ground-state densities $n_0(\boldsymbol{R})$ already fulfill (3.15b) we need another equation to determine λ^{eff} which is *not* identical to λ, the exact chemical potential. We may equally well state that a constant shift in $V^{\text{eff}}(\boldsymbol{R})$ does not change the local densities, and we need another equation to calculate this overall constant. Since $\widetilde{H}$ in (3.16a) should give the same ground-state energy as $\hat{H}$ in (3.16a), we demand that the exact ground-state energy obeys

$$E_0 = \widetilde{F}\left[n_0(\boldsymbol{R})\right] + \sum_{\boldsymbol{R}} V^{\text{eff}}(\boldsymbol{R}) n_0(\boldsymbol{R}) \,. \tag{3.16d}$$

The equations (3.16) define the equivalent one-electron problem.

For an application let us consider the one-dimensional Hubbard model. The external potential is zero, $V(\boldsymbol{R}) \equiv 0$, such that the local densities are known to be the same on all lattice sites, $\langle \hat{n}_{\boldsymbol{R}} \rangle = n$. The effective lattice potential becomes a site-independent constant which has to be determined from (3.16d). The equation (3.16c) provides no further information but only fixes λ^{eff}. The exact functional is just the ground-state energy of the Hubbard model, $F[n] = E_0(U, n)$ ($\widetilde{H}_0 \equiv \hat{T}$ describes the Fermi gas of free electrons) and the approximate functional is its ground-state energy, $\widetilde{F}[n] = E_0^{\text{Fermi gas}}(n)$. Therefore, we find

$$V^{\text{eff}}(U, n) = \left[E_0^{\text{exact}}(U, n) - E_0^{\text{Fermi gas}}(n) \right] / (Ln) \tag{3.17}$$

for the one-dimensional Hubbard model. Since the ground-state energy of the Hubbard model is a continuous function of the electron density, the effective potential $V^{\text{eff}}(U, n)$ is also a *continuous* function of n. The effective one-electron problem (3.16a) together with (3.17) gives the same ground-state energy and local densities as the one-dimensional Hubbard model.

The resulting band structure is the same as that of free fermions but the dispersion is shifted upwards in energy by an interaction and density dependent value. A "band gap", defined as discontinuity in the one-electron dispersion relation of $\widetilde{H}_0 \equiv \hat{T}$, does *not* exist. This had been obvious from the beginning since the density-functional theory can substantially change the band structure only if $V^{\text{eff}}(\boldsymbol{R})$ is site dependent.

Nevertheless, the exact density-functional theory *does* reproduce the Mott–Hubbard gap of the half-filled one-dimensional Hubbard model. The definitions (1.4) and (1.5) show that the (correlation) gap can be obtained from the exact ground-state energy, which is exactly calculated in density-functional theory. Of course, we put in this information "by hand" since we forced our effective single-electron representation (3.16a) to yield the exact ground-state energy. At half band-filling the derivative of $E_0^{\text{exact}}(U, n)$ and thus of $V^{\text{eff}}(U, n)$ with respect to the density are discontinuous, i.e.,

$$\left. \frac{\partial V^{\text{eff}}(U, n)}{\partial n} \right|_{n=1^+} = \Delta\mu(U) + \left. \frac{\partial V^{\text{eff}}(U, n)}{\partial n} \right|_{n=1^-} . \tag{3.18}$$

For a Mott insulator the density *derivative* of the effective potential has a jump discontinuity.

Although the site occupation functional theory does not give two split Hubbard bands, one might *interpret* the jump in the derivative of the effective potential in terms of a "band gap". This is the essence of the "scissor operator" approach [3.28] in which calculated "band gaps" are increased by correlation contributions to fit the experimentally observed excitation gaps.

3.2.2 Site Occupation Local Density Approximation

To set up the site occupation local density approximation we have to introduce the notion of the exchange-correlation potential. In analogy to the density-functional theory in the continuum we define the exchange-correlation function(al) $E_{\mathrm{xc}}[n(\boldsymbol{R})]$ as

$$E_{\mathrm{xc}}[n(\boldsymbol{R})] = F[n(\boldsymbol{R})] - T[n(\boldsymbol{R})] - E_{\mathrm{H}}[n(\boldsymbol{R})] \tag{3.19a}$$

with $F[n(\boldsymbol{R})]$ from (3.15a),

$$E_{\mathrm{H}}[n(\boldsymbol{R})] = \frac{1}{2}\sum_{\boldsymbol{R},\boldsymbol{R}'} n(\boldsymbol{R})V_{\mathrm{e-e}}(\boldsymbol{R}-\boldsymbol{R}')n(\boldsymbol{R}') \tag{3.19b}$$

as the Hartree interaction between the electron densities at $\boldsymbol{R}$ and $\boldsymbol{R}'$, and $T[n(\boldsymbol{R})]$ as the kinetic energy functional for non-interacting electrons at given local densities $\{n(\boldsymbol{R})\}$. The exchange-correlation potential then becomes

$$v_{\mathrm{xc}}(\boldsymbol{R}) = \left.\frac{\delta E_{\mathrm{xc}}[n(\boldsymbol{R})]}{\delta n(\boldsymbol{R})}\right|_{\{n(\boldsymbol{R})\}=\{n_0(\boldsymbol{R})\}} . \tag{3.20a}$$

The effective potential in local density approximation then reads

$$V^{\mathrm{eff}}_{\mathrm{LDA,\,exact}}(\boldsymbol{R}) = V(\boldsymbol{R}) + v_{\mathrm{H}}(\boldsymbol{R}) + v_{\mathrm{xc}}(\boldsymbol{R}) , \tag{3.20b}$$

where the Hartree potential is given by

$$v_{\mathrm{H}}(\boldsymbol{R}) = \sum_{\boldsymbol{R}'} V_{\mathrm{e-e}}(\boldsymbol{R}-\boldsymbol{R}')n(\boldsymbol{R}') . \tag{3.20c}$$

The equations (3.20) and (3.16a) with $\widetilde{H}_0 \equiv \hat{T}$ define the effective single-electron Hamilton operator for the exact site occupation local density approximation.

Obviously, the exact site occupation local density approximation does not reproduce the exact ground-state energy of the interacting system. This is evident from the fact that we did not make any difference between the Lagrange parameters λ and λ^{eff} which appeared in the exact Ritz variational procedure and in the exact density-functional theory, respectively. One of the basic problems of the local density approximation is the (incorrect) assumption that these two quantities are equal.

To further illustrate this point we study the one-dimensional Hubbard model. The local densities are constant, and both the site occupation functional theory and its LDA give the exact densities for the Hubbard model. By definition this is always the case for the site occupation functional theory; the LDA, however, gives different results, in general. For an example, see [3.23]. For the Hubbard model the exact LDA potential becomes

$$V^{\mathrm{eff}}_{\mathrm{LDA,\,exact}}(U,n) = \frac{1}{L}\frac{\partial}{\partial n}\left[E_0^{\mathrm{exact}}(U,n) - E_0^{\mathrm{Fermi\ gas}}(n)\right] . \tag{3.21}$$

Again, the band structure is merely shifted upwards in energy by an interaction and density-dependent function, i.e., no "band gap" occurs.

The expression (3.21) has to be compared to (3.17). The LDA potential (3.21) has a jump discontinuity at half band-filling, and its value for $n = 1$ is *ambiguous*. Hence, it is *impossible* to define properly a correlation gap for the LDA. Furthermore, if $v_{\mathrm{xc}}(\boldsymbol{R})$ has a jump discontinuity, then $V^{\mathrm{eff}}(\boldsymbol{R})$ and ultimately $E_0^{\mathrm{LDA,exact}}(U,n)$ will be discontinuous functions of the particle density. This is definitely wrong for the Hubbard model. In general, we do not expect such a behavior for a charge-neutral system with short-range interactions. Another example for the breakdown of the exact site occupation LDA can be found in [3.23] where a Hubbard model with an external charge-density wave potential is studied in detail.

The ultimate reason for this problem in the LDA approach becomes apparent when we look at the very definition of the effective potential in (3.20). Even if we use the exact exchange potential $v_{\mathrm{xc}}^{\mathrm{exact}}(\boldsymbol{R})$, we cannot calculate the exact gap if we ignore possible discontinuities in the overall energy shift $\lambda - \lambda^{\mathrm{eff}}$. This illustrates the fundamental "problem of the unknown derivative discontinuities" in $v_{\mathrm{xc}}(\boldsymbol{R})$ within the LDA approach [3.29]–[3.31]. The example of the one-dimensional Hubbard model indicates that $V^{\mathrm{eff}}(\boldsymbol{R})$ and thus $v_{\mathrm{xc}}(\boldsymbol{R})$ should be continuous functions of the electron density, and only their density derivatives may display a jump discontinuity.

In real band structure calculations, e.g., for semiconductors or transition-metal oxides, we have to correct for discrepancies between the predicted gaps and those actually observed. To this end we might (i) treat the size of the gap as a fit parameter [3.28], (ii) explicitly include an additional on-site interaction which we treat within an antiferromagnetic Hartree–Fock theory ("LSDA + U" calculations [3.32]), or (iii) consider many-body corrections to the band structure gap using frequency and momentum dependent self-energies [3.33]. We will not comment further on these approaches here.

3.2.3 $X\alpha$ Local Density Approximation

One way out of the dilemma in the exact LDA is to use a smooth *approximate* expression for $v_{\mathrm{xc}}(\boldsymbol{R})$. The possibly simplest way is the lattice generalization of Slater's $X\alpha$ exchange; see Sect. 2.2.3. The $X\alpha$ local density approximation is formulated as [3.31]

$$E_{\mathrm{xc}}^{\mathrm{LDA}}(n(\boldsymbol{R});a) = -aU\sum_{\boldsymbol{R}}[n(\boldsymbol{R})]^{4/3} , \tag{3.22}$$

where a is a free constant that is chosen to reproduce properly some limiting cases, e.g., the filled band ($a = 2^{-4/3}$).

This functional dependence is readily generalized to the local spin-density approximation as [3.34]

$$\begin{aligned} E_{\mathrm{xc}}^{\mathrm{LSDA}}(n_\sigma(\boldsymbol{R});a,b) &= E_{\mathrm{xc}}^{\mathrm{LDA}}(n(\boldsymbol{R});a) + \sum_\sigma E_{\mathrm{xc},\sigma}(n_\sigma(\boldsymbol{R});b) , \\ E_{\mathrm{xc},\sigma}(n_\sigma(\boldsymbol{R});b) &= -\frac{bU}{2^{4/3}-2}\sum_{\boldsymbol{R}}\left\{[2n_\sigma(\boldsymbol{R})]^{4/3} - [n(\boldsymbol{R})]^{4/3}\right\} . \end{aligned} \tag{3.23}$$

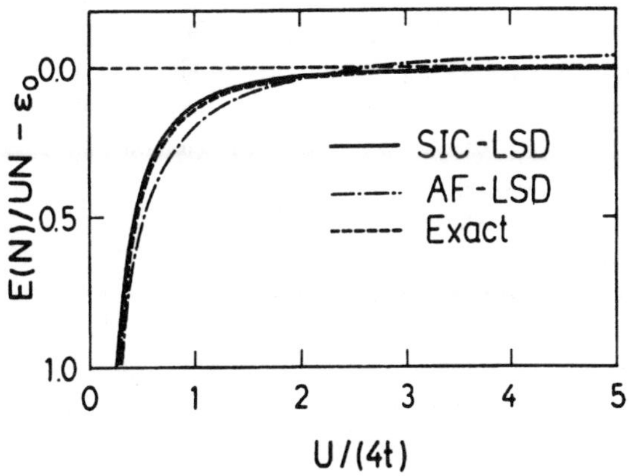

Fig. 3.2. Ground-state energy density for the one-dimensional Hubbard model at half band-filling as a function of the interaction strength. The exact result of Lieb and Wu [3.35] and the results of the local spin-density approximation with (AF-LSD) and without self-interaction corrections (SIC-LSD) are displayed. Note that the Y axis is scaled by $1/U$ (from [3.34]).

In this case the constants $a = 0.3840$, $b = 0.0705$ are chosen such that the total energy and the electron affinity of the free hydrogen atom are reproduced within the LSDA. As in Mott's example we may effectively vary the bandwidth W and keep $U = 12.9\,\mathrm{eV}$ fixed.

To assess the quality of the lattice $X\alpha$ exchange we again compare it with the one-dimensional Hubbard model for which the ground-state properties are known exactly. In particular, we are interested in the size of the Mott–Hubbard gap, and we thus restrict ourselves to half-filled bands. The densities are uniform, $n(\boldsymbol{R}) = 1$ for half band-filling. In this case the LDA reduces to the paramagnetic Hartree theory with, however, an *incorrect* prefactor for the linear term in U/t since the exchange contribution (3.22) is also linear in U. The $X\alpha$ exchange does not give rise to correlation contributions. In particular, the ground-state energy does not go to zero for large interactions.

The situation does not improve much when the local spin-density approximation in the $X\alpha$ form of (3.23) is used. The results for the ground-state energy density are displayed in Fig. 3.2. The following discrepancies can be observed.

(i) For small interaction strengths one finds [3.34, 3.36] $E_0^{\mathrm{LSDA}}(U/t \to 0) = -W/\pi + 0.116U$ while $E_0^{\mathrm{exact}}(U/t \to 0) = -W/\pi + U/4$ holds. The prefactor in front of the *linear* term is thus found to be much too small. Like the LDA the LSDA cannot reproduce the Hartree term.

(ii) For large U/t the exact ground-state energy goes to zero: the LSDA gives a diverging contribution which stems from the self-interaction of the electron. This self-interaction problem of the LSDA for the hydrogen atom has been known for some time [3.31, 3.37].

(iii) For all U/t there are unsystematic deviations from the exact ground-state energy. The LSDA does not provide any bounds on the exact ground-state energy.

(iv) A finite sublattice magnetization is predicted which is actually absent in the one-dimensional Hubbard model.

We thus see that the $X\alpha$ LSDA is not a reliable approximation scheme for the Hubbard model because it does not correctly reproduce any of the limits and provides no reliable bounds. The form of the exchange correlation potential in (3.23) is based on the $X\alpha$ exchange for the free electron gas and is not really suited for a lattice model. Any reasonable discretization must guarantee that at least the Hartree contribution is correctly reproduced.

The principally wrong behavior for $U/t \to \infty$ lies in the LSDA itself and calls for a corresponding improvement. Such corrections are included in the self-interaction corrected LSDA (SIC-LSDA) within which the self-interactions of the electron are eliminated [3.38, 3.39]. For this reason the SIC-LSDA is principally better suited for correlated electron systems than the LSDA.

3.2.4 LSDA with Self-Interaction Corrections

The SIC-LSDA allows for a localization of electrons into orbitals on a given site. Although some of the electrons remain delocalized in Bloch states, the rest of the electrons occupy orbitals which are localized around a lattice site. As shown in Fig. 3.2 and [3.34] the self-interaction corrections substantially improve on the ground-state energy, but some discrepancies remain.

(i) For small interactions one finds $E_0^{\mathrm{SIC-LSDA}}(U/t \to 0) = -W/\pi + 0.247U$. The prefactor of the *linear* term almost equals the exact Hartree contribution $U/4$.
(ii) For large U/t the exact ground-state energy goes to zero like $E_0^{\mathrm{exact}} = -4(\ln 2)t^2/U \approx -2.77t^2/U$ and the SIC-LSDA and the antiferromagnetic Hartree–Fock theory give the results $E_0^{\mathrm{SIC-LSDA}} = -2.35t^2/U$ and $E_0^{\mathrm{Slater}} = E_0^{\mathrm{Néel}} = -2t^2/U$, respectively. The SIC-LSDA thus seems to contain contributions of spin fluctuations at large U/t.
(iii) A finite sublattice magnetization is predicted which is actually absent in the one-dimensional Hubbard model.
(iv) Electrons truly localize, an effect not present in the Hubbard model.
(v) For all U/t the SIC-LSDA deviates only little from the exact ground-state energy but neither it nor the LSDA provide exact bounds for this quantity.

It is seen that the SIC-LSDA energy comes close to the exact solution for both weak and strong interactions. In this respect the SIC-LSDA solution can be considered a better description of the Hubbard model in one dimension than the bare LSDA. Since both methods do not provide exact bounds on the ground-state energy it is nevertheless impossible to judge which of the two methods is to be preferred for real systems.

The main problem of the LSDA and SIC-LSDA shows up in the energy gap. The corresponding results are displayed in Fig. 3.3. The LSDA underestimates the size of the gap, especially for small interaction strengths. The LSDA gap is vanishingly small below an interaction strength of the size of

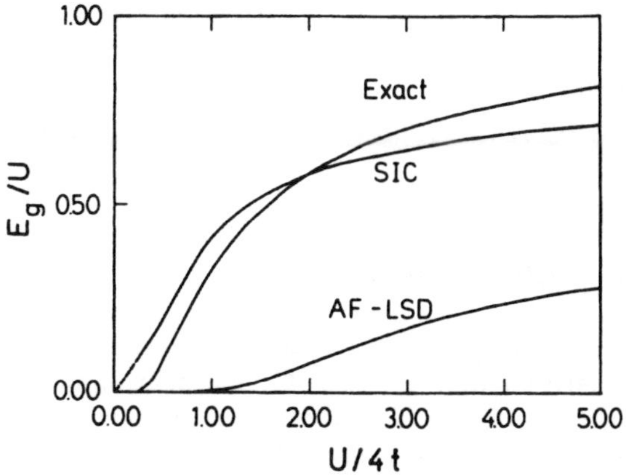

Fig. 3.3. Energy gap of the one-dimensional Hubbard model at half band-filling in the local density approximation as a function of the interaction strength. The exact result of Lieb and Wu [3.35] and the results of the local spin-density approximation with (SIC) and without (AF-LSD) self-interaction corrections are shown (from [3.34]).

the bandwidth and only reaches a fraction of the exact gap $\Delta\mu^{\text{exact}} = U - W$ for large U/W. The very presence of a gap is only a consequence of the broken translational invariance just like in the antiferromagnetic Hartree–Fock theory. Correlation contributions are completely missing.

On the other hand, the SIC-LSDA gives an energy gap that is of the right order of magnitude for large U/t, $\Delta\mu^{\text{SIC-LSDA}} = \mathcal{O}(U)$. Here the problem is that the charge excitations in the lower Hubbard band are completely immobile ($W_2 = 0$) but they delocalize too strongly in the upper Hubbard band ($W_1 > W_1^{\text{exact}}$) [3.34]. The bandwidths for the two Hubbard bands do not fit; electron–hole symmetry is violated. An even more serious problem than the behavior of the gap at large U/t is seen to occur at small interaction strengths; see Fig. 3.3. Whereas the gap in the exact solution is exponentially small for small U/W (see Sect. 4.1), the gap in the SIC-LSDA approximation strongly increases even for very small interaction strengths. Since the numerical calculations become unstable for small values of U/t an extrapolation of the behavior of the SIC-LSDA gap is indicated by dashes in Fig. 3.3. The comparison of the local moment and the sublattice magnetization clearly shows [3.36, 3.40] that some electrons localize in the SIC-LSDA approximation even at arbitrarily small U/W. Consequently, the energy gap and the sublattice magnetization increase *linearly* as a function of U/W.

For small U/W this scenario is not very likely for the metal–insulator transition in the Hubbard model in any dimension. The electrons will perform a correlated motion for $U > 0$ but we do not expect a true localization in the sense that the energy gap linearly increases with U/t. The SIC-LSDA overestimates the tendencies to electron localization in the Hubbard model. We suppose that the SIC-LSDA also predicts an insulating state for densities $n < 1$. Further investigations are certainly desirable [3.36].

The SIC-LSDA extends the scenario for possible symmetry breakings only by a rather "radical" form of breaking translational invariance, namely the localization of some of the electrons. For large interaction strengths it eliminates some pathologies of the LSDA and takes into account some localization tendencies in this limit. In this sense the SIC-LSDA may be regarded as a step forward in the description of correlated electron systems with the help of effective single-electron theories.

In the next three sections we discuss three approximate many-body approaches to the Hubbard model.

3.3 Green-Function Decoupling Method

In this section we discuss the results obtained from the Green-function decoupling scheme. We start with the derivation of the exact Green function in the atomic limit. The spectral function displays peaks at the two atomic energy levels for a single and a double occupancy. For finite hopping these levels broaden into the lower and upper Hubbard bands. The atomic limit provides the starting point for the Hubbard approximations. In the Hubbard-I solution the system is a paramagnetic insulator at half band-filling for all finite interaction strengths. The metal–insulator transition is shifted to a finite value in the Hubbard-III solution. This calculation provided the historically first example for a metal–Mott–Hubbard insulator transition. The Hubbard approximations are plagued with various deficiencies, e.g., the metallic phase is not a Fermi liquid. We discuss possible extensions and improvements at the end of this section. Some technical details are deferred to Appendix A.2.

3.3.1 Atomic Limit and Hubbard Bands

In the atomic limit the Hubbard model reduces to

$$\hat{H}_{\text{atom}} = U \sum_{\boldsymbol{R}} \hat{n}_{\boldsymbol{R},\uparrow}\hat{n}_{\boldsymbol{R},\downarrow} - \mu \sum_{\boldsymbol{R}} (\hat{n}_{\boldsymbol{R},\uparrow} + \hat{n}_{\boldsymbol{R},\downarrow}) \; . \tag{3.24}$$

We can exactly calculate the retarded one-electron Green function for this case. According to the definition in Appendix A.2 it is given by

$$\mathrm{i}G_\sigma^{\text{ret}}(\boldsymbol{R}_1, \boldsymbol{R}_2, t) = \theta(t)\Big\langle \Big[\hat{c}_{\boldsymbol{R}_1,\sigma}(t), \hat{c}^+_{\boldsymbol{R}_2,\sigma}\Big]_+ \Big\rangle \; , \tag{3.25}$$

where $\hat{X}(t) = \exp(\mathrm{i}\hat{H}t)\hat{X}\exp(-\mathrm{i}\hat{H}t)$ is the Heisenberg operator, $\theta(t)$ is the step function, and $\big[\hat{A}, \hat{B}\big]_+ = \hat{A}\hat{B} + \hat{B}\hat{A}$ is the anticommutator.

The Green function is purely local in the atomic limit, $G_\sigma^{\text{ret}}(\boldsymbol{R}_1, \boldsymbol{R}_2, t) = \delta_{\boldsymbol{R}_1,\boldsymbol{R}_2} G_\sigma^{\text{ret}}(\boldsymbol{R}_1, \boldsymbol{R}_1, t)$. Since there are only four possible configurations per site the expectation values for the Green function can readily be calculated. In the absence of a magnetic field the four contributions can be combined to

$$\mathrm{i}G_\sigma^{\text{ret}}(\boldsymbol{R}, \boldsymbol{R}, t) = \theta(t)\left\{ \mathrm{e}^{-\mathrm{i}(U-\mu)t}\langle \hat{n}_{\boldsymbol{R},-\sigma}\rangle + \mathrm{e}^{\mathrm{i}\mu t}\langle 1 - \hat{n}_{\boldsymbol{R},-\sigma}\rangle \right\} \tag{3.26a}$$

whose Fourier transformed is given by

$$G_\sigma^{\text{ret}}(\boldsymbol{R}, \boldsymbol{R}, \omega) = \frac{\langle \hat{n}_{\boldsymbol{R},-\sigma}\rangle}{\omega - (U-\mu) + \mathrm{i}\eta} + \frac{\langle 1 - \hat{n}_{\boldsymbol{R},-\sigma}\rangle}{\omega + \mu + \mathrm{i}\eta} \; . \tag{3.26b}$$

Here, $\eta = 0^+$ is a positive infinitesimal. The advanced Green function is the complex conjugate of the retarded one. According to (A.2.7a) the local spectral function is given by

$$A_\sigma(\boldsymbol{R}, \boldsymbol{R}, \omega) = \langle \hat{n}_{\boldsymbol{R},-\sigma} \rangle \delta(\omega - (U - \mu)) + \langle 1 - \hat{n}_{\boldsymbol{R},-\sigma} \rangle \delta(\omega + \mu) . \tag{3.27}$$

This result is readily understood: if we add a σ electron to the system at site $\boldsymbol{R}$, we either increase the energy by $-\mu$ if the site was empty (probability $\langle 1 - \hat{n}_{\boldsymbol{R},-\sigma} \rangle$), or by $(U - \mu)$ if the site was occupied by a $-\sigma$ electron (probability $\langle \hat{n}_{\boldsymbol{R},-\sigma} \rangle$).

The two atomic levels are the precursors of the upper and lower Hubbard bands. To see this we calculate the Green function in momentum space,

$$G_\sigma(\boldsymbol{k}, \omega) = \frac{1}{L} \sum_{\boldsymbol{R}_1, \boldsymbol{R}_2} \mathrm{e}^{-\mathrm{i}\boldsymbol{k}(\boldsymbol{R}_1 - \boldsymbol{R}_2)} G_\sigma(\boldsymbol{R}_1, \boldsymbol{R}_2, \omega) , \tag{3.28a}$$

as

$$G_\sigma^{\text{atom, ret}}(\boldsymbol{k}, \omega) = \frac{n_{-\sigma}}{\omega - (U - \mu) + \mathrm{i}\eta} + \frac{1 - n_{-\sigma}}{\omega + \mu + \mathrm{i}\eta} . \tag{3.28b}$$

The spectral function in momentum space, $A_\sigma(\boldsymbol{k}, \omega)$, displays two poles at $\omega = -\mu$ and $\omega = U - \mu$, independent of $\boldsymbol{k}$. According to Mott's general arguments these two energy levels should broaden into bands as soon as we allow for hopping of electrons between the lattice sites. Hubbard's Green-function decoupling approach was the first to substantiate these ideas, and the corresponding bands are called the upper and lower Hubbard bands, respectively.

3.3.2 Hubbard-I Approximation

The Hubbard approximations can be derived from the equation-of-motions approach to Green functions. This Green-function decoupling approach is lengthy and is carried out in Appendix A.2 for the Hubbard-I approximation. Fortunately, the Hubbard approximations can easily be understood in physical terms, as we now discuss.

Hubbard [3.41] used the atomic limit as the starting point to define a *single* self-energy to describe the *two* atomic levels, i.e., he wrote ($\omega \equiv \omega + \mathrm{i}\eta \operatorname{sgn}(\omega)$)

$$G_\sigma^{\text{atom}}(\boldsymbol{k}, \omega) = \frac{n_{-\sigma}}{\omega - (U - \mu)} + \frac{1 - n_{-\sigma}}{\omega + \mu} \tag{3.29a}$$

$$\equiv \frac{1}{\omega + \mu - \Sigma_\sigma^{\text{atom}}(\omega)} , \tag{3.29b}$$

which defines the atomic self-energy as

$$\Sigma_\sigma^{\text{atom}}(\omega) = U n_{-\sigma} + U^2 \frac{n_{-\sigma}(1 - n_{-\sigma})}{\omega + \mu - U(1 - n_{-\sigma})} . \tag{3.30}$$

At $\omega = -\mu + U(1 - n_{-\sigma})$ this self-energy has a spurious singularity that has no physical origin. It is the consequence of working with a single self-energy.

As we will argue later in this section, an approach with separate self-energies for the upper and lower Hubbard bands appears to be more natural since they are free of spurious singularities.

The Hubbard-I approximation is readily formulated as

$$\text{Hubbard-I:} \quad \Sigma_\sigma(\omega) = \Sigma_\sigma^{\text{atom}}(\omega) , \tag{3.31a}$$

such that the lattice Green function becomes

$$G_\sigma(\boldsymbol{k},\omega) = \frac{1}{\omega - (\epsilon(\boldsymbol{k}) - \mu) - \Sigma_\sigma^{\text{atom}}(\omega)} . \tag{3.31b}$$

In this approximation, the exact self-energy is replaced by the atomic self-energy. The zeros of the Green function define the quasi-particle dispersions; see Appendix A.2. They are not damped since the self-energy is purely real.

The Green function can be cast into the form

$$G_\sigma(\boldsymbol{k},\omega) = \frac{Z_\sigma^+(\boldsymbol{k})}{\omega - E_\sigma^+(\boldsymbol{k})} + \frac{Z_\sigma^-(\boldsymbol{k})}{\omega - E_\sigma^-(\boldsymbol{k})} , \tag{3.32a}$$

with the quasi-particle dispersion relations $E_\sigma^\pm(\boldsymbol{k}) \equiv E_\sigma^\pm(\epsilon(\boldsymbol{k}))$

$$E_\sigma^\pm(\boldsymbol{k}) = \frac{1}{2}\left[\epsilon(\boldsymbol{k}) + U \pm \sqrt{(\epsilon(\boldsymbol{k}) - U)^2 + 4Un_{-\sigma}\epsilon(\boldsymbol{k})}\,\right] \tag{3.32b}$$

for the upper and lower Hubbard bands. In contrast to the Hartree–Fock solution, quasi-particle states at different momenta $\boldsymbol{k}$ do not have the same, $\boldsymbol{k}$-independent spectral weight. Instead, the bands are "dynamic" with the spectral weights

$$Z_\sigma^\pm(\boldsymbol{k}) \equiv Z_\sigma^\pm(\epsilon(\boldsymbol{k})) = \pm\frac{E_\sigma^\pm(\boldsymbol{k}) - U(1 - n_{-\sigma})}{E_\sigma^+(\boldsymbol{k}) - E_\sigma^-(\boldsymbol{k})} . \tag{3.32c}$$

In particular, the number of states available in the lower (upper) band does not necessarily equal $L/2$. Instead, the number of levels per Hubbard band is given by

$$(N^\pm/L) = \int_{W_\pm^{\min}}^{W_\pm^{\max}} \mathrm{d}\epsilon D_\sigma^\pm(\epsilon) = \int_{-W/2}^{W/2} \mathrm{d}\epsilon D_{\sigma,0}(\epsilon) Z_\sigma^\pm(\epsilon) , \tag{3.33}$$

where $D_{\sigma,0}(\epsilon)$ denotes the density of states for zero interaction. The total number of states available in both Hubbard bands together equals L, as it should.

3.3.3 Advantages and Problems

The Hubbard-I solution becomes exact in the following limits:

(i) for $U = 0$ since both dispersion relations $E_\sigma^\pm(\boldsymbol{k})$ together reduce to the bare dispersion relation $\epsilon(\boldsymbol{k})$;
(ii) in the atomic limit, $\epsilon(\boldsymbol{k}) = 0$, by construction.

In addition, the Hubbard-I solution displays a variety of new, *many-particle* features compared to the Hartree–Fock description.

(i) The upper and lower Hubbard bands are "dynamic" in the sense that both the form of density of states $D_\sigma(\epsilon)$ and the number of available states per band depend on the strength of the Hubbard interaction and on the electron density. In single-electron theories one has the notion of a rigid band that can successively be filled with electrons. The Hubbard-I solution shows that this concept might no longer hold for quasi-particle bands in interacting electron systems.

(ii) The upper and lower Hubbard bands retain a finite width even in the limit $U \gg W$, in contrast to the results of Slater's Hartree–Fock theory or of the SIC-LSDA, which localize electrons on fixed lattice sites. For strong interactions we find

$$E_\sigma^+(\boldsymbol{k}) = U + n_{-\sigma}\epsilon(\boldsymbol{k}) + \mathcal{O}\left(W^2/U\right) , \tag{3.34a}$$

$$E_\sigma^-(\boldsymbol{k}) = (1 - n_{-\sigma})\epsilon(\boldsymbol{k}) + \mathcal{O}\left(W^2/U\right) . \tag{3.34b}$$

For large interactions the bandwidths for the upper and lower Hubbard bands and thus given by $W_1 = Wn_{-\sigma}$ and $W_2 = W(1 - n_{-\sigma})$.

(iii) The onset of a magnetic order is not a necessary condition for the metal–Mott insulator transition. Instead there can be a transition from a paramagnetic metal to a paramagnetic insulator.

To elucidate the last point let us consider a system whose unperturbed density of states shall be constant, $D_{\sigma,0}(\epsilon) = 1/W$ for $|\epsilon| \leq W/2$. According to (A.2.16b) the density of states of the two Hubbard bands remains constant, $D_\sigma^\pm(E) = 1/W$. At half-filling, $n_\uparrow = n_\downarrow = 1/2$, the lower of the two Hubbard bands,

$$E^\pm(\boldsymbol{k}, n = 1) = \frac{1}{2}\left[\epsilon(\boldsymbol{k}) + U \pm \sqrt{\epsilon(\boldsymbol{k})^2 + U^2}\right] , \tag{3.35}$$

is completely filled because $n_- = (W_-^{\max} - W_-^{\min})/W = 1/2$ holds. The chemical potential to add another electron is then given by $\mu^+(n = 1) = W_+^{\min} = (-W/2 + U + \sqrt{(W/2)^2 + U^2})/2$. Removing an electron from the filled lower Hubbard band requires the energy $\mu^-(n = 1) = W_-^{\max} = (W/2 + U - \sqrt{(W/2)^2 + U^2})/2$. Thus the gap for one-electron excitations is calculated to be

$$\Delta\mu^{\text{H.-I}}(n = 1) = \sqrt{(W/2)^2 + U^2} - W/2 . \tag{3.36}$$

Both Hubbard bands split for arbitrarily small interaction strength. For a constant density of states the Hubbard-I approximation predicts a metal–Mott insulator transition at $U_\text{c}^{\text{H.-I}}(n = 1) = 0^+$. As in the SIC-LSDA the gap grows with a power law for small interaction strengths.

The Hubbard-I solution is a very rough approximation to the physics of the Hubbard model. Except for the extreme cases $U = 0$ or $W = 0$ it is not exact for any other parameter value, and it is not a conserving approximation either. This can been seen for both small and large interaction strengths.

For small interactions it does *not* reproduce the Hartree–Fock solution since the quasi-particle energies go over into the Hartree–Fock expression $\epsilon_\sigma^{\mathrm{HF}}(\boldsymbol{k}) = \epsilon(\boldsymbol{k}) + Un_{-\sigma}$ for only those momenta for which $|\epsilon(\boldsymbol{k})| \gg U$ [3.42], as can easily be seen from (3.32b). For those momenta which obey $|\epsilon(\boldsymbol{k})| \ll U$ the Hubbard bands tend to their atomic limit, (3.34). This behavior is certainly not generic for the Hubbard model. It would imply that electrons with small kinetic energy are so drastically influenced by the interaction that they actually localize. In this respect the Hubbard-I approach resembles the SIC-LSDA and explains the strong enhancement of the energy gap beyond the metal–insulator transition at $U = 0^+$. For a transition at $U = 0$ a power-law behavior of the one-electron gap is not generic for the Hubbard model.

Turning to strong interactions we note that the Hubbard-I approximation lacks all antiferromagnetic correlations. The whole magnetic behavior of the electrons in the lower Hubbard band is very peculiar. Even for temperatures $U \gg k_{\mathrm{B}}T \gg W^2/U$ the magnetic susceptibility of the filled lower Hubbard band does not reduce to a Curie-law behavior of a paramagnetic band insulator [3.43]. Furthermore, ferromagnetism is impossible within the Hubbard-I solution for a constant density of states $D_{\sigma,0}(E)$. However, all statements on the possibility of ferromagnetism strongly depend on the form of the chosen density of states [3.41].

A fundamental flaw in the Hubbard-I approximation is the violation of electron–hole symmetry, which has severe consequences. For a general density of states $D_{\sigma,0}(\epsilon)$ the metal–insulator transition does not necessarily happen at half-filling, $n_{\mathrm{c}} = 1$. This is very unlikely for the Hubbard model and certainly incorrect in one dimension and in infinite dimensions; see Chaps. 4 and 5. The violation of elementary symmetries and sum rules and the poor description of magnetism in the Hubbard model lead to the conclusion that we should abandon the Hubbard-I approximation as a tool for the study of correlated electron systems, even if some features, such as the electron localization, reappear in the SIC-LSDA. The basic concepts behind the Hubbard-I approximation as discussed above became very popular since the method allowed us to describe the metal–Mott insulator transition in a rather simple "band structure picture" such that many properties of the Hubbard model and of real correlated electron systems are discussed in the terminology of the Hubbard approximations.

3.3.4 Improvements

Several of the major flaws in the Hubbard-I approximation can easily be cured if we abandon Hubbard's viewpoint that the atomic limit provides a *single* self-energy for the case of a finite bandwidth. Instead, the atomic levels broaden into *two* Hubbard bands such that we should rather work with a two-by-two matrix for the self-energy. For the sake of simplicity we neglect interband effects, which is a good approximation for large U/W. Then we are left with separate self-energies for the lower and the upper Hubbard band as independently proposed by Logan and Nozières (private communication, 1995) [3.44].

Following Hubbard we start from the atomic Green function in momentum space and write this quantity as

$$G_\sigma^{\rm atom}(\boldsymbol{k},\omega) = \frac{n_{-\sigma}}{\omega+\mu-\Sigma_\sigma^{\rm UHB,\,atom}(\omega)} + \frac{1-n_{-\sigma}}{\omega+\mu-\Sigma_\sigma^{\rm LHB\,atom}(\omega)} \;. \tag{3.37a}$$

This allows us to identify the respective self-energies for the upper and lower Hubbard bands as

$$\Sigma_\sigma^{\rm UHB,\,atom}(\omega) = U \quad , \quad \Sigma_\sigma^{\rm LHB,\,atom}(\omega) = 0 \;. \tag{3.37b}$$

We now use these two quantities to formulate an approximation for finite hopping, $\epsilon(\boldsymbol{k}) \neq 0$.

We know that the hopping of the electrons broadens the atomic levels into bands. For strong interactions, we may certainly use the *atomic* self-energies. As a reasonable approximation for large interactions we write

$$\begin{aligned} G_\sigma(\boldsymbol{k},\omega) &= \frac{n_{-\sigma}}{\omega-(\epsilon(\boldsymbol{k})-\mu)-\Sigma_\sigma^{\rm UHB,\,atom}(\omega)} \\ &\quad + \frac{1-n_{-\sigma}}{\omega-(\epsilon(\boldsymbol{k})-\mu)-\Sigma_\sigma^{\rm LHB,\,atom}(\omega)} \;. \end{aligned} \tag{3.38}$$

Obviously, the theory becomes exact in the atomic limit and for $U = 0$. Furthermore, we describe two Hubbard bands with the bandwidths $W_1 = W_2 = W$. They merge at $U_c = W$, and our approximate Green-function description in (3.38) contains a metal-to-insulator transition for half band-filling ($\mu = U/2$, $n_{-\sigma} = 1/2$). The gap opens linearly; the exponent is $\nu = 1$. Electron–hole symmetry and other basic relations are naturally fulfilled in this simple approximation.

3.3.5 Alloy-Analogy Approximation

In a subsequent paper [3.45] Hubbard extended the concepts and techniques of the Hubbard-I approximation. The corresponding Green-function decoupling scheme is very lengthy, and we present only the most important physical ideas, following Cyrot [3.46]. In his third paper on the Hubbard model [3.45] Hubbard actually introduced two approximations. We first discuss the "scattering corrections" to the Hubbard-I approximation ("alloy-analogy approximation", "Falicov–Kimball model"), and leave the "resonance broadening corrections" which lead to the (full) Hubbard-III approximation to the next subsection.

In the atomic limit the electrons do not move and either experience the potential U or no interaction on their lattice site. An itinerant σ electron encounters a sequence of empty and occupied sites during its motion through the system. The site occupancy itself dynamically changes by the motion of the $-\sigma$ electrons. Hubbard considered the simultaneous description of the dynamics of both species as a too complex problem. Hence he started out with the assumption that the $-\sigma$ electrons are fixed, $t_{-\sigma} \equiv 0$, and he considered the motion of the σ electrons. This is the Falicov–Kimball model [3.47]; see Sect. 5.4.4. Given this simplification we may still keep the assumption as in the Hubbard-I approximation that the lattice sites are occupied or unoccupied with probabilities $\langle \hat{n}_{\boldsymbol{R},-\sigma} \rangle$ and $\langle 1 - \hat{n}_{\boldsymbol{R},-\sigma} \rangle$, respectively. Since the positions $\boldsymbol{R}$ of the $-\sigma$ electrons are randomly distributed we have to deal with an alloy problem ("alloy-analogy approximation", AAA).

During the motion through the randomly positioned $-\sigma$ electrons the σ electron is frequently scattered, which results in a energy-dependent "scattering correction" $\Delta_\sigma(\omega)$ for the unperturbed energy levels at $\omega = -\mu$ and $\omega = U - \mu$. The equation (3.30) for the atomic self-energy is now supposed to hold for energies $\omega - \Delta_\sigma(\omega)$,

$$\begin{aligned} \text{AAA:} \quad \Sigma_\sigma(\omega) &= \Sigma_\sigma^{\text{atom}}(\omega - \Delta_\sigma(\omega)) \\ &= U n_{-\sigma} + U^2 \frac{n_{-\sigma}(1 - n_{-\sigma})}{\omega - \Delta_\sigma(\omega) + \mu - U(1 - n_{-\sigma})} \, . \end{aligned} \tag{3.39a}$$

This self-energy then enters the lattice Green function via the Dyson equation

$$G_\sigma(\boldsymbol{k}, \omega) = \frac{1}{\omega - (\epsilon(\boldsymbol{k}) - \mu) - \Sigma_\sigma(\omega)} \, . \tag{3.39b}$$

The local Green function is found as the average over all momenta,

$$\begin{aligned} G_\sigma(\omega) &= \frac{1}{L} \sum_{\boldsymbol{k}} \frac{1}{\omega - (\epsilon(\boldsymbol{k}) - \mu) - \Sigma_\sigma(\omega)} \\ &= \int_{-W/2}^{W/2} \frac{\mathrm{d}\epsilon D_{\sigma,0}(\epsilon)}{\omega - (\epsilon - \mu) - \Sigma_\sigma(\omega)} \, . \end{aligned} \tag{3.40}$$

Thus far, the scattering corrections $\Delta_\sigma(\omega)$ remained undetermined. Hubbard now demanded that the local Green function still obeys (3.29b) for the shifted energy levels, i.e.,

$$G_\sigma(\omega) = \frac{1}{\omega - \Delta_\sigma(\omega) + \mu - \Sigma_\sigma(\omega)} \, . \tag{3.41}$$

The equations (3.40) and (3.41) have to be solved self-consistently and constitute Hubbard's alloy-analogy approximation ("scattering corrections only" [3.45]).

The self-energy $\Sigma_\sigma(\omega)$ is precisely the "coherent potential" of the "coherent potential approximation" (CPA) [3.48] which was therefore already anticipated by Hubbard [3.45]; see also [3.49, Chap. 9.4]. The CPA gives the *exact* result for the one-electron Green function for the alloy problem in the limit of infinitely many neighboring sites [3.50]. This limit is obtained in infinite dimensions or for a Bethe lattice with infinite coordination number. As a consequence of this observation the CPA cannot give unphysical results, e.g., a violation of sum rules is excluded. For a thorough discussion, see [3.51].

The alloy-analogy approximation provides the exact solution of the Falicov–Kimball model in infinite dimensions. To see this we write (3.41) as

$$G_\sigma(\omega) = \frac{1 - n_{-\sigma}}{(\omega - \Delta_\sigma(\omega)) + \mu} + \frac{n_{-\sigma}}{(\omega - \Delta_\sigma(\omega)) + \mu - U} \,. \tag{3.42}$$

Hence, the scattering correction $\Delta_\sigma(\omega)$ can be viewed as the effect of an external field on the atomic problem. As shown in Sect. 5.4.4 the solution of the atomic problem in the presence of an arbitrary external field is enough to solve exactly the Falicov–Kimball model in infinite dimensions.

In general, the self-consistency equations (3.40) and (3.41) can only be solved numerically. Hubbard [3.45] realized that they can be rewritten for half-filling, $n_\sigma = 1/2$, in the form of a third-order algebraic equation if a special density of states is chosen. The density of states shall be given by the "Hubbard semi-ellipse",

$$D_{\sigma,0}(\epsilon) = \frac{4}{\pi W}\sqrt{1 - \left(\frac{2\epsilon}{W}\right)^2} \quad \text{for} \quad |\epsilon| \leq W/2 \,. \tag{3.43}$$

This density of states becomes exact for a Bethe lattice with infinite coordination number; see, e.g., [3.5, 3.51]. With the abbreviation $E = \omega + \mu$, (3.40) becomes

$$G_\sigma(\omega) = \frac{8}{W^2}\left[E - \Sigma_\sigma(\omega) - \sqrt{[E - \Sigma_\sigma(\omega)]^2 - (W/2)^2}\right] \tag{3.44}$$

under the condition of a positive square root for $|E - \Sigma_\sigma(\omega)| > W/2$. Furthermore, we find from (3.41) that

$$\Delta_\sigma(\omega) = \left(\frac{W}{4}\right)^2 G_\sigma(\omega) \,. \tag{3.45}$$

On a Bethe lattice with infinite coordination number the self-consistency equations for the paramagnetic phase of the Falicov–Kimball model at half band-filling ($\mu = U/2$) thus reduce (3.42) to

$$2G_\sigma(\omega) = \frac{1}{\omega + U/2 - (W/4)^2 G_\sigma(\omega)} + \frac{1}{\omega - U/2 - (W/4)^2 G_\sigma(\omega)} \,. \tag{3.46a}$$

We denote $\widetilde{G}_\sigma(\omega) = W G_\sigma(\omega)/4$, $\widetilde{U} = 4U/W$, and $\widetilde{\omega} = 4\omega/W$, and finally find

$$[\widetilde{G}_\sigma(\omega)]^3 - 2\widetilde{\omega}[\widetilde{G}_\sigma(\omega)]^2 + \widetilde{G}_\sigma(\omega)\left(1 + \widetilde{\omega}^2 - (\widetilde{U}/2)^2\right) - \widetilde{\omega} = 0 \,. \tag{3.46b}$$

If (3.46b) has complex solutions, the density of states is finite.

The width of the upper and lower Hubbard bands for strong coupling can easily be determined since, for large U, the upper and lower Hubbard bands are centered around $\omega = \pm U/2$. The cubic equation (3.46b) reduces to a quadratic equation which has complex solutions for $|\omega \pm U/2| \leq \sqrt{2}W/4$. Consequently, the bandwidth of the Hubbard bands is $W_{1,2} = W/\sqrt{2}$ for large interactions. The upper and lower Hubbard bands do not display the full bandwidth because we made the approximation that one spin species is frozen while the other spins move.

Electron–hole symmetry at half band-filling means the insulating state ceases to exist when the density of states becomes finite at $\omega = 0$. At $\omega = 0$ the Green function from (3.46b) is given by $\widetilde{G}_\sigma^2(\omega = 0) = (\widetilde{U}/2)^2 - 1$, and the density of states at $\omega = 0$ is finite for $\widetilde{U} \leq 2$. It is seen that there is a Mott transition from a paramagnetic metal to a paramagnetic insulator at a *finite* critical value $U_{\mathrm{c}}^{\mathrm{AAA}}(n = 1) = W/2$. A more detailed analysis of the cubic equation [3.5] shows that the gap increases like $\Delta\mu^{\mathrm{AAA}}(n = 1) \sim (U/U_{\mathrm{c}} - 1)^{3/2}$. The exponent for the gap function in the alloy-analogy approximation thus is $\nu^{\mathrm{AAA}} = 3/2$. This is the exact result for the Falicov–Kimball model on a Bethe lattice with infinite coordination number.

3.3.6 Hubbard-III Approximation

Thus far we have assumed the $-\sigma$ electrons to be fixed. In fact, this is not the case so we have to supply $\Delta_\sigma(\omega)$ with two "resonance broadening corrections" [3.45]. These come from the physically allowed processes that a σ electron arrives at a site but a $-\sigma$ electron leaves the site (correction $\Delta_{-\sigma}(\omega)$). Alternatively, a σ electron may arrive at an empty site and a $-\sigma$ hole is scattered out, which, because of electron–hole symmetry, results in the correction $[-\Delta_{-\sigma}(U - \omega)]$. The (full) Hubbard-III approximation then becomes [3.45, 3.46]

$$\Delta_\sigma^{\mathrm{tot}}(\omega) = \Delta_\sigma(\omega) + \Delta_{-\sigma}(\omega) - \Delta_{-\sigma}(U - \omega) , \tag{3.47a}$$

where $\Delta_\sigma^{\mathrm{tot}}(\omega)$ now changes the self-energy to

$$\text{Hubbard-III:} \quad \Sigma_\sigma(\omega) = \Sigma_\sigma^{\mathrm{atom}}(\omega - \Delta_\sigma^{\mathrm{tot}}(\omega)) \tag{3.47b}$$
$$= Un_{-\sigma} + U^2 \frac{n_{-\sigma}(1 - n_{-\sigma})}{\omega - \Delta_\sigma^{\mathrm{tot}}(\omega) + \mu - U(1 - n_{-\sigma})} .$$

The self-consistency equations (3.40) and (3.41) formally stay the same. However, since $\Delta_\sigma(\omega)$ and $\Delta_\sigma^{\mathrm{tot}}(\omega)$ are different in general, the Green function can no longer be written in the form of (3.42).

To illustrate the (full) Hubbard-III solution we concentrate on the paramagnetic half-filled band for which we have $\Delta_\sigma^{\mathrm{tot}}(\omega) = 3\Delta_\sigma(\omega)$. For the semi-elliptic density of states equation (3.45) still holds, and the self-consistency equations again reduce to a cubic equation,

$$3[\widetilde{G}_\sigma(\omega)]^3 - 4\widetilde{\omega}[\widetilde{G}_\sigma(\omega)]^2 + \widetilde{G}_\sigma(\omega)\Big(3 + \widetilde{\omega}^2 - \big(\widetilde{U}/2\big)^2\Big) - \widetilde{\omega} = 0 , \tag{3.48}$$

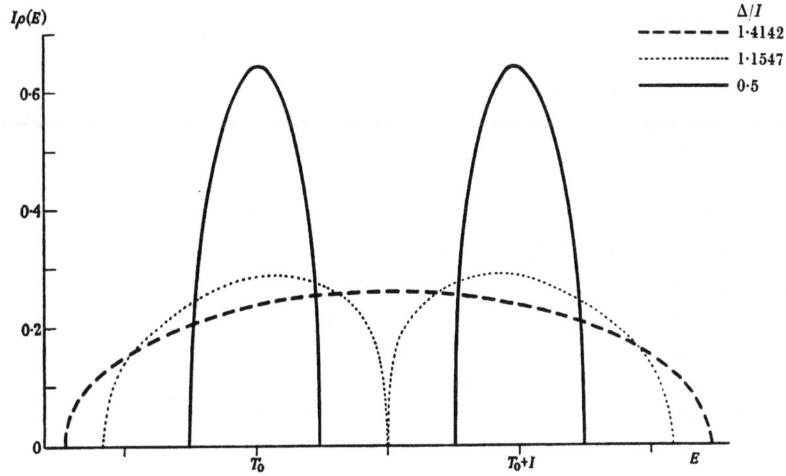

Fig. 3.4. One-electron density of states $UD_\sigma(E)$ in the Hubbard-III approximation as a function of energy for various interaction strengths and half band-filling $n = 1$. The notation is $\Delta \equiv W$, $I \equiv U$, $T_0 \equiv -U/2$. The lower and upper Hubbard bands split at $(U_{\rm c,P}^{\rm H.\text{-}III}(n=1))^{-1} = (2/\sqrt{3})W^{-1} \approx 1.1547W^{-1}$ at which the metal–Mott insulator transition occurs (from [3.45]).

where we used the abbreviations $\widetilde{G}_\sigma(\omega) = WG_\sigma(\omega)/4$, $\widetilde{U} = 4U/W$, and $\widetilde{\omega} = 4\omega/W$. The final result for the density of states is displayed in Fig. 3.4.

The same analysis as in the last subsection shows that the upper and lower Hubbard bands now have the full bandwidth, $W_{1,2} = W$, for large interaction strengths. The metal–Mott insulator transition is shifted to the critical value $U_{\rm c,P}^{\rm H.\text{-}III}(n=1) = \sqrt{3}W/2 \approx 0.866W$. The exponent for the opening of the gap, however, does not change from its value for the Falicov–Kimball model, $\nu^{\rm H.\text{-}III} = 3/2$. The Hubbard-III approximation is further analyzed in [3.52].

3.3.7 Advantages and Problems

Like the Hubbard-I solution Hubbard's "improved" solution is exact in the limits $U = 0$ and $t = 0$. In the Hubbard-III solution the Hubbard bands split at a finite strength of the interaction. This is the "classical" metal–Mott–Hubbard insulator transition: independent of the formation of a magnetic (long-range) order there is a transition from a metal to an insulator at a finite interaction strength $U = \mathcal{O}(W)$. In contrast to the Hubbard-I solution, the upper and lower Hubbard bands now display the full non-interacting bandwidth for large interaction strengths, in agreement with Mott's general argument: the charge excitations should move freely, if the influence of the spin background could be ignored.

Unfortunately, the fundamental problems of the Hubbard-I approximation remain.

(i) For small U/t the Hartree solution cannot be reproduced.
(ii) For large U/t antiferromagnetic correlations are missing and the magnetic behavior of the electrons in the lower Hubbard band does not reproduce a Curie law of band-electrons for temperatures $U \gg k_B T \gg W^2/U$ [3.43]. Recent studies by Vollhardt et al. [3.53] show that neither ferromagnetism nor antiferromagnetism are possible within the Hubbard-III solution.
(iii) The Hubbard bands do not precisely contain one state per atom and, in general, violate electron–hole symmetry and sum rules.
(iv) The metallic state for $U < U_{\mathrm{c,P}}^{\mathrm{H.\text{-}III}}(n = 1)$ is not a Fermi liquid [3.54, 3.55] because the self-energy does not vanish at the Fermi energy. The insulating state of the atomic limit dominates the Hubbard-III approximation too strongly to allow for a correct description of the metallic state below the transition [3.56].

Similar to the Hubbard-I solution the importance of the Hubbard-III solution lies in the facts that (i) basic ideas are introduced and used for the first time (Falicov–Kimball model and CPA), and (ii) a metal–insulator transition at a finite critical interaction strength $U_{\mathrm{c,P}}(n = 1)$ is predicted to be possible. It is driven by (charge) correlations, and there is no onset of magnetic order at the transition. The Hubbard-III solution thus provides an instructive approximate description of a metal–Mott–Hubbard insulator transition.

3.3.8 Improvements

The shortcomings of the Hubbard solutions are of a different type. First, they do not allow for magnetism in the Hubbard model and violate sum rules. The reason for that is the *arbitrariness in the decoupling* of higher-order Green functions to close the system of equations of motion. Second, any Green-function decoupling is based on the *alloy-analogy approximation* in which it is assumed that the dynamics of the σ and the $-\sigma$ electrons can be treated separately. While the dynamics of the σ electrons is studied, the $-\sigma$ electrons are kept fixed and act as centers for impurity scattering. This approach always results in a solution which cannot be a Fermi liquid because the imaginary part of the self-energy is finite at the Fermi energy, a standard result for electrons scattered off impurities.

As a consequence of this twofold failure of the theory there are two directions to improve the alloy-analogy approximation: the formulation of a thermodynamically consistent approximation on one hand, and the inclusion of Fermi-liquid behavior in the metallic phase on the other hand.

Thermodynamic Consistency. It is clear that one has to find a more systematic decoupling scheme than the rather intuitive decoupling used by

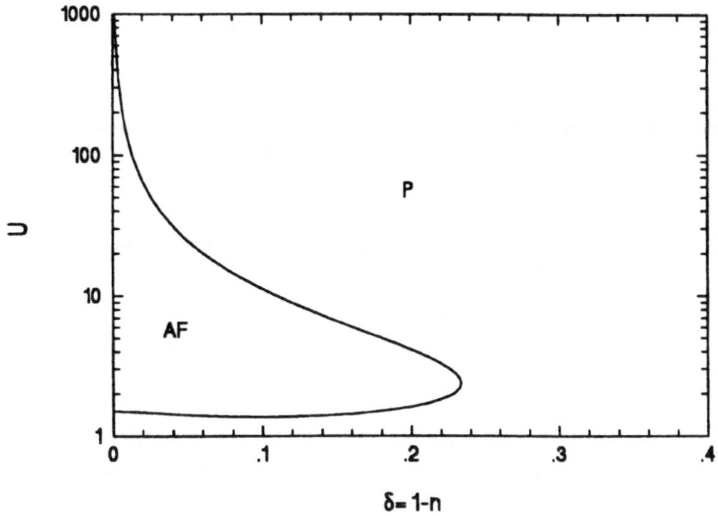

Fig. 3.5. Phase diagram of the Hubbard model in the thermodynamically consistent alloy-analogy approximation as a function of the hole concentration $\delta = 1 - n$ and the interaction strength $U/(W/4)$. Note the logarithmic scale of the Y axis (from [3.53]).

Hubbard. This can be achieved by a decoupling scheme that conserves the lowest-order moments of the spectral function, i.e., the moments $M_\sigma(\boldsymbol{k}, n) = \int \mathrm{d}E E^n A_\sigma(\boldsymbol{k}, E)$ are calculated exactly for the Hubbard model for $n \leq 2$. As a result of this "method of spectral moments" [3.57]–[3.59] the "atomic levels" are determined self-consistently; they were fixed at $E^1_{\mathrm{at}} = 0, E^2_{\mathrm{at}} = U$ in Hubbard's work. The idea of variable atomic levels is also the basis for the improvements suggested by Cyrot [3.60, 3.61] and Kawabata [3.43] who was the first to identify this shortcoming in the Green-function decouplings of the Hubbard-III approximation. Within the alloy-analogy approximation with self-consistently determined atomic levels it is now possible to obtain magnetic solutions [3.43], [3.60]–[3.62].

Janiš and Vollhardt [3.53] provided a comprehensive approach to the Hubbard model within the alloy-analogy approximation. They used the exact solution of the Falicov–Kimball model in infinite dimensions to construct a thermodynamically consistent mean-field theory for the Hubbard model. For the semi-elliptic density of states (3.43) this approach describes a transition from a paramagnetic metal to an antiferromagnetic Mott–Heisenberg insulator at $U_{\mathrm{c,AF}}^{\mathrm{cons.\ AAA}}(n = 1) \approx 0.75(W/2)$ instead of the transition into a paramagnetic Mott–Hubbard insulator. If we (artificially) suppressed the antiferromagnetic solution in the self-consistency equations there would be a transition into the Mott–Hubbard insulator at $U_{\mathrm{c,P}}^{\mathrm{cons.\ AAA}}(n = 1) = W/2$, as derived previously. As well as the possibility for antiferromagnetism the thermodynamically consistent evaluation of the alloy-analogy approximation for the Hubbard model also allows the evaluation of two-electron quantities within the same approximation scheme; see [3.53]. This had not been feasible within Hubbard's original approach.

The whole phase diagram in the thermodynamically consistent alloy-analogy approximation is shown in Fig. 3.5. In contrast to the Hartree–Fock phase diagram there is no ferromagnetic phase. Even at half-filling, antiferromagnetism exists only above the threshold value of the interaction $U_{\mathrm{c,AF}}^{\mathrm{cons.\ AAA}}(n=1) \approx 0.75(W/2)$. A "perfect nesting" between lattice and dispersion relation would not play a role in this approach because the metallic state is not a Fermi liquid. For the Hubbard model with the "perfect nesting" property this is unlikely since we may indeed expect a transition into the antiferromagnetic state at $U_{\mathrm{c}}(n=1)=0^{+}$; see Sect. 4.4. In contrast to Hartree–Fock theory, which overestimates the importance of magnetic phases, the thermodynamically consistent alloy-analogy approximation (Fig. 3.5) produces a phase diagram that indicates that magnetism may actually play a minor role in the physics of the Hubbard model. The theory based on the alloy-analogy approximation, however, tends to underestimate the importance of (ferro-)magnetism in the model since we know that there are tendencies to ferromagnetism in the Hubbard model for large U/t; see Sect. 2.5.

Inclusion of Fermi-Liquid Behavior. The thermodynamically consistent version of the alloy-analogy approximation is still plagued by the fact that the "metallic" state is not a Fermi liquid. It is more difficult to improve on this shortcoming of the Hubbard-III approximation than on thermodynamic consistency because the alloy-analogy approximation has to be abandoned. The basic idea is to restore the Fermi-liquid properties by allowing a recoil of the scattering centers. The σ electrons are then moving no longer in a static but in a "time-dependent" effective field of the $-\sigma$ electrons [3.63]. However, it is difficult to design an improved version of the alloy-analogy approximation "by hand" to generate the desired new feature without destroying the positive aspects of the old approximation. For example, Kawabata proposed an improvement to the Hubbard-III approximation which is not exact in the atomic limit [3.55].

The limit of infinite dimensions offers a systematic way to go beyond the alloy-analogy approximation. This is the subject of Sect. 5.4 and will be discussed in the framework of the Hubbard model in infinite dimensions. Here we just mention that it is indeed possible within this framework to restore the Fermi-liquid property into the Hubbard-III approximation. For example, Edwards [3.64] constructed an approximation that is exact to order $(U/t)^2$ for the Hubbard model in infinite dimensions. The range of the Fermi-liquid phase, however, does not reach up to the transition into the (paramagnetic) insulator. Instead there is another transition to a paramagnetic, non-Fermi-liquid state before the insulating state is reached.

One of the characteristic features of the Hubbard approximations is the site-independence of the proper self-energy. It is therefore natural to extend the approximation by the inclusion of cluster corrections which also appear in the theory of disordered systems. Preliminary results [3.65, 3.66] indicate that the critical value for the Mott–Hubbard transition is shifted to higher values of the interaction strength [3.65] and that local antiferromagnetic correlations become more important [3.66] since the "perfect nesting" property of the lattice is partially taken into account by the cluster corrections.

Some older approaches to improve on the Hubbard approximations have not been listed here. The recent developments outlined in this subsection show that now the Hubbard-III approximation is understood in so much detail that it can be formulated and extended in a well-controlled manner. Our understanding of the physics contained in correlated variational wave functions for the Hubbard model has reached the same level of sophistication. The controlled and systematic evaluation of these wave functions is subject of the next section.

3.4 Variational Wave Functions

In this section we examine the variational approach to the metal–insulator transition. After the introduction of the class of Gutzwiller-correlated variational wave functions we give a definition of an insulator based on a variational description. In one dimension the simplest of all these variational states, the Gutzwiller-correlated paramagnetic Fermi sea ("Gutzwiller wave function"), can be evaluated exactly. In addition, an approximation-free evaluation of general Gutzwiller-correlated wave functions is feasible in the limit of infinite dimensions, and a $1/d$ expansion can be carried out. The "Gutzwiller approximation", a semi-classical technique for the calculation of expectation values, gives the exact results for the Gutzwiller wave function in infinite dimensions, within which, at half band-filling and above a finite critical interaction strength, all electrons are localized. This "Brinkman–Rice transition" is actually absent in all finite dimensions but serves as an important instructive example for a metal–Mott–Hubbard insulator transition. Finally, we present the variational phase diagram, and discuss possible extensions and problems of Gutzwiller-correlated wave functions.

3.4.1 Gutzwiller-Correlated Wave Functions

Variational wave functions are an educated guess for the true ground state of a system. Therefore, statements based on variational arguments are not exact but the trial state may still contain the "essential physics" of the system under investigation. The impressive success of the BCS wave function for the explanation of superconductivity may serve as an illustrative example.

Furthermore, Jastrow–Feenberg variational wave functions play an important role in the calculation of low-temperature properties of the Helium quantum liquids [3.67].

For correlated electrons with a purely local interaction Gutzwiller [3.68, 3.69] proposed the following class of variational wave functions,

$$\text{Gutzwiller-correlated wave functions:} \quad |\Psi\rangle_g = g^{\hat{D}}|\Psi_0\rangle \ , \tag{3.49}$$

where $\hat{D} = \sum_{\boldsymbol{R}} \hat{n}_{\boldsymbol{R},\uparrow}\hat{n}_{\boldsymbol{R},\downarrow} = \sum_{\boldsymbol{R}} \hat{D}_{\boldsymbol{R}}$ is the operator for double occupancies and $|\Psi_0\rangle$ denotes a one-electron product wave function, typically a Hartree–Fock state. If we choose the paramagnetic Fermi sea for $|\Psi_0\rangle$, we obtain the "Gutzwiller wave function" (GWF)

$$|\text{GWF}\rangle_g = g^{\hat{D}}|\text{Fermi sea}\rangle = g^{\hat{D}} \prod_{\epsilon(\boldsymbol{k})\leq E_{\text{F}},\sigma} \hat{c}^{+}_{\boldsymbol{k},\sigma}|\text{vacuum}\rangle \ . \tag{3.50}$$

The operator $g^{\hat{D}}$ is called the "Gutzwiller correlator". For $g = 0$ it reduces to the "Gutzwiller projector" onto the subspace without double occupancies, $\hat{P}_{D=0}$,

$$g^{\hat{D}} = \exp\Big(\ln(g)\sum_{\boldsymbol{R}} \hat{D}_{\boldsymbol{R}}\Big) = \prod_{\boldsymbol{R}} \Big[1-(1-g)\hat{D}_{\boldsymbol{R}}\Big] \ , \tag{3.51a}$$

$$\hat{P}_{D=0} = \lim_{g\to 0} g^{\hat{D}} = \prod_{\boldsymbol{R}} \Big[1-\hat{D}_{\boldsymbol{R}}\Big] \ . \tag{3.51b}$$

Equation (3.51a) shows that the Gutzwiller correlator reduces the number of double occupancies for $g < 1$ whereby density fluctuations are globally suppressed in $|\Psi_0\rangle$.

Our task is to evaluate these variational wave functions, i.e, to calculate expectation values of operators $\hat{O}$,

$$\langle\hat{O}\rangle\,(g,\{\Psi_0\}) = \frac{{}_g\langle\Psi|\hat{O}|\Psi\rangle_g}{{}_g\langle\Psi|\Psi\rangle_g} \ . \tag{3.52}$$

These expectation values depend both on g and on variational parameters contained in $|\Psi_0\rangle$. They are determined by the minimization of the variational ground-state energy $E_0^{\text{var}}(g,\{\Psi_0\}) = \langle\hat{H}\rangle(g,\{\Psi_0\})$.

For $U = 0$ the Hubbard model describes a paramagnetic metal with the Fermi sea as its ground state which completely lacks correlations between $\uparrow$ and $\downarrow$ electrons; see Sect. 1.4. For $U > 0$ double occupancies are energetically unfavorable. The Fermi sea is a product state in momentum space. In position space it is a sum over configurations with $D = 0,1,2,\ldots,N_\uparrow N_\downarrow$ double occupancies, the phase of these configurations being given by Slater determinants. Gutzwiller's central assumption is that only the *weight* of the position space configurations with D double occupancies has to be reduced by a factor g^D ($g < 1$) to obtain a good variational ground state for the Hubbard model. The *phases* between these configurations, however, are not modified.

3.4.2 Variational Characterization of an Insulator

When we work with variational wave functions we have to define how to characterize a "metallic" or an "insulating" state. As discussed in Sect. 1.1, we should study the low-energy charge excitations to distinguish between the two cases. By construction, however, the variational state is only an approximation to the true ground state. Hence, $|\Psi\rangle_g$ is a complicated mixture of the true ground state and its excitations. In fact, a variational calculation results in a finite conductivity at *negative* excitation energies [3.70] and a further interpretation of the results is required. However, there are various other ways to characterize an insulator based on variational wave functions, as we now discuss.

Overlap with the Fermi Sea. The Gutzwiller wave function is constructed from the Fermi sea. Therefore, it seems natural to argue that the Gutzwiller-correlated Fermi sea $|\mathrm{GWF}\rangle_g$ describes a correlated metal for all $0 < g < 1$ since only the weights and not the phases of the configurations of the Fermi sea in position space are modified by the correlator $g^{\hat{D}}$. Therefore, the overlap between the Gutzwiller wave function and the Fermi sea,

$$\langle \text{Fermi sea}|\mathrm{GWF}\rangle_g = \langle \text{Fermi sea}|g^{\hat{D}}|\text{Fermi sea}\rangle > 0 \text{ for } 0 < g < 1, \quad (3.53)$$

should be finite in the metallic phase. Both states are only quantitatively distinct but are qualitatively similar with respect to their physical properties.

A metal–insulator transition in the Gutzwiller wave function should appear only if the variation procedure fixes g at its limiting value $g = 0$. The same line of argument should hold true for all Gutzwiller-correlated wave functions: if the uncorrelated wave function $|\Psi_0\rangle$ does not already describe an insulating state (e.g., an antiferromagnetic Slater insulator), a metal–insulator transition could happen only if $g = 0$ is the result of the variation with respect to g. Then there are no more double occupancies in the Gutzwiller-projected state. For less than half band-filling there are still holes left even for $g = 0$ when double occupancies are absent. For $n < 1$ the remaining holes will serve for (hypothetical) low-energy charge excitations even at $g = 0$, and the Gutzwiller wave function describes a metallic ground state for all $(n < 1, g)$. Apparently, an insulating state for Gutzwiller-correlated wave functions is limited to the parameter values $(n = 1, g = 0)$.

Jump Discontinuity in the Momentum Distribution. For a Fermi liquid the one-electron spectral function $A_\sigma(\boldsymbol{k}, E)$ (see Appendix A.2) contains a coherent part at the Fermi energy, $A_\sigma^{\mathrm{coh.}}(\boldsymbol{k}_\mathrm{F}, E_\mathrm{F}) = Z_\sigma(\boldsymbol{k}_\mathrm{F})\delta(E_\mathrm{F} - E_\sigma(\boldsymbol{k}_\mathrm{F}))$ [3.71]. This feature results in a jump discontinuity in the momentum distribution $n_\sigma(\boldsymbol{k})$ of size $Z_\sigma(\boldsymbol{k}_\mathrm{F})$ at the Fermi wave vector $\boldsymbol{k}_\mathrm{F}$. Hence it appears natural to conclude that a discontinuity in the momentum distribution at the Fermi energy is a unique signal for a Fermi liquid and thus for a metallic ground state.

A jump in $n_\sigma(\boldsymbol{k}_\mathrm{F})$ is not unique to Fermi liquids. This is not too surprising since it is the quasi-particle peak in the momentum- and frequency-dependent spectral function that truly specifies a Fermi liquid. It contains more information than the momentum distribution. As discussed in Sect. 4.4.1, the chiral Luttinger liquid also has a jump discontinuity in the momentum distribution at the Fermi energy and describes a metallic ground state. However, it displays no quasi-particle peak away from the Fermi energy but has algebraic singularities at different velocities for charge and spin excitations.

In turn, the absence of a discontinuity in $n_\sigma(\boldsymbol{k})$ do not necessarily imply that the system is not metallic. For example, one-dimensional Luttinger liquids (see Sect. 4.4.1) have gapless charge excitation but a continuous momentum distribution. Hence, a jump discontinuity in $n_\sigma(\boldsymbol{k}_\mathrm{F})$ may be taken as an indication for a metallic ground state but its absence does not prove the contrary.

Jump in the Chemical Potential. Since the momentum distribution provides rather detailed information on the variational ground state we prefer to extend the definition of a Mott insulator on p. 34 to variational wave functions [3.72]. Consequently, a "variational" insulator displays a gap for charge excitations $\Delta\mu^\mathrm{var}$ defined by

$$\begin{aligned}\Delta\mu^\mathrm{var}(N) &= \mu^{\mathrm{var},+}(N) - \mu^{\mathrm{var},-}(N) \\ &= (E_0^\mathrm{var}(N+1) - E_0^\mathrm{var}(N)) - (E_0^\mathrm{var}(N) - E_0^\mathrm{var}(N-1)) \ .\end{aligned} \tag{3.54}$$

As usual this equation for the gap can be written as the difference between right and left partial derivatives. The variational gap defined in this way is not a variational bound on the true gap since it involves energy differences between variational estimates. Nevertheless, we obtain an educated guess for the size of the gap and the exponent with which it opens as a function of the interaction strength.

3.4.3 Exact Statements

Gutzwiller-correlated wave functions are obviously exact in two limits, namely (i) for small interactions ($U \to 0$), and (ii) for vanishing hopping (atomic limit, $\epsilon(\boldsymbol{k}) \equiv 0$). Although these limiting cases are of minor interest for us we recall that they are not properly described by all approximate theories. For example, the limit of small interactions and the reproduction of the Hartree-Fock solution posed major problems for Hubbard's Green-function decoupling scheme (Sect. 3.3), and the correct atomic limit could not be obtained within the local density approximation without self-interaction corrections (Sect. 3.2).

One of the main advantages of the variational approach lies in its upper bound property [3.10]. Hartree–Fock theory (Sect. 3.1) also gives exact upper bounds for the true ground-state energy and is exact to lowest-order perturbation theory in U/t. Since the Gutzwiller correlator in $|\Psi\rangle_g$ typically acts on Hartree–Fock product states, the Gutzwiller-correlated wave functions go beyond Hartree–Fock theory. For example, one can readily derive upper bounds for the correlation energy in the Hubbard model up to and including second-order $\mathcal{O}(U^2/t)$ [3.73].

To make full use of the advantages of these correlated wave functions we have to calculate the corresponding expectation values for the ground-state energy, the momentum distribution, and correlation functions. For this purpose we may employ the electron–hole transformations of Sect. 2.5 to derive useful relations for the expectation values which also allows us to restrict the parameter range for our calculations to $n \leq 1$ and $g \leq 1$.

Unfortunately, general statements on the behavior of the variational energy are impossible for large U/t since the evaluation of these wave functions poses a many-particle problem of its own. It can be exactly solved for the Gutzwiller wave function in one dimension (see Sect. 3.4.4) and for the general Gutzwiller-correlated wave function in the limit of infinite dimensions (see Sect. 3.4.5). In $d = 2, 3$ dimensions complete analytical solutions do not exist but perturbation expansions in the parameter $\ln(g)$ [3.73] and $1/d$ expansions can be carried out; see Sect. 3.4.7. In addition, it can be shown [3.74] that the Gutzwiller wave function does not predict a metal–insulator transition for the Hubbard model for finite interaction strengths in all finite dimensions. Within the Gutzwiller wave function a localization transition at finite interaction strength, the celebrated Brinkman–Rice transition [3.75], exists only in infinite dimensions; see Sect. 3.4.6.

3.4.4 Gutzwiller Wave Function in One Dimension

The task posed by (3.52) is a true many-body problem which is unsolvable in general. In close analogy to the treatment of Green functions [3.14] we may set up a many-particle diagrammatic perturbation theory for the problem. The advantage of variational functions lies in the fact that the graphs involve only space coordinates and are independent of time or frequency. Furthermore, the expansion parameter of the perturbation series around $g = 1$ is not fixed but can be chosen in a suitable way. For the calculation of the corresponding lowest-order terms of a perturbation theory in U/t an expansion in $\ln(g) \sim U$ [3.73] is recommended from which we can derive the (variational) correlation energy to order U^2/t.

For the exact evaluation of the Gutzwiller wave function in one dimensions a series expansion in $(g^2 - 1)$ is more advantageous. The structure of the corresponding graphs [3.76] shows that the expansion parameter actually is the product of the interaction parameter and the density, i.e, the true expansion parameter is $n(g^2 - 1)$. If we then apply the electron–hole symmetries to the variational problem (see Sect. 2.5) we can *exactly* calculate *all* coefficients of the perturbation series and sum it up. In this way we can evaluate not only the expectation values for the ground-state energy, the momentum distribution, and the mean double occupancy [3.76], but also all (static) two-electron correlation functions for the Gutzwiller wave function in one dimension [3.77].

For small U/t the Gutzwiller wave function is lower in energy than all Hartree–Fock states. This does not come as a surprise since the latter are lacking all contributions to the correlation energy to order $\mathcal{O}(U^2/t)$. For large U/t the Gutzwiller wave function displays totally incorrect behavior,

$$E_0^{\text{var}}(U \gg t, n = 1) \sim -\frac{t^2}{U} \frac{1}{\ln(U/t)} . \tag{3.55}$$

Instead of an antiferromagnetic coupling proportional to t^2/U there is an additional logarithmic dependence for $U \gg t$. It does not exists in the exact solution and yields only a rather unsatisfactory upper bound for the exact ground-state energy in one dimension. In the limit of strong interactions the paramagnetic Gutzwiller wave function is higher in energy than the antiferromagnetic Hartree–Fock (Slater) state. The reason for this unsatisfactory behavior is the missing correlation between double occupancies and holes in the limit of strong interactions [3.78]. On the other hand the spin correlations in the Gutzwiller wave function in one dimension are in excellent agreement with the results for the Heisenberg chain [3.77]. This result is the basis for possible improvements; see Sect. 3.4.9.

For all $U/t < \infty$ the variational ground-state energy is minimized for values $0 < g < 1$. Thus the variational theory predicts a metal–insulator transition in one dimension in the paramagnetic phase at $U_{\text{c,P}}^{\text{GWF, } d=1} = \infty$. In contrast to that there is a metal–insulator transition in the exact solution of the one-dimensional Hubbard model at $U_{\text{c}}^{\text{exact, } d=1} = 0^+$. Consequently, we have to improve on the Gutzwiller wave function, e.g., by an inclusion of antiferromagnetism in $|\Psi_0\rangle$. Unfortunately, "obviously" better choices for variational wave functions cannot be evaluated exactly even in one dimension, in general.

In any case we are most interested in dimensions larger than one since the physics in one spatial dimension is very peculiar; see Sect. 4.4. The *exact* evaluation of general Gutzwiller-correlated wave functions in $d > 1$ dimensions is necessary to make use of the genuine advantages of the variational approach. *Uncontrolled* approximations in the evaluation of a variational wave function lead to new uncertainties on its physical contents and may even lead to unphysical results. For example, negative occupation probabilities in momentum space can be produced [3.79].

3.4.5 Variational Wave Functions in Infinite Dimensions

In the limit of infinite dimensions an approximation-free evaluation of the Gutzwiller wave function is also possible [3.80]. With the method developed in [3.81] we can exactly evaluate the whole class of Gutzwiller-correlated wave functions in infinite dimensions and even carry out a $1/d$ expansion.

To this end we use a new representation of the wave function as

$$|\Psi\rangle_g = g^{\hat{D}}|\Psi_0\rangle = g^{\hat{K}}|\Phi_0\rangle \,, \tag{3.56a}$$

where the correlators $\hat{K}$ and $\hat{D}$ differ by one-electron operators only,

$$\hat{K} = \hat{D} - \sum_{\boldsymbol{R},\sigma} \mu_{\boldsymbol{R},\sigma} \hat{n}_{\boldsymbol{R},\sigma} + \sum_{\boldsymbol{R}} \eta_{\boldsymbol{R}} \,. \tag{3.56b}$$

$|\Psi_0\rangle$ and $|\Phi_0\rangle$ are *both* one-electron product wave functions. The parameters $\mu_{\boldsymbol{R},\sigma}$ and $\eta_{\boldsymbol{R}}$ can now be chosen conveniently. Instead of an expansion of expectation values in powers of (g^2-1) we use the expansion parameter $(G = 1 - g^2)$

$$x_{\boldsymbol{R}} = \frac{-1 + G(n_{\boldsymbol{R},0} - 2\overline{d}_{\boldsymbol{R},0}) - 1 + \sqrt{1 - G[n_{\boldsymbol{R},0}(2-n_{\boldsymbol{R},0}) + g^2 m^2_{\boldsymbol{R},0}]}}{2G\overline{d}_{\boldsymbol{R},0}(1-n_{\boldsymbol{R},0})} \,, \tag{3.57}$$

where $n_{\boldsymbol{R},0}$, $m_{\boldsymbol{R},0}$, and $\overline{d}_{\boldsymbol{R},0}$ denote the local density, magnetization, and mean double occupancy in the uncorrelated state $|\Phi_0\rangle$. The main advantage of this method is that the self-energy of the corresponding diagrammatic expansion vanishes identically in infinite dimensions. Even if one succeeded in deriving the general results with the help of $|\Psi_0\rangle$ instead of $|\Phi_0\rangle$, all formulae would be much more complicated [3.82].

In infinite dimensions the variational ground-state energy for general Gutzwiller-correlated wave function becomes [3.81]

$$\begin{aligned} \langle\hat{H}\rangle(g,\{\Phi_0\}) &= \sum_{\boldsymbol{R}_1\neq\boldsymbol{R}_2,\sigma} t(\boldsymbol{R}_1-\boldsymbol{R}_2)\sqrt{q_{\boldsymbol{R}_1,\sigma}}\sqrt{q_{\boldsymbol{R}_2,\sigma}}\langle\hat{c}^+_{\boldsymbol{R}_1,\sigma}\hat{c}_{\boldsymbol{R}_2,\sigma}\rangle_0 \\ &\quad + U\sum_{\boldsymbol{R}}\overline{d}_{\boldsymbol{R}} \,, \end{aligned} \tag{3.58a}$$

where $\langle\ldots\rangle_0$ denotes expectation values in $|\Phi_0\rangle$, and

$$\sqrt{q_{\boldsymbol{R},\sigma}} = \frac{\sqrt{(1-n_{\boldsymbol{R},0}+\overline{d}_{\boldsymbol{R}})(n_{\boldsymbol{R},\sigma,0}-\overline{d}_{\boldsymbol{R}})} + \sqrt{\overline{d}_{\boldsymbol{R}}(n_{\boldsymbol{R},-\sigma,0}-\overline{d}_{\boldsymbol{R}})}}{\sqrt{n_{\boldsymbol{R},\sigma,0}(1-n_{\boldsymbol{R},\sigma,0})}} \,, \tag{3.58b}$$

$$g^2 = \frac{\overline{d}_{\boldsymbol{R}}(1-n_{\boldsymbol{R},0}+\overline{d}_{\boldsymbol{R}})}{(n_{\boldsymbol{R},\uparrow,0}-\overline{d}_{\boldsymbol{R}})(n_{\boldsymbol{R},\downarrow,0}-\overline{d}_{\boldsymbol{R}})} \,. \tag{3.58c}$$

Equation (3.58c) allows us to express the variational parameter g by the physical quantity $\overline{d}_{\boldsymbol{R}}$, the average double occupancy per lattice site. We can immediately generalize these equations to multi-band systems [3.83]; applications can be found in [3.84].

The interpretation of the result (3.58a) is very simple. The hopping between two lattice sites $\boldsymbol{R}_1$ and $\boldsymbol{R}_2$ is reduced by two factors $\sqrt{q_{\boldsymbol{R}_1,\sigma}}\sqrt{q_{\boldsymbol{R}_2,\sigma}}$.

Equation (3.58c) describes a law of mass action between the variational parameter g^2 and the probabilities for singly occupied sites against double occupancies and holes [3.85]. These simple results are valid if we express expectation values $\langle \ldots \rangle_0$ in terms of $|\Phi_0\rangle$. They do *not* hold if we work with $|\Psi_0\rangle$; see (3.56).

3.4.6 Brinkman–Rice Transition

The result (3.58) appears to be of classical rather than quantum-mechanical origin. In fact, it can be obtained from semi-classical counting arguments on the correlated motion of electrons [3.85, 3.79] where, however, the two representations of $|\Psi\rangle_g$, (3.56a), cannot be distinguished. For the translationally invariant Gutzwiller wave function the operator $\hat{K} - \hat{D}$, (3.56b), gives rise only to an irrelevant normalization. In this case the results in (3.58a) become in fact identical with those of the "Gutzwiller approximation" for the Gutzwiller wave function [3.68, 3.69]. Because of its simple and intuitive structure the Gutzwiller approximation has become very popular and has frequently been applied to many correlated electron systems [3.86]–[3.88]. Equation (3.58) implies that the Gutzwiller approximation on the Gutzwiller wave function becomes exact in infinite dimensions. For the translationally invariant case and the Gutzwiller-correlated Hartree–Fock (Slater) state this was first shown in [3.76, 3.80] via a different method; see Sect. 5.4.

Brinkman and Rice [3.75] recognized that the Gutzwiller approximation describes a metal–Mott insulator transition. The equations (3.58) are exact in infinite dimensions for the Gutzwiller wave function. For half-filling ($n_\sigma = 1/2$) the minimization of the variational ground-state energy $\langle \hat{H} \rangle(g)$ gives

$$q(n=1,\overline{d}) = 1 - \left(\frac{U}{U_{\rm BR}}\right)^2 , \tag{3.59a}$$

$$\overline{d} = \frac{1}{4}\left(1 - \frac{U}{U_{\rm BR}}\right) , \tag{3.59b}$$

$$[E_0^{\rm var}(U, n=1)]/L = -|\overline{\epsilon}_0| \left(1 - \frac{U}{U_{\rm BR}}\right)^2 , \tag{3.59c}$$

$$U_{\rm BR} = 8|\overline{\epsilon}_0| \equiv U_{\rm c,P}^{{\rm GWF},\, d=\infty} , \tag{3.59d}$$

for $U \leq U_{\rm BR}$, where $\overline{\epsilon}_0 = \langle \hat{T} \rangle_0 / L$ denotes the average kinetic energy of the Fermi sea per lattice site, and $U_{\rm BR}$ is the critical interaction strength for the transition. Above the Brinkman–Rice transition the q factor, the mean double occupancy, and the variational ground-state energy are zero.

From (3.59b) we can infer that all double occupancies vanish beyond the transition, and only singly occupied sites remain since the number of double occupancies and holes is equal at half band-filling. This rather drastic concept for a metal–Mott insulator transition, the localization of all carriers, is called the “Brinkman–Rice” transition, the “Brinkman–Rice insulator” being a special case of a Mott–Hubbard insulator. The paramagnetic Gutzwiller wave function in infinite dimensions obviously lacks all antiferromagnetic correlations. Beyond the Brinkman–Rice transition there are no fluctuations in the local electron density and all exchange interactions between the localized spins are thus missing. The ground-state energy vanishes accordingly for $U > U_{\mathrm{BR}}$ and it lacks all contributions proportional to $(-t^2/U)$. We see that the Brinkman–Rice transition is *solely* caused by charge correlations and spin correlations do not play any role.

In this respect the Brinkman–Rice transition is similar to the transitions described by the (paramagnetic) Hubbard solutions. Whereas the Gutzwiller wave function properly describes the metallic phase, the (thermodynamically consistent) Hubbard-III approximation provides a fair description of the insulating phase if we can safely exclude antiferromagnetism; see Sect. 3.3. Both approaches together provide a complementary description for the metallic and the insulating state of the paramagnetic phases of the Hubbard model. The merits (and failures) of these two approaches are united in the “iterated perturbation theory” on the Hubbard model in infinite dimensions; see Sect. 5.6.

The Brinkman–Rice insulator is an insulator also according to our various definitions in Sect. 3.4.2. First, the transition happens at half band-filling when $\overline{d} = 0$, i.e., $g = 0$; see (3.58c). Second, the momentum distribution is constant inside and outside of the Fermi surface with the jump discontinuity given by $q(n, \overline{d})$ [3.85]. Hence, the discontinuity vanishes at the transition as required for an insulator. Third, the variational gap as defined in (3.54) opens at the transition. We thus see that in infinite dimensions the Gutzwiller wave function describes a transition from a paramagnetic correlated metal into a non-magnetic Mott–Hubbard insulator.

To obtain a variational estimate for the gap we evaluate (3.54) in the following. Using electron–hole symmetry we find that the chemical potential at half band-filling becomes

$$\mu^{\mathrm{var},+}(n=1) = U - \mu^{\mathrm{var},-}(n=1) , \tag{3.60a}$$

$$\mu^{\mathrm{var},-}(n=1) = \left. \frac{\mathrm{d}E_0^{\mathrm{var}}(\overline{d}, n)}{\mathrm{d}n} \right|_{n=1^-} , \tag{3.60b}$$

where we took the total derivative with respect to n to account for the density dependence of the average double occupancy. This had not been necessary since $\overline{d}$ acts as independent variational parameter (instead of g). Owing to the minimization condition

$$\frac{\partial E_0^{\mathrm{var}}(\overline{d}, n)}{\partial \overline{d}} = 0 \tag{3.61}$$

we may replace the total derivative in (3.60b) by a partial one. Because the average kinetic energy of the Fermi sea is extremal as a function of the density at half band-filling we thus find

$$\mu^{\text{var},-}(n=1) = \overline{\epsilon}_0 \left. \frac{\partial q(\overline{d}, n)}{\partial n} \right|_{n=1^-} . \tag{3.62}$$

In this calculation the average density of double occupancies $\overline{d}$ is a thermodynamically large quantity even if it becomes small. The deviation from half band-filling $\delta = 1 - n$, however, vanishes in the thermodynamic limit since $\delta = 1/L$. Hence, we have to calculate in the limit $\delta \ll \overline{d}$. Then we find

$$q(\overline{d}, \delta) = 8\overline{d}(1 - 2\overline{d}) + 4\delta(1 - 4\overline{d}) + \mathcal{O}\left(\delta^2\right) , \tag{3.63a}$$

such that

$$\mu^{\text{var},-}(n=1) = \frac{U_{\text{BR}}}{2}\left(1 - 4\overline{d}\right) . \tag{3.63b}$$

With the help of (3.59b) it follows that the chemical potential is continuous below the Brinkman–Rice transition, $\mu^{\text{var}}(n=1) = U/2 = \mu^{\text{exact}}(n=1)$. The variational gap opens for $U > U_{\text{BR}}$ whose size is given by

$$\Delta\mu^{\text{var}}(n=1) = U - U_{\text{BR}} . \tag{3.64}$$

The gap opens *linearly* as a function of $U - U_{\text{BR}}$, hence the critical exponent is given by $\nu = 1$. We may also *interpret* the chemical potential as effective "bandwidths" of the corresponding Hubbard bands, i.e, $\mu^{\text{var},-}(n=1) = U_{\text{BR}}/2 = W_2/2$, $U - \mu^{\text{var},+}(n=1) = U_{\text{BR}}/2 = W_1/2$; see Sect. 1.4. Hence, the Brinkman–Rice energy $U_{\text{BR}} = 8|\epsilon_0|$ gives a variational estimate for the width of the upper and lower Hubbard bands. For an application of the theory to real systems, see [3.85, 3.89].

3.4.7 Expansion Around the Limit of Infinite Dimensions

The simple structure of the solutions in infinite dimensions allows us to carry out a $1/d$ expansion explicitly [3.81]. Such an expansion is particularly interesting for the Gutzwiller wave function since the Brinkman–Rice transition in one dimension occurs only at infinite interaction strengths whereas we just saw that the critical interaction strength in infinite dimensions is finite.

We can carry out the $1/d$ expansion at half band-filling including second order without too much effort [3.81]. Even in dimensions as low as $d = 1$ many expectation values, e.g., for the momentum distribution and the mean double occupancy, can quantitatively be reproduced. The results suggest, however, that a Brinkman–Rice transition should happen at a finite value. For finite d the transition would then just be shifted from U_{BR}, (3.59d), to higher critical values. This is actually *not* the case since we can generally show that the Brinkman–Rice transition does not exist in any finite dimension [3.74],

$$\begin{aligned} U_{\text{c,P}}^{\text{GWF},\, d} &= \infty , \\ U_{\text{c,P}}^{\text{GWF},\, d=\infty} &= U_{\text{BR}} < \infty . \end{aligned} \tag{3.65}$$

The example of the paramagnetic Gutzwiller wave function already indicates that the limit $d \to \infty$ may produce peculiar effects which remain in a series expansion in $1/d$, although they actually do not exist in any finite dimension. The limit of infinite dimensions is very singular in many respects; see Chap. 5.

As for the Hubbard solutions we emphasize the conceptual importance of the Brinkman–Rice transition. It had been clear from the beginning that a state with all electrons completely localized cannot be a conceivable ground state for the Hubbard model because it completely neglects the itinerant exchange between the electrons. However, the Brinkman–Rice scenario nicely visualizes Mott's idea of a metal–Mott–Hubbard insulator transition at half band-filling as a quantum phase transition. The transition at a finite interaction strength is caused solely by charge correlations and no ordering of the magnetic moments occurs.

3.4.8 Variational Phase Diagram in Infinite Dimensions

The approximation-free evaluation of general Gutzwiller-correlated wave functions [3.80, 3.81] allows for the calculation of a variational phase diagram for an infinite-dimensional hyper-cubic lattice [3.72]. The ground-state wave functions for the paramagnetic, antiferromagnetic, and ferromagnetic phases are the corresponding Gutzwiller-correlated Hartree–Fock variational wave functions. The resulting phase diagram is shown in Fig. 3.6 where only the stability regions for the homogeneous phases are displayed.

Antiferromagnetism is stable at half band-filling for all finite interactions $U/t > 0$ because of the perfect nesting property. The charge correlations force $g < 1$ for all $U/t > 0$ but we always find $g > 0$ for all $U < \infty$. For this reason there is no Brinkman–Rice transition even in infinite dimensions within the Gutzwiller-correlated wave functions. The spin correlations force a transition into an antiferromagnetically ordered state at $U_{\mathrm{c,AFM}} = 0^+$, and an electron localization as in the Brinkman–Rice scenario does not happen. The quantum phase transition is "hidden" behind the thermodynamic phase transition into the antiferromagnetically ordered phase.

Beyond half-filling one further obtains *phase separation.* The Néel-type antiferromagnetic state with pure bipartite structure does not mix with the ferromagnetic and paramagnetic states and domain walls are formed instead. This may be taken as an indication that the variational wave functions have to be improved. It might also imply that phase separation is a genuine effect of the Hubbard model in high dimensions away from half band-filling [3.90].

The ferromagnetic phase for large interactions is stable up to a critical density $n_{\mathrm{c}} \approx 0.67$. Below that value the paramagnetic state is energetically favorable, in agreement with the general properties of the Hubbard model at low electron densities; see Sect. 2.5.

The phase diagram in Fig. 3.6 would be further modified if we considered Gutzwiller-correlated wave functions with other starting wave functions $|\Psi_0\rangle$, or totally different variational wave functions. Nonetheless, the variational

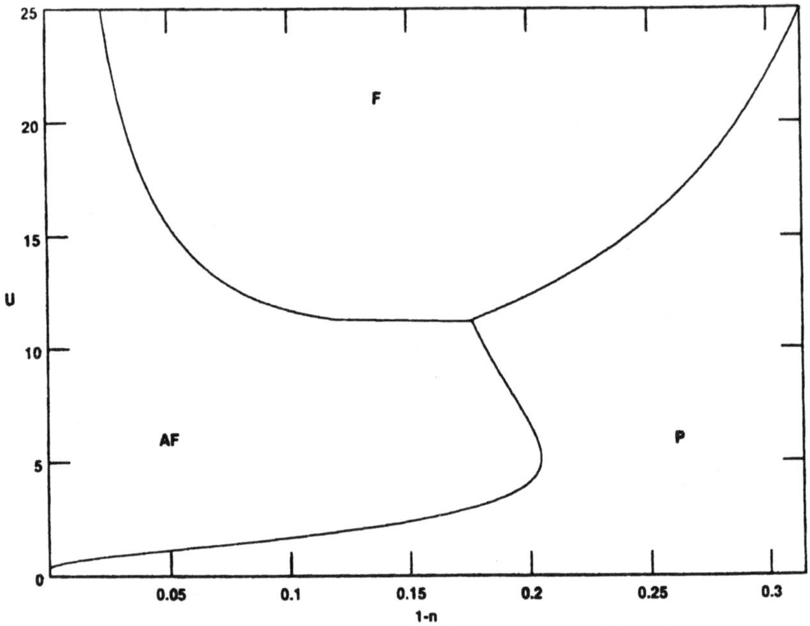

Fig. 3.6. Simplified variational phase diagram of the Hubbard model for an infinite-dimensional hyper-cubic lattice as a function of the hole concentration $\delta = 1 - n$ and the interaction strength, calculated with Gutzwiller-correlated wave functions ($U_{\mathrm{BR}} = 16/\sqrt{2\pi} \approx 6.38$). Only the stability regions for the homogeneous phases are shown. In fact, a pure antiferromagnetic phase exists only for $n = 1$ for all U/t and phase separation between the antiferromagnetic (AF) and ferromagnetic (F) or paramagnetic (P) phases occurs for $n < 1$ (from [3.72]).

phase diagram shows the essential phases that are expected for the Hubbard model. Moreover, it indicates that the regions with magnetic order are strongly reduced in the presence of correlations as compared to the Hartree–Fock phase diagram; see Sect. 3.1.4. If we allow for local moments due to correlations, the demand for long-range order to minimize the ground-state energy is strongly reduced.

3.4.9 Extensions and Improvements

At half band-filling and $U \gg W$ we can express the exact ground state of the Hubbard model with the help of the exact ground state of the Heisenberg model; see Sect. 2.5. The perturbatively exact result for the ground-state wave function for the half-filled Hubbard model to order $\mathcal{O}(t/U)$ can be written as

$$|\Psi^{\mathrm{Hubb}}\rangle = \left(1 + \frac{\alpha}{U}\hat{T}\right)|\Psi^{\mathrm{Heis}}\rangle \,. \tag{3.66}$$

Since the operator for the kinetic energy $\hat{T}$ precisely creates one hole and one double occupancy in the Heisenberg wave function $|\Psi^{\mathrm{Heis}}\rangle$ we have a strong spatial correlation between that hole and its double occupancy. For a pure nearest-neighbor hopping in $\hat{T}$ these two objects are also nearest neighbors. A good variational wave function for $n = 1$ and $U \gg W$ must contain these charge correlations. In addition, it must reproduce the spin–spin correlations of the Heisenberg model in this limit.

One Dimension. We can show that the Gutzwiller-projected Fermi sea is an excellent trial state for the one-dimensional Heisenberg model [3.77]. Correspondingly, the missing spatial correlations between double occupancies and holes in the Gutzwiller wave function lead to a poor ground-state energy in one dimension, (3.55). To improve the wave function for large U/W we have to incorporate the missing charge correlations into the wave function. Kaplan, Horsch, and Fulde [3.91] proposed to introduce another Jastrow correlator which suppresses configurations in which double occupancies and holes are not nearest neighbors. Unfortunately, an analytical evaluation of this wave function is not possible.

Baeriswyl [3.92] made another proposal to improve the Gutzwiller wave function. In analogy to (3.66) we may set

$$|\mathrm{BWF}\rangle_b = b^{\hat{T}}|\mathrm{GWF}\rangle_{g=0} = b^{\hat{T}}\hat{P}_{D=0}|\text{Fermi sea}\rangle \tag{3.67}$$

with $0 \leq b \leq 1$. For large U/t we will have $(1-b) \ll 1$ and the expansion shows that double occupancies and holes are correlated just as in the exact wave function. Since the spin–spin correlations of the one-dimensional Heisenberg model are well approximated by the Gutzwiller-projected Fermi sea, the Baeriswyl wave function $|\mathrm{BWF}\rangle_b$ is an excellent trial state for the one-dimensional Hubbard model for large U/t. For $U = 0$ we have $b = 0$ and the trial state reproduces the Fermi sea as the exact ground state. Unfortunately, we cannot reach small U/t since, in general, expectation values cannot be calculated exactly for the Baeriswyl wave function for all b.

It is obvious that the combination of the Gutzwiller and Baeriswyl wave functions will give a particularly good variational bound on the ground-state energy for the one-dimensional Hubbard model. The "Gutzwiller–Baeriswyl" wave function, first considered in [3.93], is defined as

$$|\mathrm{GBWF}\rangle_{g,b} = b^{\hat{T}}g^{\hat{D}}|\text{Fermi sea}\rangle\ , \tag{3.68a}$$

and the corresponding "Baeriswyl–Gutzwiller" wave function [3.94] reads

$$|\mathrm{BGWF}\rangle_{b,g} = g^{\hat{D}}b^{\hat{T}}\hat{P}_{D=0}|\text{Fermi sea}\rangle\ . \tag{3.68b}$$

Unfortunately, these wave functions cannot be evaluated exactly, with the exception of the one-dimensional $1/r$ Hubbard model [3.94, 3.95]; see below.

Higher Dimensions. None of the above extensions of the Gutzwiller wave function can be evaluated in higher dimensions. There, however, the Gutzwiller-projected Fermi sea at half filling is not a satisfactory trial state for the antiferromagnetic Heisenberg model. Hence, we would have to take into account antiferromagnetic order in the uncorrelated basis state that is not contained in the paramagnetic Fermi sea.

Although it is fairly obvious how to write down better trial states it is very difficult to derive exact expressions for expectation values. Thus far only Gutzwiller-correlated one-electron product states as defined in (3.49) can be exactly evaluated, in the limit of infinite dimensions.

3.4.10 Advantages and Problems

The main advantage of the variational approach lies in its "upper bound property". Each variational wave function allows for the calculation of an upper bound for the exact ground-state energy, which in turn permits a comparison of variational states. This "energy criterion" for the assessment of the quality of a trial state is very important in principle but not too helpful in practice because two wave functions may be very close in energy but may completely differ in the physics they describe; see, e.g., [3.53]. Furthermore, a good variational energy does not necessarily imply that the physics of the model is indeed properly reproduced.

An explicit example for this dilemma is given in [3.94, 3.95]. For the exactly solvable $1/r$ Hubbard model (see Sect. 4.2) the Gutzwiller, Baeriswyl, Gutzwiller–Baeriswyl, and Baeriswyl–Gutzwiller wave functions can be evaluated exactly. Whereas the Gutzwiller wave function at half-filling leads to an unsatisfactory variational energy for large U/t [see (3.55)] the Baeriswyl wave function lacks the correlation energy $\mathcal{O}(U^2/t)$ for small U/t and half band-filling. On the other hand the Gutzwiller–Baeriswyl and Baeriswyl–Gutzwiller wave functions provide a variational energy that differs from the exact value by at most half a percent and become *exact* both for weak *and* strong interactions.

Unfortunately, however, none of these wave functions is able to reproduce correctly the metal–insulator transition in the $1/r$ Hubbard model at $U_{\mathrm{c}}^{\mathrm{exact}} = W$ [3.96, 3.97]. The Baeriswyl–Gutzwiller wave function describes an insulator for all $U > 0$ since $b > 0$ is the result of the variational procedure for all $U > 0$. The Gutzwiller–Baeriswyl wave function on the other hand describes a metallic state for all $U < \infty$ since $g > 0$ holds for all $U < \infty$. This example indicates that variational parameters tend to take their limiting values only if a model parameter becomes extreme ($U = 0$ or $U = \infty$). Of course, we could use the energy criterion to locate the Mott transition [3.94]. Since the Gutzwiller–Baeriswyl wave function is lower in energy for $U \leq W$ than the Baeriswyl–Gutzwiller wave function, we would properly locate the transition at $U_c^{\mathrm{exact}} = W$. However, the variational ground-state energy would then displays a kink ("first-order transition"), and other physical quantities such as the mean double occupancy would discontinuously change as a function of U/t since the variational wave function itself abruptly changes its nature, in contrast to the exact ground state. Hence, this route is not very satisfactory either.

This example makes us sceptical whether variational wave functions in finite dimensions will be able to locate correctly a metal–insulator transition and properly reproduce its nature. Since a transition at finite U/t is supposed to be the canonical case in the absence of perfect nesting we believe that variational wave functions are only of limiting value for the description of the metal–Mott insulator transition.

In view of these principle problems the Brinkman–Rice transition in the Gutzwiller wave function in infinite dimensions deserves special attention. From a mathematic point of view the transition is only the result of a limit ($d = \infty$) on a variational, i.e., approximate description of the ground state of the Hubbard model. The significance of the Brinkman–Rice concept lies in its basic agreement with Mott's general physical ideas on the metal–Mott–Hubbard insulator transition as a quantum phase transition: it occurs at half band-filling, symmetries or conservation laws are not violated, and a magnetic order need not occur. The ground-state energy is continuous and differentiable at $U = U_{\mathrm{BR}}$ ("second-order transition"), and the energy gap continuously opens at the transition according to a power law (exponent $\nu = 1$) as expected for short-range interactions. The Brinkman–Rice transition is due to the charge correlations; spin correlations play no role and are, in fact, totally ignored. This is a very crude approximation and results in a very poor variational energy. Nevertheless, the major principle ingredients of the Mott–Hubbard transition are properly described in the Brinkman–Rice scenario.

3.5 Slave Boson Approach

In this section we discuss two Slave Boson methods which are very popular in many-particle condensed matter theory: the Kotliar–Ruckenstein Slave Boson approach on one hand, and the Barnes–Coleman or Read–Newns Slave Boson approach on the other hand. The latter is mainly applied to multi-band models since these bosons are most suitable for the limit of infinitely strong Hubbard interaction. Both variants, their evaluation on the mean-field level, and their main advantages and problems are investigated after a brief look at their relation to functional integrals.

3.5.1 Slave Bosons and Functional Integrals

The quantum mechanics of a single particle can be formulated in terms of path integrals. In many particle physics we have to perform a functional integration over many degrees of freedom rather than a single path integral but the two notions are often used synonymously [3.98].

The functional integral representation of models for correlated electrons allows us to implement elegantly approximation schemes like the alloy-analogy approximation. An early example for the Hubbard model was given by Cyrot [3.60, 3.61]; for more recent applications, see [3.63, 3.99, 3.100]. The partition function can be written as a functional integral over Grassmann variables $\psi_{\boldsymbol{R},\sigma}(\tau)$ representing the fermions at site $\boldsymbol{R}$ and (imaginary) time τ in the time slice $(\mathrm{d}\tau)$,

$$\mathcal{Z} = \int \mathrm{d}\left\{\psi_\uparrow \psi_\downarrow\right\} \exp\left(-S_0 - S_{\mathrm{int}}\right) , \tag{3.69a}$$

where the interacting and the free parts of the action read

$$S_{\mathrm{int}} = U \int_0^\beta \mathrm{d}\tau \sum_{\boldsymbol{R}} \psi^*_{\boldsymbol{R},\uparrow}(\tau)\psi^*_{\boldsymbol{R},\downarrow}(\tau)\psi_{\boldsymbol{R},\downarrow}(\tau)\psi_{\boldsymbol{R},\uparrow}(\tau) , \tag{3.69b}$$

$$\begin{aligned} S_0 = & \int_0^\beta \mathrm{d}\tau \Bigg[\sum_{\boldsymbol{R},\sigma} \psi^*_{\boldsymbol{R},\sigma}(\tau) \left(\partial_\tau - \mu\right) \psi_{\boldsymbol{R},\sigma}(\tau) \\ & - \sum_{\boldsymbol{R}_1 \neq \boldsymbol{R}_2,\sigma} t(\boldsymbol{R}_1 - \boldsymbol{R}_2) \psi^*_{\boldsymbol{R}_1,\sigma}(\tau)\psi_{\boldsymbol{R}_2,\sigma}(\tau) \Bigg] . \end{aligned} \tag{3.69c}$$

Although the two-fermion terms in S_0 could be exactly integrated, the four-fermion terms naturally make an exact treatment impossible.

The four-fermion term of the Hubbard interaction can formally be decoupled with the help of a Hubbard–Stratonovich transformation [3.98, 3.101]. To make this decoupling an exact transformation it is necessary to introduce a new set of bosonic fields. These Hubbard–Stratonovich fields typically represent fluctuating spin and/or charge densities. For example [3.102], we may write

$$\begin{aligned} \exp\left(-(\mathrm{d}\tau) U \psi^*_\uparrow \psi^*_\downarrow \psi_\downarrow \psi_\uparrow\right) = & \frac{(\mathrm{d}\tau)}{\pi U} \int \mathrm{d}\left\{\tilde{\Delta}_{\mathrm{c}} \tilde{\Delta}_{\mathrm{s}}\right\} \exp\Bigg[-\frac{(\mathrm{d}\tau)}{U} \left(\tilde{\Delta}_{\mathrm{c}}^2 + \tilde{\Delta}_{\mathrm{s}}^2\right) \\ & + \mathrm{i}(\mathrm{d}\tau)\tilde{\Delta}_{\mathrm{c}} n + (\mathrm{d}\tau)\tilde{\Delta}_{\mathrm{s}} \sigma^z \Bigg] , \end{aligned} \tag{3.70}$$

where $n = \psi^*_\uparrow \psi_\uparrow + \psi^*_\downarrow \psi_\downarrow$, and $\sigma^z = \psi^*_\uparrow \psi_\uparrow - \psi^*_\downarrow \psi_\downarrow$. $\tilde{\Delta}_{\mathrm{c,s}}(\boldsymbol{r},\tau)$ are the (real) Hubbard–Stratonovich fields for charge and spin excitations which have to be integrated over. From a physical point of view we describe electrons moving in the presence of (all possible) charge and spin configurations. For a fixed configuration the electron problem can (formally) be exactly solved, i.e., we may exactly integrate out the fermionic degrees of freedom.

In practice, only a restricted class of bosonic configurations can be taken into account, and the resulting electronic one-particle problem must still be solved for these cases. Nevertheless, the functional integral representation is a formally exact reformulation of the model and allows for an elegant introduction of approximations (e.g., "coherent potential approximation", "saddle-point approximation"). For example, in the saddle-point approximation one includes only the contribution of the supposedly dominant Hubbard–Stratonovich field configuration, and the electrons move in the corresponding background field which has to be determined self-consistently. At least in principle one could then go beyond and systematically calculate corrections to the saddle-point solution ("fluctuations around the saddle point").

These dynamic fields are historically the first examples for the application of Slave Bosons. Whereas the Hubbard–Stratonovich decoupling still plays an important role for quantum Monte-Carlo techniques [3.103] and field-theoretical approaches to the Hubbard model [3.102, 3.104] Slave Bosons may replace the Hubbard–Stratonovich fields on the Hamilton operator level. This offers the advantage that considerable technical problems in the formulation and application of the functional integral formalism can be avoided when we are interested in the saddle-point approximation only.

3.5.2 Kotliar–Ruckenstein Slave Bosons

Kotliar and Ruckenstein [3.105] enlarged the Hilbert space of the Hubbard model by four bosonic operators,

$$\begin{aligned} \hat{d}_{\boldsymbol{R}} \quad &: \quad \text{double occupancy,} \\ \hat{s}_{\boldsymbol{R},\sigma} \quad &: \quad \text{single } \sigma \text{ electron, } \sigma = \uparrow, \downarrow, \text{ and} \\ \hat{e}_{\boldsymbol{R}} \quad &: \quad \text{empty site or “hole”} \; . \end{aligned} \tag{3.71}$$

The physical Hilbert space of the Hubbard model is recovered if we impose the following *local* constraints, which have to be fulfilled *exactly*,

$$\hat{n}^e_{\boldsymbol{R}} + \hat{n}^s_{\boldsymbol{R},\uparrow} + \hat{n}^s_{\boldsymbol{R},\downarrow} + \hat{n}^d_{\boldsymbol{R}} = 1 \, , \tag{3.72a}$$

$$\hat{n}^s_{\boldsymbol{R},\sigma} + \hat{n}^d_{\boldsymbol{R}} = \hat{c}^+_{\boldsymbol{R},\sigma}\hat{c}_{\boldsymbol{R},\sigma} = \hat{n}^c_{\boldsymbol{R},\sigma} \, . \tag{3.72b}$$

Equation (3.72a) states that only one of the four physical possibilities on a lattice site is allowed, either a double occupancy, a single occupancy, or a hole. Equation (3.72b) expresses the fact that the presence of a physical electron implies either a double or a single occupancy.

The main advantage of this approach is that we are now in the position to write the Hubbard interaction as a bosonic occupation number operator. The operator for the kinetic energy on the other hand becomes much more complicated since the motion of a physical electron changes the occupation numbers for the Slave Bosons on both lattice sites involved. The Hubbard Hamiltonian becomes

$$\hat{H} = \sum_{\boldsymbol{R}_1 \neq \boldsymbol{R}_2, \sigma} t(\boldsymbol{R}_1 - \boldsymbol{R}_2)\hat{z}^+_{\boldsymbol{R}_1,\sigma}\hat{z}_{\boldsymbol{R}_2,\sigma}\hat{c}^+_{\boldsymbol{R}_1,\sigma}\hat{c}_{\boldsymbol{R}_2,\sigma} + U\sum_{\boldsymbol{R}} \hat{d}^+_{\boldsymbol{R}}\hat{d}_{\boldsymbol{R}} \, , \tag{3.73a}$$

where we introduced the short-hand notation

$$\hat{z}_{\boldsymbol{R},\sigma} = \hat{e}^+_{\boldsymbol{R}}\hat{s}_{\boldsymbol{R},\sigma} + \hat{s}^+_{\boldsymbol{R},-\sigma}\hat{d}_{\boldsymbol{R}} \tag{3.73b}$$

for the boson hopping operators. In addition to the fact that the kinetic energy operator has become much more complicated we must also deal with the enlarged Hilbert space. As long as we exactly fulfill the constraints (3.72) we may, for example, substitute the boson hopping operators by [3.105]

$$\hat{z}_{\boldsymbol{R},\sigma} \mapsto \widetilde{z}_{\boldsymbol{R},\sigma} = \frac{1}{\sqrt{1-\hat{n}^d_{\boldsymbol{R}}-\hat{n}^s_{\boldsymbol{R},\sigma}}}\hat{z}_{\boldsymbol{R},\sigma}\frac{1}{\sqrt{1-\hat{n}^e_{\boldsymbol{R}}-\hat{n}^s_{\boldsymbol{R},-\sigma}}} \tag{3.74}$$

without changing the Hamiltonian. However, as soon as we introduce any approximations it becomes of utmost importance whether we use $\widetilde{z}_{\boldsymbol{R},\sigma}$ or rather prefer to work with $\hat{z}_{\boldsymbol{R},\sigma}$ in (3.73a). On the one hand, this freedom allows for an "optimum choice" whereas, on the other hand, it gives way to an uncontrolled approximation. In any case it is by no means clear how far an approximation will violate the constraints. Typically, the local constraints (3.72) for the operators are replaced by their time- and space-independent expectation values and (3.72) is only fulfilled "on average".

3.5.3 Kotliar–Ruckenstein Approximation

One can show [3.106] that the choice of (3.74) is the best possible in the sense that it reproduces both the exact solution for $U = 0$ and the exact result for the fully ferromagnetically polarized system after the saddle-point approximation to the functional integral has been performed. This "mean-field" approximation can also be formulated on the Hamilton operator level without referring to functional integrals, whereby technical problems in this formalism are avoided.

The Kotliar–Ruckenstein mean-field approximation is easily formulated after the best choice for the boson hopping operators has been found. We substitute the Bose operators by their time-independent averages which may be interpreted as "condensation of the bosons". We thus replace

$$\hat{d}^+_{\boldsymbol{R}}, \hat{d}_{\boldsymbol{R}} \quad\mapsto\quad \langle \hat{d}^+_{\boldsymbol{R}}\rangle_0 = \langle \hat{d}_{\boldsymbol{R}}\rangle_0 \equiv \sqrt{\overline{d}_{\boldsymbol{R}}}\,, \tag{3.75a}$$

$$\hat{s}^+_{\boldsymbol{R},\sigma}, \hat{s}_{\boldsymbol{R},\sigma} \quad\mapsto\quad \langle \hat{s}^+_{\boldsymbol{R},\sigma}\rangle_0 = \langle \hat{s}_{\boldsymbol{R},\sigma}\rangle_0 \equiv \sqrt{n_{\boldsymbol{R},\sigma,0}-\overline{d}_{\boldsymbol{R}}}\,, \tag{3.75b}$$

$$\hat{e}^+_{\boldsymbol{R}}, \hat{e}_{\boldsymbol{R}} \quad\mapsto\quad \langle \hat{e}^+_{\boldsymbol{R}}\rangle_0 = \langle \hat{e}_{\boldsymbol{R}}\rangle_0 \equiv \sqrt{1-n_{\boldsymbol{R},0}+\overline{d}_{\boldsymbol{R}}}\,. \tag{3.75c}$$

In the last expressions we used the mean-field versions of the local constraints (3.72) and the notation of Sect. 3.4. In particular we see that

$$\langle \widetilde{z}_{\boldsymbol{R},\sigma}\rangle_0 = \sqrt{q_{\boldsymbol{R},\sigma}} \tag{3.76}$$

holds; see (3.58b). In the Kotliar–Ruckenstein mean-field theory we simply replace the Hubbard Hamilton operator by the single-electron effective Hamiltonian

$$\hat{H}^{\mathrm{KR}} = \sum_{\boldsymbol{R}_1\neq\boldsymbol{R}_2,\sigma} t(\boldsymbol{R}_1-\boldsymbol{R}_2)\sqrt{q_{\boldsymbol{R}_1,\sigma}}\sqrt{q_{\boldsymbol{R}_2,\sigma}}\hat{c}^+_{\boldsymbol{R}_1,\sigma}\hat{c}_{\boldsymbol{R}_2,\sigma} + U\sum_{\boldsymbol{R}}\overline{d}_{\boldsymbol{R}}\,. \tag{3.77}$$

Since we still have to fulfill the constraints (3.72) the number of parameters to be determined reduces to the average double occupancy $\overline{d}_{\boldsymbol{R}}$ and the average local densities $n_{\boldsymbol{R},\sigma,0}$. As in the Hartree–Fock theory these quantities are determined by the minimization of the free energy f^{KR} of the effective single-electron Hamiltonian $\hat{H}^{\mathrm{KR}}$,

$$f^{\mathrm{KR}}(\overline{d}_{\boldsymbol{R}}, n_{\boldsymbol{R},\sigma,0}) = -\frac{1}{\beta}\ln \mathcal{Z}^{\mathrm{KR}}(\overline{d}_{\boldsymbol{R}}, n_{\boldsymbol{R},\sigma,0}) \, , \tag{3.78a}$$

$$\mathcal{Z}^{\mathrm{KR}}(\overline{d}_{\boldsymbol{R}}, n_{\boldsymbol{R},\sigma,0}) = \mathrm{Tr}\left\{\exp\left[(-\beta)(\hat{H}^{\mathrm{KR}} - \sum_{\sigma}\mu_\sigma \hat{N}_\sigma)\right]\right\} \, . \tag{3.78b}$$

The minimization equations are nothing but the saddle-point equations of the corresponding functional integral formalism; see [3.105]. For zero temperature the saddle-point equations are *identical* to (3.58a), which we deduced from the general Gutzwiller-correlated wave functions in infinite dimensions [3.81].

Owing to the complete equivalence between Gutzwiller-correlated wave functions in infinite dimensions and the Kotliar–Ruckenstein saddle-point solution at $T = 0$ [3.81, 3.83] the latter theory also describes the Brinkman–Rice transition in the paramagnetic phase. Consequently, (3.64) for the gap at half-filling also holds within the Slave Boson mean-field theory. In [3.107] the limits $\delta \to 0$ and $\overline{d} \to 0$ were exchanged. The proper result for half-filling within the Slave Boson approach is also given by (3.64).

The Kotliar–Ruckenstein Hamilton operator naturally describes a Fermi liquid. The explicit expression for the ground-state energy as a function of the interaction strength, magnetization, and electron density allows us to determine the Landau parameters for the corresponding Fermi-liquid theory; see Sect. 1.4, and [3.85]. In this way we do not only make direct contact to experiments on static quantities but also derive the low-energy, small-momentum behavior of dynamic correlation functions [3.108]. This link to Fermi-liquid theory allows us to deduce dynamic quantities from the Gutzwiller variational theory. They may also be calculated within the saddle-point approximation [3.107, 3.109], whereby the results of the variational Fermi-liquid approach are fully confirmed.

Beyond the region of validity of Fermi-liquid theory there are substantial problems with the excitations described by the Kotliar–Ruckenstein Hamilton operator. We will comment further on that in the following subsection.

3.5.4 Advantages and Problems

In the Kotliar–Ruckenstein method, correlation effects are incorporated into an effective single-electron Hamilton operator in a very simple and elegant manner. Consequently, the technique has found many applications, for example for three-band Hubbard models for high-temperature superconductivity [3.110], for the Peierls–Hubbard Hamilton operator [3.111], for the single-impurity Anderson model [3.112], and for the description of magnetic phases with spiral order in the two-dimensional Hubbard model [3.113]. Furthermore, the method can be formulated in a manifestly spin-rotationally invariant form and can be used to calculate dynamic correlation functions and the $T^3 \ln T$ contribution to the specific heat of normal fluid ^{3}He [3.109].

The difficulties of the approach are hidden in the functional integral formulation of the saddle-point approximation, which is basically uncontrolled. For example, we have no control within the approach whether the local constraints, (3.72), are really obeyed, and not only on average. For this reason the free energy and the ground-state energy according to (3.77) do not obey a variational principle, in general. It is because of only the equivalence between Gutzwiller-correlated wave functions and the Kotliar–Ruckenstein saddle-point equations at temperature $T = 0$ that the ground-state energy obeys a variational principle in infinite dimensions and that the local constraints are exactly fulfilled in this case. Hence the results of the saddle-point theory cannot be totally unphysical for zero temperature since explicit wave functions for the ground state can be given.

The situation changes for finite temperatures. For $T > 0$ the free energy and partition function of Kotliar and Ruckenstein, (3.78), are not well-defined quantities. In the corresponding expression for the entropy states are incorrectly counted and weighed at finite temperatures, an effect that worsens with increasing temperature. This well-known entropy problem of saddle-point solutions to the functional integral [3.105, 3.114] results in a completely unphysical transition from a metal at low temperatures to an insulator at high temperatures [3.105].

To illustrate this point we construct the free energy and partition function of Kotliar and Ruckenstein, (3.78), for the translationally and spin-rotationally invariant case using the results for Gutzwiller-correlated wave functions, a genuine zero-temperature approach. In Sect. 3.4 we derived the variational energy for an arbitrary Gutzwiller-correlated wave function,

$$|\Psi_m\rangle_g = g^{\hat{D}}|\Psi_m\rangle \;, \tag{3.79}$$

where $|\Psi_m\rangle$ has to be a one-electron product wave function, e.g., an exact eigenstate of the kinetic energy operator $\hat{T}$. Since the 4^L states $|\Psi_m\rangle$ span the full Hilbert space we may indeed construct a "variational partition function" from the correlated energies (3.58):

$$E_m(\overline{d}, n) = q(\overline{d}, n)\langle\hat{T}\rangle_m + UL\overline{d} \;, \tag{3.80}$$

where $\overline{d}$ replaces g as variational parameter and $q(\overline{d}, n)$ is given in (3.58b). We then have

$$\mathcal{Z}^{\text{var}}(\overline{d}, n) = \sum_m \left\{\exp\left[-\beta(E_m(\overline{d}, n) - \mu nL)\right]\right\} \;, \tag{3.81}$$

which is identical to (3.78b) for our example.

From this construction it becomes immediately evident that the Kotliar–Ruckenstein Slave Boson theory may at best be applied to the region of the validity of Fermi-liquid theory [3.81, 3.83]. In analogy to Landau's ideas there may be a one-to-one correspondence between the low-lying excitations of the Fermi gas (states $|\Psi_m\rangle$) and the correlated electron system (states $|\Psi_m\rangle_g$) but this cannot hold for the whole spectrum. For example, for large enough interactions, charge excitations involving double occupancies are projected onto spin excitations which are thus severely overcounted. This entropy problem is ultimately responsible for the spurious metal-to-insulator transition on raising the temperature.

The same problems with the description of high-energy excitations become apparent when one tries to calculate explicitly the upper and lower Hubbard bands for the translationally invariant case since the excitations found do not fulfill elementary sum rules [3.115]. Hence, all statements on excitation energies and temperature ranges beyond the validity of the Fermi-liquid picture must be taken with great care.

In addition to problems at finite temperatures and excitation energies, the calculation of "fluctuation corrections" around the saddle point poses new problems of its own. On one hand, the treatment of the functional integral is technically involved and prone to errors [3.116]. On the other hand the calculation of corrections reveals the weakness and arbitrariness [3.116, 3.117] of the bosonic representation of the kinetic energy operator chosen for the optimum saddle point; see (3.74). The recent investigations of Arrigoni and Strinati [3.116] indicate that previous results of "fluctuation corrections" around the saddle point have to be taken with care. This does *not* apply to those results that can alternatively be obtained from Gutzwiller-correlated wave functions and thus from a well-controlled independent approach. At the moment it seems as if considerable difficulties have to be overcome before one can go substantially beyond the Kotliar–Ruckenstein saddle-point approximation and thus beyond the results for Gutzwiller-correlated wave functions.

3.5.5 Barnes–Coleman Slave Bosons and Approximation

Barnes [3.118] was the first to introduce the Slave Boson method explicitly. It was subsequently used and extended by Coleman [3.119], and Read and Newns [3.120]. This version of the Slave Boson method was mainly applied to problems of valence fluctuations and heavy fermions [3.121]–[3.123], and to three-band models for high-temperature superconductivity [3.124].

A necessary condition for the applicability of the method is the limit $U = \infty$ for the Hubbard interaction. Then we have the physical constraint that at most one electron can occupy a lattice site, $\hat{n}_{\boldsymbol{R}} \leq 1$. The lattice site is thus either empty or singly occupied. The inconvenient inequality can be transformed into an equality by the introduction of a single Slave Boson $\hat{b}_{\boldsymbol{R}}$ for the hole,

$$\hat{n}_{\boldsymbol{R},\uparrow} + \hat{n}_{\boldsymbol{R},\downarrow} + \hat{b}^{+}_{\boldsymbol{R}}\hat{b}_{\boldsymbol{R}} = 1 \, . \tag{3.82}$$

For the Hubbard model at $U = \infty$ ($\hat{T}$ model) we thus find

$$\hat{P}_{D=0}\hat{H}\hat{P}_{D=0} = \sum_{\boldsymbol{R}_1 \neq \boldsymbol{R}_2,\sigma} t(\boldsymbol{R}_1 - \boldsymbol{R}_2)\hat{b}_{\boldsymbol{R}_1}\hat{c}^+_{\boldsymbol{R}_1,\sigma}\hat{c}_{\boldsymbol{R}_2,\sigma}\hat{b}^+_{\boldsymbol{R}_2} , \tag{3.83}$$

since every hopping of an electron from site $\boldsymbol{R}_2$ to site $\boldsymbol{R}_1$ creates a hole at $\boldsymbol{R}_2$ and annihilates another one at $\boldsymbol{R}_1$, if we exactly fulfill the constraint (3.82).

In analogy to the Kotliar–Ruckenstein Slave Bosons we replace the boson operators by their (spatially independent) averages to arrive at the Barnes–Coleman mean-field approximation. Again, these averages have to be determined self-consistently. Within the theory one cannot judge how strongly the local constraint is violated. For the $\hat{T}$ model (3.83) the translationally invariant Barnes–Coleman mean-field theory describes a Fermi liquid for which the electron hopping amplitude is reduced by a factor

$$q^{\mathrm{BC}} = \langle \hat{b}^+_{\boldsymbol{R}} \rangle_0 \langle \hat{b}_{\boldsymbol{R}} \rangle_0 = 1 - n . \tag{3.84}$$

The factor q^{BC} expresses the fact that on average an empty site for a hopping process is available with probability $(1 - n)$.

It is more interesting to use the Barnes–Coleman Slave Boson mean-field theory for the study of the (periodic) Anderson model at $U = \infty$. There the strongly interacting f electrons cannot directly move but only via hybridization with the conduction electrons; see Sect. 2.4, (2.25). Within the mean-field theory one recovers a formally non-interacting effective Hamiltonian with renormalized hybridization matrix elements $V_\sigma^{\mathrm{eff}} = \sqrt{q^{\mathrm{BC}}} V_\sigma$ [3.125, 3.126]. For the Kotliar–Ruckenstein mean-field theory [3.83] the corresponding q factors can be found in (3.58b).

3.5.6 Advantages and Problems

As we mentioned above, the Barnes–Coleman theory is applied mostly to models with N_f-fold degenerate f levels where N_f is often assumed to be large such that $1/N_f$ appears as a natural expansion parameter. Unfortunately, for the Hubbard model one cannot use the inverse band-degeneracy as an expansion parameter [3.127]. Hence, we will discuss the advantages and problems of the Slave Boson method for the Anderson models below.

For finite temperatures the same restrictions apply for all Slave Boson mean-field theories. Both the Barnes–Coleman and the Kotliar–Ruckenstein version are only applicable in the Fermi-liquid regime. These conclusions have earlier been reached for the single-impurity Anderson in [3.128].

Single-Impurity Anderson Model. If we apply the method to the Anderson model with a single, N_f-fold degenerate impurity, we can show that the results of the Barnes–Coleman mean-field theory become *exact* for the model to lowest order in $\mathcal{O}\left((1/N_f)^0\right)$. The same results can be obtained from a variational approach [3.129]; see [3.128] for a comprehensive review on $1/N_f$ expansions.

For a single impurity it is even possible to calculate the lowest order of a $1/N_f$ expansion from the fluctuations around the saddle point [3.128, 3.130], where the derivation again shows that the manipulations of the functional integral require great care [3.131].

Periodic Anderson Model. For the case of the periodic Anderson or Varma–Yafet model we can show in general [3.83] that the Barnes–Coleman Slave Boson mean-field theory is variationally controlled at zero temperature to order $\mathcal{O}\left((1/N_f)^0\right)$. The difference between the Barnes–Coleman and the Kotliar–Ruckenstein Slave Boson mean-field theory at $U = \infty$ lies in the fact that the former is variationally controlled in the limit $N_f \to \infty$, the latter in the limit $d \to \infty$. In these limiting cases the constraints are actually exactly fulfilled within the framework of the variational approach [3.83].

The results for the periodic and the single-impurity Anderson models are identical within the Barnes–Coleman mean-field theory since in both cases the renormalized hybridizations are given by (3.84). The limit $N_f \to \infty$ completely suppresses lattice coherence effects. This is not surprising because the motion of an f electron from one lattice site to another requires at least two hybridizations with the lattice electrons. Since the lattice electron keeps its orbital quantum number during its motion, f-electron transport is less likely by a factor of $1/N_f$ than a purely local hybridization process.

The limit $N_f \to \infty$ also suppresses exchange effects due to the Pauli principle since the probability that an electron encounters another electron at some site with the same orbital quantum numbers is a $1/N_f$ effect only. This is in contrast to the results of a generalized Kotliar–Ruckenstein mean-field theory for high dimensions for which the hybridization strength for the lth orbital is scaled by $(n_l = n/N_f)$

$$q_l^{\mathrm{KR}} = \frac{1-n}{1-n_l} \tag{3.85}$$

(see (3.58b) and [3.83]). The additional orbital density dependence in the denominator in q_l^{KR} is the decisive difference between the two Slave Boson approaches. q_l^{KR} reflects the transition probability for $U = \infty$ *relative* to $U = 0$. Even for $U = 0$ we have the Pauli blocking, which reduces the transition probability by a factor $(1 - n_l) = (1 - n/N_f)$ in any case. Hence we find the *relative* factor q_l^{KR}, and not q^{BC}. Of course, both factors agree for $N_f = \infty$. For $N_f = 2$ ($l \equiv \sigma = \uparrow, \downarrow$) we find a strong tendency towards magnetic order in the Kotliar–Ruckenstein theory. This magnetic exchange effect and the lattice coherence effect mentioned above are underestimated in the Barnes–Coleman mean-field theory.

Local Constraints. There are similar problems in all techniques that introduce additional Slave particles. In the Schwinger Boson method for spin systems spin degrees of freedom are overcounted [3.132, 3.133] which applies equally well to the "Majorana" representation of the periodic Anderson model [3.134] and to the Slave Fermion method for the Hubbard model [3.135]. The fundamental problem lies in the fact that the local constraints cannot be fulfilled exactly such that the additional degrees of freedom of the enlarged Hilbert space reappear in the physical Hilbert space. It is principally unclear how to formulate approximations for lattice problems within the Slave Boson approach in such a way that the constraints can be exactly fulfilled.

The Kotliar–Ruckenstein and the Barnes–Coleman mean-field approximations offer the advantage of being variationally controlled at $T = 0$ in the limits $d \to \infty$ and $N_f \to \infty$, respectively. Hence they indeed fulfill the constraints in the respective limits. Statements beyond the validity of the variational theory (regime of Fermi-liquid theory) are, in general, uncontrolled.

4. One-Dimensional Hubbard Models

In this chapter we present exact results for one-dimensional modified Hubbard models with different dispersion relations. If not specified otherwise, we use the lattice constant a as the unit length from now on: $a = 1$.

In the first section we discuss the "standard" Hubbard model with cosine dispersion (nearest-neighbor hopping only) which was solved exactly by Lieb and Wu via the Bethe ansatz. We outline their solution which shows that the model describes a Mott insulator for half band-filling for all values $U > 0$. Strictly speaking there is no metal–insulator transition in the half-filled model because the metallic phase does not exist for finite interaction strengths.

In the second section we present the results for the $1/r$ Hubbard model. In this model the hopping amplitude between two lattice sites at distance r decays like $1/r$ and is thus long-ranged. The dispersion relation is linear within the Brillouin zone. The complete thermodynamics of the model can be deduced from the solution of the two-electron problem and numerical diagonalizations of small systems. At half band-filling the model describes a quantum phase transition from a metal to a Mott–Hubbard insulator at a finite interaction strength, $U_c = W$, where W is the bandwidth.

In the third section we discuss the $1/\sinh r$ Hubbard model. The electron hopping is now of variable range, $t(r) \sim \sinh(\kappa)/\sinh(\kappa r)$, where κ controls the hopping range. The model includes the "standard" Hubbard model for $\kappa = \infty$ and the $1/r$ Hubbard model for $\kappa = 0$, respectively. It can be solved exactly with the help of the asymptotic Bethe ansatz. For all $\kappa > 0$ the model posses the "perfect nesting" property. As a consequence, just like in the "standard" Hubbard model, the critical coupling for the transition into an insulating state at half band-filling is $U_c(\kappa > 0) = 0^+$.

In the fourth section we address the low-energy physics of one-dimensional models via the general g-ology Hamiltonian. Umklapp scattering for systems with perfect nesting leads to a metal–Mott insulator transition already at $U_c^{d=1} = 0^+$. Nesting is absent in the $1/r$ Hubbard model such that it describes a transition at a finite interaction strength. We compare the analytical results for our one-dimensional modified Hubbard models with the basic theoretical concepts for the Mott transition developed in Chap. 1. We find that Mott's views are widely supported by the exact solutions and discrepancies are due to the peculiarities of low-dimensional systems.

4.1 Hubbard Model with Nearest-Neighbor Hopping

In this section we discuss the one-dimensional Hubbard model with electron transfer between nearest neighbors. We begin with the solution of the two-electron problem on two and L lattice sites. The form of the two-electron wave function is simple enough to allow for its generalization to the N-electron problem ("Bethe ansatz"). We briefly outline the necessary steps to show that this "educated guess" indeed gives exact eigenstates of the Hubbard model. The analysis of the resulting Lieb–Wu equations at half band-filling gives explicit expressions for the ground-state energy and the Mott–Hubbard gap: the model describes an insulator for arbitrarily small Hubbard interaction. Near half band-filling, the Drude weight indicates a hole-like conductivity that vanishes because of carrier depletion when the lower Hubbard band becomes gradually filled.

4.1.1 Two-Electron Problem for Two Lattice Sites

The two-electron problem for two lattice sites already reveals some fundamental features of the Hubbard model. For $L = 2$ only the case of "half band-filling" with one spin $\uparrow$ and one $\downarrow$ electron ($S^z = 0$) is non-trivial since interaction is possible only in the presence of both spin species. The analysis of the remaining subspace of four states is simple [4.1].

One eigenstate belongs to the spin triplet ($S = 1$, $S^z = 0$) and charge singlet ($C = C^z = 0$) (see Sect. 2.5) and thus has the energy $E_{S=1}(U) = 0$ for all U. Another state is a spin singlet ($S = S^z = 0$) and charge triplet ($C = 1$, $C^z = 0$) and thus has the energy $E_{C=1}(U) = U$. The remaining two states are not simultaneous eigenstates of the kinetic and the potential energy operators. In position space the diagonal elements of the Hamilton matrix for the two remaining states are zero and U, respectively, whereas the non-diagonal elements are given by $-4t$ for periodic boundary conditions. The corresponding eigenenergies thus become

$$(E^{\pm}/L)(N = L = 2, U) = \frac{U}{4} \pm \sqrt{\left(\frac{U}{4}\right)^2 + (2t)^2} \, . \tag{4.1}$$

The spin singlet with energy $E^-(U)$ is the ground state for all U.

For zero interaction the "Fermi sea" is the ground state with energy density $(E^-/L)(N = L = 2, U = 0) = -2t$. The energy correction to the ground-state energy density in first order is given by $(U/4)$, the Hartree term for half band-filling. The correlation energy in second order is found to be $-U^2/(64t) = -0.015625\, U^2/t$.

For strong interactions the four states are grouped in two states with one double occupancy ($E_{C=1}(U) = U$; $E^+(U) = U + 4t^2/U$) and two states without a double occupancy ($E_{S=1}(U) = 0$; $E^-(U) = -4t^2/U$). One recognizes the two fundamental energy scales for a spin excitation from the ground state, $J \sim t^2/U$, and for charge excitations which require an energy U. The ground-state energy for large U/t nicely displays the energy gain due to the antiferromagnetic itinerant exchange interaction J.

4.1.2 General Two-Electron Problem in One Dimension

The general two-electron problem on L lattice sites can always be solved in any dimension for a translationally invariant Hamiltonian. The conservation of the total momentum allows us to reduce the task to the solution of an effective one-electron problem [4.2]. In the following we will analyze the one-dimensional two-electron problem in the language of the Bethe-ansatz wave function. The discussion follows the excellent review by Andrei [4.3]; for further reading, see the reprint volume edited by Korepin and Eßler [4.4].

We have to solve the Schrödinger equation in second quantization, $\hat{H}|\psi\rangle = E|\psi\rangle$, for the two-electron state

$$|\psi\rangle = \sum_{x_1,x_2} \psi(x_1,x_2)\hat{c}^+_{x_1,\uparrow}\hat{c}^+_{x_2,\downarrow}|\text{vacuum}\rangle \ . \tag{4.2}$$

As in the last subsection, we first restrict ourselves to the spin singlet. For the triplet the scattering matrix is unity since the electrons cannot interact with each other. For total spin $S=0$ the spin wave function is antisymmetric such that the space part has to be symmetric, $\psi(x_1,x_2)=\psi(x_2,x_1)$. For the space part the Schrödinger equation in first quantization becomes

$$\begin{aligned}(E-U\delta_{x_1,x_2})\psi(x_1,x_2) &= (-t)\Big[\psi(x_1+1,x_2)+\psi(x_1-1,x_2)\\ &\quad +\psi(x_1,x_2+1)+\psi(x_1,x_2-1)\Big] \ .\end{aligned} \tag{4.3}$$

Since the Hubbard interaction is purely local we expect that *scattering solutions* should, at least asymptotically, become plane-wave states for $x_1 \ll x_2$ and $x_1 \gg x_2$, respectively. Since we have electron transfer between nearest neighbors only, we may make the ansatz of plane-wave states in both sectors $x_1 < x_2$ and $x_1 > x_2$, respectively. Our "educated guess" is to set

$$\psi(x_1 \le x_2) = A\mathrm{e}^{\mathrm{i}(k_1x_1+k_2x_2)} - B\mathrm{e}^{\mathrm{i}(k_2x_1+k_1x_2)} \ . \tag{4.4}$$

Here, k_1 and k_2 are quasi-momenta and A, B are the amplitudes for the incoming and the scattered wave, respectively. In particular,

$$s = (-B)/A = -\mathrm{e}^{\mathrm{i}\theta(k_1,k_2)} \tag{4.5}$$

is the (scalar) scattering matrix for spin $S=0$ and $\theta(k_1,k_2)$ is the scattering phase shift. The scattering process is purely local and electrons are transferred between nearest neighbors only. Hence the wave function above is also valid at the boundary of the sectors for $x_1 = x_2$.

The Schrödinger equation within the sector $x_1 \le x_2 - 1$ gives

$$E = \epsilon(k_1) + \epsilon(k_2) = (-2t)\left(\cos(k_1)+\cos(k_2)\right) \ . \tag{4.6}$$

It is important to note that k_1 and k_2 are not given by multiples of $2\pi/L$ but are U dependent in general. We find that the two sectors are not coupled for all coordinates within a sector. Such a coupling between sectors appears only when the Schrödinger equation is solved at the boundary, $x_1 = x_2$, where terms from both sectors contribute. The interaction thus appears as a boundary condition problem for the individual sectors. For $x_1 = x_2$

$$(E-U)(A-B) = (-2t)\left[A(\mathrm{e}^{\mathrm{i}k_2}+\mathrm{e}^{-\mathrm{i}k_1}) - B(\mathrm{e}^{\mathrm{i}k_1}+\mathrm{e}^{-\mathrm{i}k_2})\right] \tag{4.7}$$

has to be fulfilled. Together with (4.6) we thus find that

$$s = -\frac{B}{A} = \frac{\mathrm{i}(\sin(k_1)-\sin(k_2))+U/(2t)}{\mathrm{i}(\sin(k_1)-\sin(k_2))-U/(2t)} \ . \tag{4.8}$$

For the total scattering matrix between electrons with arbitrary spin we obviously have $S^{12} = (I+P^{12})/2 + (I-P^{12})s/2$ since the scattering matrix between electrons with total spin $S=1$ is unity. Here, P^{12} is the spin exchange operator ($P^{\eta_1,\eta_2}_{\sigma_1,\sigma_2} = \delta^{\eta_2}_{\sigma_1}\delta^{\eta_1}_{\sigma_2}$ in matrix notation), I is the unit operator in spin space ($I^{\eta_1,\eta_2}_{\sigma_1,\sigma_2} = \delta^{\eta_1}_{\sigma_1}\delta^{\eta_2}_{\sigma_2}$), and $(I\pm P^{12})/2$ are the operators for projection onto the spin subspaces with $S=0$ and $S=1$, respectively. The total two-electron scattering matrix thus reads

$$S^{12} = \frac{(\sin(k_1)-\sin(k_2))I + \mathrm{i}U/(2t)P^{12}}{\sin(k_1)-\sin(k_2)+\mathrm{i}U/(2t)} \ . \tag{4.9}$$

For systems that can be solved with the help of the Bethe ansatz the information contained in the two-electron scattering matrix is sufficient to determine the eigenenergies of the Bethe-ansatz wave functions. In general, however, the scattering states do not exhaust the Hilbert space. It was only recently that Eßler, Korepin, and Schoutens constructed all eigenstates of the Hubbard model [4.5, 4.6] with the help of its $SO(4)$ symmetry, which has been discussed in Sect. 2.5.

To quantize the quasi-momenta we have to apply boundary conditions whose effect can most easily be discussed with the help of the scattering phase shift. From its relation to the scattering matrix in the singlet channel, (4.5), it is found from (4.8) as

$$\theta(k_1,k_2) = (-2)\arctan\left[\frac{\sin(k_2)-\sin(k_1)}{U/(2t)}\right] \ . \tag{4.10}$$

If we take the electron with the quasi-momentum k_1 around the chain that was closed into a ring (periodic boundary conditions) we find the total phase shift. It is given by the contribution Lk_1 and the contribution $\theta(k_1,k_2)$ due to the scattering from the other electron with quasi-momentum k_2. For periodic boundary conditions we thus have to demand that

$$2\pi I_1 = Lk_1 + \theta(k_1,k_2) \ , \tag{4.11a}$$

$$2\pi I_2 = Lk_2 + \theta(k_2,k_1) \ , \tag{4.11b}$$

where I_1, I_2 are half-odd integer due to the spin symmetry of the wave function. Since $\theta(k_1, k_2)$ is antisymmetric in its arguments the conserved total momentum $Q = k_1 + k_2$ is an integer multiple of $2\pi/L$.

For a better comparison to the N-electron problem we rewrite (4.11) as

$$\begin{aligned} k_j L &= 2\pi I_j + \Theta\big(2\sin(k_j) - 2\Lambda\big) , \qquad j = 1, 2 , &(4.12a)\\ 0 &= \sum_{j=1}^{2} \Theta(2\sin(k_j) - 2\Lambda) , &(4.12b)\\ \Theta(x) &= (-2)\arctan\Big(\frac{x}{U/(2t)}\Big) , \qquad -\pi \le \Theta(x) < \pi . &(4.12c) \end{aligned}$$

Equation (4.12b) has the obvious solution $\Lambda = (\sin(k_1) + \sin(k_2))/2$, and we immediately recover (4.11).

4.1.3 Bethe Ansatz for the N-Electron Problem

The analysis of the two-electron problem showed that the electrons do not interact within a sector. Thus, we may represent them there as plane waves. The interaction appears only as a boundary condition problem at the borders of the sectors. It is obvious that we should also try to construct the solution of the general N-electron problem in terms of plane waves. Our "educated guess" for the N-electron state thus reads

$$|\psi\rangle = \sum_{\sigma_1 \ldots \sigma_N} \sum_{x_1 \ldots x_N} \psi_{\sigma_1 \ldots \sigma_N}(x_1, \ldots, x_N) \prod_{j=1}^{N} \hat{c}^{+}_{x_j,\sigma_j} |\text{vacuum}\rangle , \qquad (4.13a)$$

with the Bethe-ansatz wave function

$$\begin{aligned} \psi_{\sigma_1 \ldots \sigma_N}(x_1, \ldots, x_N) &= \sum_{\mathcal{P}} (-1)^{\mathcal{P}} \exp\Big(\mathrm{i} \sum_j k_{\mathcal{P}(j)} x_j\Big) \\ &\quad \sum_{\mathcal{Q}} A_{\sigma_{\mathcal{P}(1)} \ldots \sigma_{\mathcal{P}(N)}}(\mathcal{Q}) \theta_{\mathrm{S}}(x_{\mathcal{Q}}) . \qquad (4.13b) \end{aligned}$$

Here, $\mathcal{P}$ and $\mathcal{Q}$ are permutations of the numbers $1, \ldots, N$ into $\mathcal{P}(1), \ldots, \mathcal{P}(N)$ and $\mathcal{Q}(1), \ldots, \mathcal{Q}(N)$, respectively. If $\mathcal{P}$ can be decomposed into an odd (even) number of permutations of two numbers (transpositions) we set $(-1)^{\mathcal{P}} = -1$ $(+1)$. The function $\theta_{\mathrm{S}}(x_{\mathcal{Q}})$ is unity if the electrons are found in the sector $\mathcal{Q}$ $(x_{\mathcal{Q}(1)} \le x_{\mathcal{Q}(2)} \cdots \le x_{\mathcal{Q}(N)})$ and zero otherwise. Whereas the permutations $\mathcal{P}$ take care of the antisymmetry of the wave function ("Slater determinant") the permutations $\mathcal{Q}$ guarantee that we sum over all sectors.

The solution of the Schrödinger equation within the sectors immediately gives the total energy and the total momentum as a function of the N quasi-momenta as

$$E = \sum_{j=1}^{N} \epsilon(k_j) \quad ; \quad P = \sum_{j=1}^{N} k_j . \qquad (4.14)$$

These two quantities are conserved for all translationally invariant and time-independent Hamiltonians. In our Bethe-ansatz wave function we demand also that the *individual* quasi-momenta k_j are constants of motion. Our ansatz can be successful only if the system is indeed integrable such that it has $L \to \infty$ many conserved quantities. For the Hubbard model an infinite number of conserved quantities was constructed by Shastry [4.7].

For the two-electron problem it is obvious that the quasi-momenta are the same for both sectors since the total energy and the total momentum have to be the same. As stated above it implies additional conservation laws for $N \geq 3$. Within the Bethe-ansatz wave function the ansatz of equal quasi-momenta in all sectors results in consistency equations for the amplitudes $A(\mathcal{Q})$. To derive them we consider two permutations $\mathcal{Q}_1$ and $\mathcal{Q}_2$ which differ only by a transposition of two indices. In particular, if $\mathcal{Q}_1$ and $\mathcal{Q}_2$ differ only by a transposition of two neighboring indices $[i = \mathcal{Q}_1(l+1) = \mathcal{Q}_2(l)$; $j = \mathcal{Q}_1(l) = \mathcal{Q}_2(l+1)]$ we may interpret this as the result of a scattering process between neighboring electrons. Hence we have $A(\mathcal{Q}_1) = S^{ij} A(\mathcal{Q}_2)$. If the two permutations are not neighbors we can generate them from each other by subsequent transpositions of neighboring index pairs (ij), (jk), (kl), e.g., $A(\mathcal{Q}_1) = S^{ij} S^{jk} S^{kl} A(\mathcal{Q}_2)$. For three electrons we can go from the permutation $\mathcal{Q}_2 = (123)$ to the permutation $\mathcal{Q}_1 = (321)$ on two different paths: namely via the sequence $(123) \mapsto (132) \mapsto (312) \mapsto (321)$ or the sequence $(123) \mapsto (213) \mapsto (231) \mapsto (321)$. Since the result is claimed to be independent of the sequence we have to fulfill the following condition:

$$S^{12} S^{13} S^{23} = S^{23} S^{13} S^{12} \, . \tag{4.15a}$$

This is the Yang–Baxter equation [4.8, 4.9]. It states that the three-electron scattering matrix can be represented by the product of two-electron scattering matrices and that there is no ambiguity in the corresponding partition of the three-electron scattering matrix. This is formally expressed in (4.15a). In addition, however, this has to be true for the full N-electron scattering matrix. For $N = 2$ and $N = 4$ this results in the additional conditions

$$S^{ij} S^{ji} = I \, , \tag{4.15b}$$

$$S^{ij} S^{kl} = S^{kl} S^{ij} \, . \tag{4.15c}$$

All three conditions of (4.15) are fulfilled by the scattering matrix in (4.9). In the Hubbard model a scattering process at some site can at most involve a bound state (double occupancy) at this side and two scattered waves. Hence, the conditions (4.15) on the N-electron scattering matrices for $N \leq 4$ are sufficient to guarantee a path-independent factorization of the general N-electron scattering matrix into two-electron scattering matrices [4.10].

Next we have to implement periodic boundary conditions to quantize the quasi-momenta. Consequently, the following N equations should hold,

$$\psi_{\sigma_1 \ldots \sigma_N}(x_1, \ldots, x_j = 0, \ldots, x_N) = \psi_{\sigma_1 \ldots \sigma_N}(x_1, \ldots, x_j = L, \ldots, x_N) \, , \quad j = 1, \ldots, N \, . \tag{4.16}$$

With the help of the scattering matrix this can be turned into a condition for the amplitudes within the same sector. In explicit matrix notation for the spin indices (4.16) reads

$$\sum_{\eta_1\dots\eta_N} (Z_j)^{\eta_1\dots\eta_N}_{\sigma_1\dots\sigma_N} A_{\eta_1\dots\eta_N}(\mathcal{Q}) = \mathrm{e}^{-\mathrm{i}k_j L} A_{\sigma_1\dots\sigma_N}(\mathcal{Q}) \,, \tag{4.17a}$$

where the operator $(Z_j)^{\eta_1\dots\eta_N}_{\sigma_1\dots\sigma_N}$ in spin space is given by

$$(Z_j)^{\eta_1\dots\eta_N}_{\sigma_1\dots\sigma_N} = (S^{jj-1}\dots S^{j1}S^{jN}\dots S^{jj+1})^{\eta_1\dots\eta_N}_{\sigma_1\dots\sigma_N} \,. \tag{4.17b}$$

The form of the operator becomes evident when we recall the fact that the electron j can be shifted to the beginning of the chain with the sequence $S^{1j}\dots S^{j-1j}$ and can be shifted to the end of the chain with the sequence $S^{jN}\dots S^{jj+1}$. Both sequences may differ by only a phase factor $\mathrm{e}^{\mathrm{i}k_j L}$ because we effectively took the electron around the ring.

The remaining task is the diagonalization of the spin Hamiltonians Z_j which commute with each other due to the Yang–Baxter equations. For that purpose we have to set up another Bethe-ansatz wave function which will result in "nested" Bethe-ansatz equations for the quasi-momenta k_j and those of the spin problem Λ_α. The spin problem is equivalent to the six-vertex model [4.11] of classical statistical mechanics. The formidable task of its solution was accomplished by Yang [4.8] and Gaudin [4.9]. The calculations are complicated and lengthy; see [4.12] for details.

A more elegant and powerful technique for the solution of the problem is the "quantum inverse scattering" or algebraic Bethe-ansatz method [4.13]. For the six-vertex model it is compactly presented in [4.3] and will not be repeated here. The formulation of the quantum inverse scattering approach for the full Hubbard model is a rather complex problem, and substantial progress has been made only fairly recently [4.14, 4.15].

From the solution of the spin problem one finally obtains the nested Bethe-ansatz equations:

$$\mathrm{e}^{\mathrm{i}k_j L} = \prod_{\alpha=1}^{N_\downarrow} \frac{\Lambda_\alpha - \sin(k_j) - \mathrm{i}U/(4t)}{\Lambda_\alpha - \sin(k_j) + \mathrm{i}U/(4t)} \,, \quad j = 1,\dots,N \,, \tag{4.18a}$$

$$-\prod_{\beta=1}^{N_\downarrow} \frac{\Lambda_\beta - \Lambda_\alpha + \mathrm{i}U/(2t)}{\Lambda_\beta - \Lambda_\alpha - \mathrm{i}U/(2t)} = \prod_{j=1}^{N} \frac{\Lambda_\alpha - \sin(k_j) - \mathrm{i}U/(4t)}{\Lambda_\alpha - \sin(k_j) + \mathrm{i}U/(4t)} \,, \quad \alpha = 1,\dots,N_\downarrow \,, \tag{4.18b}$$

where $N_\downarrow$ is the number of electrons with spin $\downarrow$. These equations were first derived by Yang [4.8] and Gaudin [4.9] for the continuum problem with contact interaction. Lieb and Wu realized [4.16] that one obtains the Bethe-ansatz equations for the Hubbard model on the lattice, (4.18), from those of Yang and Gaudin by the replacement $k^2 \mapsto -2\cos(k)$ and $k \mapsto \sin(k)$.

We take the logarithm on both sides of (4.18) and thereby obtain the Lieb–Wu equations:

$$\begin{aligned} k_j L &= 2\pi I_j + \sum_{\alpha=1}^{N_\downarrow} \Theta\big(2\sin(k_j) - 2\Lambda_\alpha\big) , \\ & j = 1, \ldots, N , \end{aligned} \tag{4.19a}$$

$$\begin{aligned} -\sum_{\beta=1}^{N_\downarrow} \Theta(\Lambda_\alpha - \Lambda_\beta) &= -2\pi J_\alpha + \sum_{j=1}^{N} \Theta\big(2\sin(k_j) - 2\Lambda_\alpha\big) , \\ & \alpha = 1, \ldots, N_\downarrow . \end{aligned} \tag{4.19b}$$

Here, $\Theta(x) = -2\arctan\big[x/(U/(2t))\big]$ is the function that we introduced in (4.12c). The quantum numbers I_j are half odd-integer when $N_\downarrow$ is odd and integer otherwise. The values of $2\pi I_j/L$ are restricted to the first Brillouin zone ($N \leq L$). The quantum numbers J_α are half odd-integer when $N - N_\downarrow - 1$ is odd and integer otherwise. Furthermore, $2\pi J_\alpha \leq \pi(N - N_\downarrow)$. A particularly elegant method for the analysis of the nested Bethe-ansatz equations has been developed by Klümper, Schadschneider, and Zittartz [4.17]; see [4.4] for a review and [4.6] for the inclusion of the $SO(4)$ symmetry.

In the following two subsections we discuss the ground-state properties and the charge excitations at half band-filling.

4.1.4 Ground-State Energy at Half Band-Filling

In the thermodynamic limit at fixed electron density n_σ the numbers k_j, Λ_α are continuously distributed with densities $\rho(k) - 1/[(k_{j+1} - k_j)L]$ for $|k| \leq Q \leq \pi$ and $\sigma(\Lambda) = 1/[(\Lambda_{\alpha+1} - \Lambda_\alpha)L]$ for $|\Lambda| \leq B$. Their boundaries are determined by the conditions

$$\int_{-Q}^{Q} \mathrm{d}k\rho(k) = N/L , \tag{4.20a}$$

$$\int_{-B}^{B} \mathrm{d}\Lambda\sigma(\Lambda) = N_\downarrow/L . \tag{4.20b}$$

The total energy and momentum then become

$$E/L = -2t \int_{-Q}^{Q} \mathrm{d}k\rho(k)\cos(k) , \tag{4.21a}$$

$$P/L = \int_{-Q}^{Q} \mathrm{d}k\rho(k)k . \tag{4.21b}$$

In the ground state the quantum numbers k_j and Λ_α are real, and symmetrically distributed around zero. The corresponding quantum numbers I_j and J_α are subsequent numbers distributed symmetrically around zero [4.16].

We subtract the corresponding equations for I_{j+1} and I_j, and $J_{\alpha+1}$ and J_α $[(I_j - I_{j-1})/L = \Delta I = 1/L$, $(\Lambda_\alpha - \Lambda_{\alpha-1})/L = \Delta\Lambda = 1/L]$, respectively, and expand the Lieb-Wu equations in ΔI and $\Delta\Lambda$; see Appendix A.3. This results in the integral equations for the ground state

$$2\pi\rho_0(k) = 1 + \cos(k)\int_{-B_0}^{B_0} \mathrm{d}\Lambda\sigma_0(\Lambda)K\left[\sin(k) - \Lambda; U/(4t)\right] , \quad (4.22a)$$

$$2\pi\sigma_0(\Lambda) = \int_{-Q_0}^{Q_0} \mathrm{d}k\rho_0(k)K\left[\sin(k) - \Lambda; U/(4t)\right] - \int_{-B_0}^{B_0} \mathrm{d}\Lambda'\sigma_0(\Lambda')K\left[\Lambda - \Lambda'; U/(2t)\right] , \quad (4.22b)$$

with $K[x;y] = 2y/(x^2 + y^2) = \int_{-\infty}^{\infty} \mathrm{d}p \exp(-y|p|)\exp(\mathrm{i}px)$. In the absence of an external magnetic field the ground state is a singlet, in particular $N_\downarrow = N/2$. The equations (4.22) then give $B_0 = \infty$, which is easily verified using (4.22b) and (4.20). With the help of a Fourier transformation we can explicitly solve for the density $\sigma_0(\Lambda)$ in terms of $\rho_0(k)$; see Appendix A.3.

For half band-filling ($N = L$) we have $Q_0 = \pi$. In this case we can analytically solve the remaining integral equation for $\rho_0(k)$. The result for half band-filling is [4.16]

$$\rho_0(k) = \frac{1}{2\pi} + \frac{\cos(k)}{\pi}\int_0^\infty \mathrm{d}p\frac{J_0(p)\cos(p\sin(k))}{1 + \exp(pU/(2t))} , \quad (4.23a)$$

$$\sigma_0(\Lambda) = \frac{1}{2\pi}\int_0^\infty \mathrm{d}\omega \cos(\omega\Lambda)\frac{J_0(\omega)}{\cosh(\omega U/(4t))} . \quad (4.23b)$$

The ground-state energy density can then be written as

$$(E_0/L)(N = L, U) = -4t\int_0^\infty \mathrm{d}\omega\frac{J_0(\omega)J_1(\omega)}{\omega[1 + \exp(\omega U/(2t))]} , \quad (4.24)$$

where $J_0(\omega)$ and $J_1(\omega)$ are Bessel functions. For all densities the ground-state energy is a differentiable and monotonously increasing function of the interaction strength. For large U we obtain the result for the Heisenberg model, $(E_0/L)(N = L, U \to \infty) = -2\ln(2)(2t^2/U) + \mathcal{O}(t^3/U^2)$ [4.18]. The result for two electrons on two lattices sites, $(E_0/L)(N = L = 2, U \to \infty) = -2t^2/U$, comes very close to the result in the thermodynamic limit. For small interaction strengths we obtain $(E_0/L)(N = L, U \to 0) = -(4t/\pi) + U/4 - 7\zeta(3)/(16\pi^3)U^2/t + \mathcal{O}(U^3/t^2)$ [4.19, 4.20] where $\zeta(x)$ is the Riemann zeta function. The numerical value for the correlation energy to second order is $-0.01696\ U^2/t$. Again, the numerical value for the correlation energy for two electrons on two lattice sites, $(E_0/L)(N = L = 2, U \to 0) = -0.015625\ U^2/t$, obviously comes very close to its value in the thermodynamic limit. A series expansion in U/t, however, is an asymptotic series only. Terms of the order $\mathcal{O}(\mathrm{e}^{-t/U})$ which are present in the exact expression are also missing [4.20].

4.1.5 Metal–Insulator Transition at Half Band-Filling

To gain some insight into the conducting properties at half band-filling we have to investigate whether or not there is a gap for charge excitations. Due to electron–hole symmetry [see Sect. 2.5, (2.32)] the ground-state energy for more than half band-filling obeys

$$E_0(2-n,U) = E_0(n) + UL(1-n) \ . \tag{4.25a}$$

Hence, the chemical potential at more than half-filling can be deduced from its values for $n \leq 1$,

$$\mu(2-n) = U - \mu(n) \ . \tag{4.25b}$$

In particular, we find for a half-filled band that

$$\mu^+ = U - \mu^- \ . \tag{4.25c}$$

Therefore, we need to study only the properties of a system with $L-1$ electrons on L sites to derive the gap $\Delta\mu = \mu^+ - \mu^- = U - 2\mu^-$ at half band-filling.

It is natural to interpret the state with $L-1$ electrons as an excitation of the half-filled band. Since an electron carries both charge and spin, we have to look for those excitations with minimum excitation energy to find the gap. The spin excitations are always gapless [4.3]. Therefore, it is sufficient to determine the gap for the charge excitations. A real electron with minimum excitation energy can be imagined as a charge excitation with energy $\epsilon_{\mathrm{c}}^{\mathrm{QP}}(k)$ and spin excitation(s) of vanishing energy. The distance between the energetically lowest state in the upper Hubbard band $\epsilon_{\mathrm{UHB}}(k) = U - \epsilon_{\mathrm{c}}^{\mathrm{QP}}(k)$ and the energetically highest state in the lower Hubbard band $\epsilon_{\mathrm{LHB}}(k) = \epsilon_{\mathrm{c}}^{\mathrm{QP}}(k)$ will then determine the size of the gap $\Delta\mu$,

$$\Delta\mu = \min_{\{k\}} \left(U - 2\epsilon_{\mathrm{c}}^{\mathrm{QP}}(k) \right) \ . \tag{4.26}$$

Recall that k is a quasi-momentum and not the true momentum of the excitation.

There are several ways to calculate the quasi-particle dispersion $\epsilon_{\mathrm{c}}^{\mathrm{QP}}(k)$. In the traditional approach one creates a hole in the sequence of the real quasi-momenta k_j and calculates its energy and momentum relative to the ground state at half band-filling [4.4, 4.21] (this is the "holon" excitation in Andrei's terminology [4.3]). This approach is used in [4.3] where all the other elementary excitations are explicitly calculated. The various quasi-particles in the Hubbard model interact with each other but their scattering matrices naturally obey the Yang–Baxter equations since the model is integrable. It is a highly non-trivial task to calculate the scattering matrices between the elementary excitations. It was first accomplished in [4.6] and was extended to general band-fillings in [4.3].

A more elegant method to calculate the quasi-particle dispersions is based on the direct formulation of the model properties in terms of the elementary excitations [4.22, 4.23]. In this approach the coupled integral equations for the charge and spin excitations of the ground state read

$$\tilde{\epsilon}_{\mathrm{c}}^{\mathrm{QP}}(k) = \epsilon_{\mathrm{c}}^{0}(k) + \int_{-B_0}^{B_0} \frac{\mathrm{d}\Lambda}{2\pi} K[\sin(k) - \Lambda; U/(4t)]\tilde{\epsilon}_{\mathrm{s}}^{\mathrm{QP}}(\Lambda) , \tag{4.27a}$$

$$\begin{aligned} \tilde{\epsilon}_{\mathrm{s}}^{\mathrm{QP}}(\Lambda) &= \int_{-Q_0}^{Q_0} \frac{\mathrm{d}k}{2\pi} \cos(k) K[\sin(k) - \Lambda; U/(4t)]\tilde{\epsilon}_{\mathrm{c}}^{\mathrm{QP}}(k) \\ &\quad - \int_{-B_0}^{B_0} \frac{\mathrm{d}\Lambda'}{2\pi} K[\Lambda - \Lambda'; U/(2t)]\tilde{\epsilon}_{\mathrm{s}}^{\mathrm{QP}}(\Lambda') , \end{aligned} \tag{4.27b}$$

where $\epsilon_{\mathrm{c}}^{0}(k) = -\mu + \epsilon(k) = -\mu - 2t\cos(k)$, and k is a quasi-momentum. In addition,

$$\tilde{\epsilon}_{\mathrm{s}}^{\mathrm{QP}}(B_0) = \tilde{\epsilon}_{\mathrm{s}}^{\mathrm{QP}}(-B_0) , \tag{4.28a}$$

$$\tilde{\epsilon}_{\mathrm{c}}^{\mathrm{QP}}(Q_0) = \tilde{\epsilon}_{\mathrm{c}}^{\mathrm{QP}}(-Q_0) . \tag{4.28b}$$

The chemical potential μ^- is derived from the condition

$$0 = \max_{\{k\}} \left[\tilde{\epsilon}_{\mathrm{c}}^{\mathrm{QP}}(k)\right] . \tag{4.29}$$

Since the chemical potential enters the definition of $\tilde{\epsilon}_{\mathrm{c}}^{\mathrm{QP}}(k)$ we now have $\epsilon_{\mathrm{LHB}}(k) = \mu^- + \tilde{\epsilon}_{\mathrm{c}}^{\mathrm{QP}}(k)$, $\epsilon_{\mathrm{UHB}}(k) = U - (\mu^- + \tilde{\epsilon}_{\mathrm{c}}^{\mathrm{QP}}(k)) = \mu^+ - \tilde{\epsilon}_{\mathrm{c}}^{\mathrm{QP}}(k)$.

The ground-state energy can be calculated with the help of the quasi-particle dispersion as

$$(E_0 - \mu N)/L = \int_{-Q_0}^{Q_0} \frac{\mathrm{d}k}{2\pi} \tilde{\epsilon}_{\mathrm{c}}^{\mathrm{QP}}(k) . \tag{4.30}$$

The approach via the quasi-particle dispersions also allows for a direct study of further model properties, especially for the calculation of critical exponents [4.22, 4.24] and the interpretation of the results in terms of generalized Landau parameters ("Landau–Luttinger liquid" [4.25]).

As shown in Appendix A.3 the equations can be analytically solved at half band-filling with the help of Fourier transformations. The result for the charge excitations becomes

$$\tilde{\epsilon}_{\mathrm{c}}^{\mathrm{QP}}(k) = \epsilon_{\mathrm{c}}^{0}(k) - 4t \int_0^\infty \mathrm{d}\omega \frac{J_1(\omega)}{\omega} \frac{\cos\left[\omega \sin(k)\right]}{1 + \exp(\omega U/(2t))} , \tag{4.31}$$

where $J_1(\omega)$ is a Bessel function. For $U = 0$ we have $\tilde{\epsilon}_{\mathrm{c}}^{\mathrm{QP}}(k) = 2\epsilon_{\mathrm{c}}^{0}(k)$ for $|k| \leq \pi/2$ and $\tilde{\epsilon}_{\mathrm{c}}^{\mathrm{QP}}(k) = 0$ for $|k| > \pi/2$. For $U \gg W = 4t$ we find $\tilde{\epsilon}_{\mathrm{c}}^{\mathrm{QP}}(k) = \epsilon_{\mathrm{c}}^{0}(k) - 4t\ln(2)/U + \mathcal{O}(t^3/U^2)$. For large U/t we obtain an upper and a lower Hubbard band of bandwidth $W = 4t$, centered around the energies zero and U, in full agreement with the general ideas of Mott and Hubbard. A more detailed discussion, including figures for the dispersion relations of the elementary excitations, can be found in [4.17].

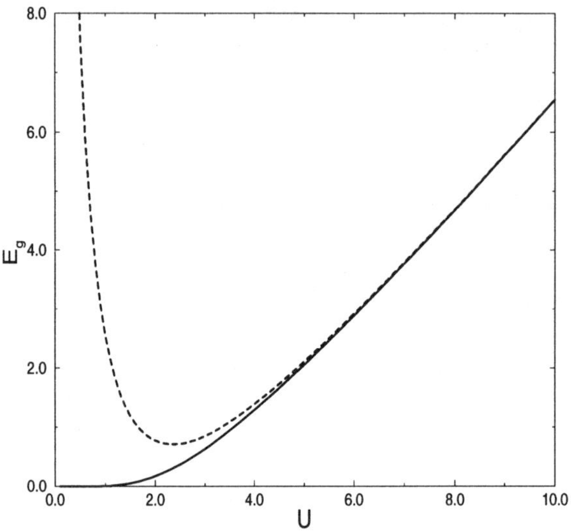

Fig. 4.1. Mott–Hubbard gap $E_{\mathrm{g}} \equiv \Delta\mu$ for the one-dimensional Hubbard model at half band-filling as a function of the interaction strength U/t. The *dashed line* is the result of the large-U expansion in (4.33) (from [4.27]).

At half band-filling we have $k_{\max} = Q_0 = \pi$ in (4.29), and the chemical potential follows from (4.31) as

$$\mu^-/t = 2 - 4\int_0^\infty \mathrm{d}\omega \frac{J_1(\omega)}{\omega} \frac{1}{1+\exp(\omega U/(2t))} \, . \tag{4.32a}$$

With the help of the series representation

$$\frac{1}{1+\mathrm{e}^x} = \sum_{n=0}^{\infty} (-1)^n \mathrm{e}^{-x(n+1)} \quad \text{for} \quad x > 0 \tag{4.32b}$$

this can be transformed into the Lieb–Wu result [4.16]

$$\mu^-/t = 2 - 4\sum_{n=1}^{\infty} (-1)^{n+1} \left[\sqrt{1+\left(\frac{nU}{2t}\right)^2} - \frac{Un}{2t} \right] . \tag{4.32c}$$

This representation, however, converges very slowly for small U/t. Ovchinnikov [4.26] expressed that series in terms of a contour integral whose integration path can be deformed such that it agrees with the imaginary axis. The expression for the energy gap can then be written as

$$\Delta\mu = U - 2\mu^- = \frac{16t}{U} \int_1^\infty \frac{\mathrm{d}y\sqrt{y^2-1}}{\sinh(2\pi t y/U)} \, . \tag{4.32d}$$

The Mott–Hubbard gap for the one-dimensional Hubbard model at half band-filling as a function of the interaction strength is shown in Fig. 4.1.

The gap as expressed in (4.32d) allows us to evaluate the limiting cases for weak and strong interaction strengths as

$$\Delta\mu = \begin{cases} (8t/\pi)\sqrt{U/t}\exp(-2\pi t/U) & \text{for} \quad U \lesssim W/2 = 2t \\ U - 4t + 8\ln(2)t^2/U & \text{for} \quad U \gtrsim W = 4t \end{cases} \, . \tag{4.33}$$

The upper and lower Hubbard bands are separated for *all* $U > 0$. The half-filled one-dimensional Hubbard model describes an *insulator* with critical $U_c = 0^+$. Equation (4.33) and Fig. 4.1 also show that the gap is only of an appreciable size for $U \gtrsim W/2$. Apparently, there is a contribution to the Mott–Hubbard gap from charge correlations which opens around $U = \mathcal{O}(W)$, as predicted by Mott (see Sect. 1.4). In addition, the tendency to magnetic order in systems with perfect nesting results in a Mott–Hubbard gap $\Delta\mu \sim \exp\left(-\mathcal{O}(J/t)\right)$ for arbitrarily small interaction strength.

Since the metallic phase is actually confined to the singular point $U = 0$, a real metal–insulator transition in the model does not exist at half band-filling. The point $(U = 0, n = 1, T = 0)$ in the parameter space of the Hubbard model displays the features of a critical point. For example, the Drude weight for finite lattice systems obeys a (hyper-)scaling function [4.28]–[4.30].

4.1.6 Drude Weight Below Half Band-Filling

The half-filled Hubbard model describes an insulator for all $U > 0$ since the one-electron excitations always display an energy gap for finite interaction strength. On the other hand, it is easy to show (see, e.g., [4.3]) that the system is an ideal metal for all $n < 1$ since gapless charge excitations ("holon–antiholon excitations" [4.3]) exist for all U. Consequently, we may study the Drude weight for all densities $n < 1$, and, in particular, for $n \to n_c = 1$.

Kohn [4.31] was apparently the first to point out that the Drude weight can be obtained from the ground-state energy that depends on an external vector potential. The discussion follows [4.32]. In the presence of a potential $\mathcal{A} = \mathcal{A}^0 \exp(-\mathrm{i}\omega t)$ the electron transfer matrix element t is modified by the Peierls phase factor to give $t(\Phi) = t\exp(\pm\mathrm{i}\Phi/L)$ where the sign depends on the hopping direction. In the following we measure the total flux through the ring of length La (a is the lattice constant) in 2π-fold units of the elementary flux quantum hc/e, $\Phi = La\mathcal{A}e/(\hbar c)$. If we treat Φ as a small parameter we may use second-order perturbation theory to obtain the standard result of linear response theory for the real part of the conductivity

$$\mathrm{Re}\big[\sigma(\omega)\big] = \frac{2\pi e^2}{\hbar^2} D\delta(\omega) + \text{reg. terms} = \mathcal{D}_c\delta(\omega) + \text{reg. terms}\ , \qquad (4.34a)$$

$$D = \frac{La}{2}\left.\frac{\partial^2 E_0(\Phi)}{\partial^2\Phi^2}\right|_{\Phi=0} . \qquad (4.34b)$$

Here, $\mathcal{D}_c$ denotes the Drude weight which is thereby given by

$$\mathcal{D}_c = \frac{\pi e^2 La}{\hbar^2}\left.\frac{\partial^2 E_0(\Phi)}{\partial^2\Phi^2}\right|_{\Phi=0} . \qquad (4.34c)$$

The calculation of the Drude weight $\mathcal{D}_c$ is thus reduced to the evaluation of the ground-state energy in the presence of the phase Φ. After a redefinition of the Fermi operators the phase appears as a "twist" of the boundary conditions; see [4.33] for details.

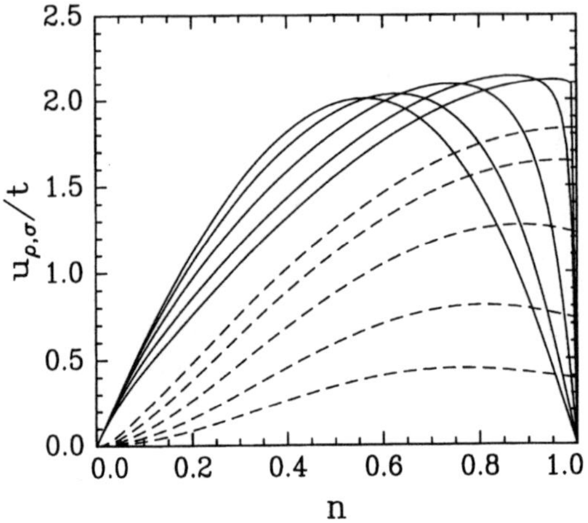

Fig. 4.2. Velocities of the charge and spin excitations in the one-dimensional Hubbard model as a function of the density for interaction strengths $U/t = 1, 2, 4, 8, 16$. *Full lines* correspond to the charge velocity $v_c \equiv u_\rho$ (increasing interaction strength from bottom to top in the left part of the figure); *dashed lines* correspond to the spin velocity $v_s \equiv u_\sigma$ (increasing interaction strength from top to bottom) (from [4.34]).

In a simple Boltzmann transport theory for non-interacting electrons with quadratic dispersion and (impurity) scattering time τ [4.35, Chap. 1] one finds for the real part of the conductivity $\mathrm{Re}[\sigma(\omega)] = (n_e e^2/m)[\tau/(1+\omega^2\tau^2)]$, where $n_e = N/(La) = n/a$ is the number of electrons per unit length. The DC conductivity becomes $\sigma^{\mathrm{DC}} = \mathrm{Re}[\sigma(\omega=0)] = n_e e^2\tau/m$. The Drude weight is obtained for the case of vanishing scattering, $\tau^{-1} \to 0$, as $\mathcal{D}_c = \pi n_e e^2/m$. The same result can be obtained from (4.34c) since all momenta are just shifted by the constant value $\Phi/(La)$.

For non-interacting electrons with cosine dispersion $\epsilon(k) = -2t\cos(ka)$ we find from (4.34b) that $D^0 = \hbar v_F/\pi$ with Fermi velocity $v_F = 2ta\sin(k_F a)/\hbar$ and Fermi momentum $k_F = n_e\pi/2$. For non-interacting tight-binding electrons the Drude weight thus becomes

$$\mathcal{D}_c^0 = (e^2/\hbar)\,2v_F\,. \tag{4.35}$$

In the presence of interactions the quasi-momenta obtain more than the trivial phase shift $\Phi/(La)$, and more subtle arguments are needed.

According to (4.34b), changes in the boundary conditions result in energy corrections of the order of $1/L$. Hence, the twist of boundary conditions can be interpreted as the generation of elementary excitations. Consequently, for the calculation of the Drude weight for the one-dimensional Hubbard model it is sufficient to find the proper elementary excitations that correspond to the twisting of the boundary conditions.

For the Hubbard model such a calculation was first carried out by Frahm and Korepin [4.22]; see also [4.24, 4.25]. In the absence of an external magnetic field one finds

$$\mathcal{D}_c = (e^2/\hbar)\,\xi^2(Q_0)v_c(Q_0)\,. \tag{4.36}$$

The velocity of the charge excitations is given by

$$\hbar v_c(Q_0) = \frac{\tilde{\epsilon}_c^{\mathrm{QP}\prime}(Q_0)}{2\pi\rho(Q_0)}\,, \tag{4.37}$$

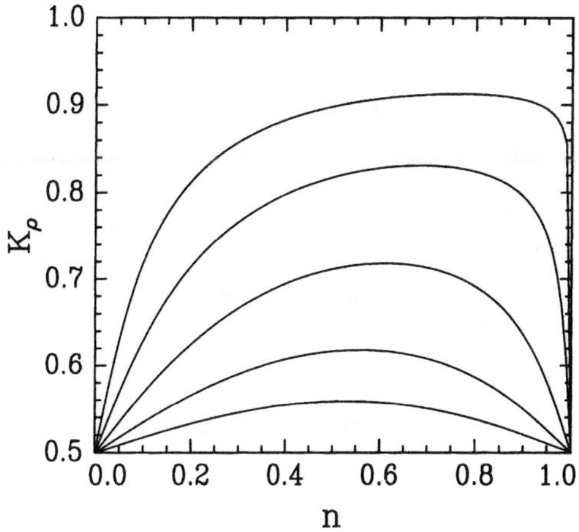

Fig. 4.3. Correlation exponent $K_\rho = \xi^2(Q_0)/2$ for the $4k_F$-oscillations in the Hubbard model as a function of the density for various interaction strengths ($U/t = 1, 2, 4, 8, 16$, increasing from top to bottom). For $U = 0$, $K_\rho(n) = 1$ (from [4.34]).

where $\rho(Q_0)$ and $\tilde{\epsilon}_c^{\mathrm{QP}\prime}(Q_0)$ are the values for the quasi-particle density and the derivative of the quasi-particle dispersion at the border of the quasi-particle density in the ground state, $\rho(k) = 0$ for $|k| > Q_0$; see [4.22] and Appendix A.3. For $U = 0$ we had $v_c^0(Q_0) \equiv v_F = 2ta\sin(k_F a)/\hbar$ while for $U = \infty$ we find $v_c^\infty(Q_0) = v_F^h = 2ta\sin(2k_F a)/\hbar$. For $U = \infty$ the system behaves as if the number of electrons had doubled compared to the number for $U = 0$. As far as the number of charge carriers is concerned, the system behaves like the fully polarized ferromagnet for which we have a gas of free holes at density $n_h = 1 - n$. For all finite U/t the calculation of $v_c(Q_0)$ has to be performed numerically. The result for the charge and spin velocities is shown in Fig. 4.2.

In (4.36) $\xi^2(Q_0)$ denotes an element of the "dressed charge matrix", which describes the scattering between the quasi-particles. For its derivation one has to analyze an integral equation [4.22], which can be explicitly solved for small and large interaction strengths and/or densities. Alternatively one can utilize the following relation between $\xi^2(Q_0)$ and the isothermal compressibility κ_T at $T = 0$ [4.22, 4.34],

$$\xi^2(Q_0) = \pi\hbar v_c(Q_0)(n_e^2\kappa_T) \ . \tag{4.38a}$$

In turn, the compressibility at $T = 0$ can be calculated from the variation of the ground-state energy density $e_0(n_e) = E_0(n_e)/(La)$ with the electron density $n_e = n/a$,

$$\left(n_e^2\kappa_T\right)^{-1} = \frac{\partial^2 e_0(n_e)}{\partial n_e^2} \ . \tag{4.38b}$$

The variation can be performed numerically [4.34] and gives the result for $\xi^2(Q_0) \equiv 2K_\rho$ that is displayed in Fig. 4.3. For all (U, n) one finds $1/2 \le K_\rho(U) \le 1$, $K_\rho(U = 0) = 1$, $K_\rho(U = \infty) = 1/2$.

One can show [4.22, 4.24, 4.34] that the density–density correlation function at zero temperature contains a part that oscillates with $\cos(4k_F r)$ as a function of distance r and algebraically decays proportional to r^{-4K_ρ}. In the absence of an external magnetic field this correlation exponent determines all critical exponents for the Green function and the correlation functions. In particular, this implies that the Hubbard model at low excitation energies belongs to the generic class of Luttinger liquids [4.36] for which even the spectral functions have been calculated [4.37, 4.38]. The optical conductivity at half band-filling and small interaction strengths was calculated in [4.39]. Thus, the low-energy properties of the Hubbard model are well understood. The optical conductivity for very large interactions was studied in [4.40].

4.1.7 Metal–Insulator Transition Approaching Half Filling

There is a metal–insulator transition for all finite interaction strengths if the electron density approaches the critical value $n_{\rm c}$ of half band-filling, i.e., for $U > 0, n \to n_{\rm c} = 1$. We point out, however, that this transition is *conceptually different* from the "canonical" Mott–Hubbard transition at half band-filling, i.e., for $U = U_{\rm c}, n = 1$. In the latter case the upper and lower Hubbard bands split apart such that the charge excitations acquire a gap and in the former case a whole class of excitations ceases to exist.

The interpretation of the lower Hubbard band being gradually filled up can be substantiated. For low hole densities, $n_{\rm h} = 1 - n$, one can find an analytical expression for the Drude weight [4.24]:

$$
\begin{aligned}
\frac{\mathcal{D}_{\rm c}}{\mathcal{D}_{\rm c}^0} &= \frac{n_{\rm h}}{2}\frac{B}{A^2}\,, \\
A &= \int_1^\infty {\rm d}z \left[\sqrt{z^2-1}\sinh(2\pi t z/U)\right]^{-1}\,, \\
B &= \frac{\pi^3 t}{4U}\int_1^\infty {\rm d}z \sqrt{z^2-1}\,\frac{2\coth^2(2\pi t z/U)-1}{\sinh(2\pi t z/U)}\,.
\end{aligned}
\tag{4.39a}
$$

In the limits of small and large interaction strengths this reduces to

$$
\frac{\mathcal{D}_{\rm c}}{\mathcal{D}_{\rm c}^0} = \frac{n_{\rm h}}{2}\left\{\begin{array}{ccc} \pi t/[2\Delta\mu(U/t\to 0)] & \text{for} & U/t\to 0 \\ \pi/2 & \text{for} & U/t\to\infty \end{array}\right.\,,
\tag{4.39b}
$$

where $\Delta\mu(U/t \to 0)$ is the energy gap for the case of a half-filled band for small U/t; see (4.33). For small U/t, however, the region in $n_{\rm h}$ for which (4.39) may be used is exponentially small. $\mathcal{D}_{\rm c}$ stays finite for $U/t \to 0$ but displays an infinite slope as a function of $n_{\rm h}$. The Drude weight for various interaction strengths according to [4.24] is shown in Fig. 4.4; see also [4.34].

Figure 4.4 shows that the Drude weight vanishes in the limit of half band-filling as is expected for a metal–insulator transition. In analogy to non-interacting electrons the Drude weight is written as $\mathcal{D}_{\rm c} = \pi e^2(n/m)^*$ with an effective particle density n^* and an effective mass m^*. We may ask how to interpret the metal–insulator transition for $U > 0, n \to 1$. A vanishing $\mathcal{D}_{\rm c}$

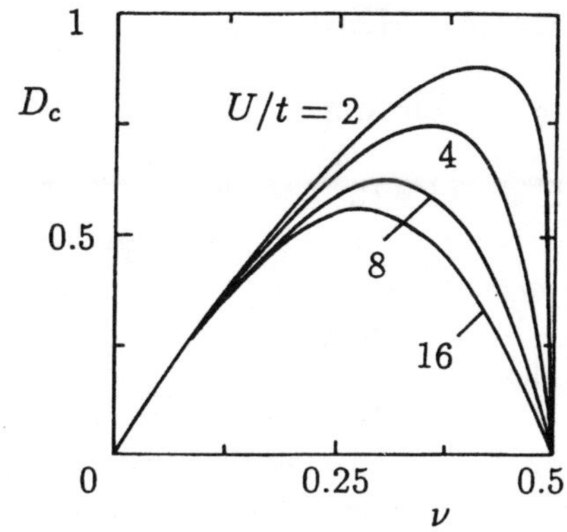

Fig. 4.4. Drude weight ($\mathcal{D}_c = 4tae^2/\hbar^2 D_c$) in the Hubbard model as a function of the density $\nu = n/2$ for interaction strengths $U/t = 2, 4, 8, 16$. For $U = 0^+$ there is a jump discontinuity at $n = 1$ (from [4.24]).

may individually be caused by the divergence of the effective mass or by the disappearance of the charge carriers. In interacting electron systems, however, it may not be justified to view the Drude weight as the ratio of two separate contributions from the carrier density and the effective mass, and one can indeed provide arguments for both points of view [4.24, 4.41].

One might argue that both the bandwidth of the holon band [4.41] and the velocity of the charge quasi-particles v_c go to zero for small hole densities n_h (see Fig. 4.2). Such behavior is expected for a diverging effective mass. In addition, the product of the Drude weight and the compressibility is finite for all (U, n); see (4.36) and (4.38a). Correspondingly, the compressibility of the system diverges for $n \to 1$ which is generically interpreted as a diverging effective mass. The transfer of spectral weight between the two Hubbard bands as a function of the density at strong coupling is discussed in [4.42].

On the other hand, according to (4.39), the Drude weight is proportional to the density of charge carriers, the hole density $n_h = 1 - n$. If we adopt a simple band structure picture, it implies that the effective concentration of charge carriers vanishes at the transition, $n^* = 1 - n \to 0$. This picture is supported by the fact that the "holon–antiholon" excitations exist only for $n < 1$. At half band-filling this corresponds to a filled, and thus inert, band in the case of non-interacting electrons. In principle, comparisons with non-interacting electrons could be misleading because the form of the quasi-particle bands depends on (U, n) and, in particular, on the density of quasi-particle excitations which in fact interact with each other. In the case of the Hubbard model this caveat does not apply for the charge excitations in the limit $U = \infty$. There, they can be described by those of a spinless *Fermi gas* at density $n_h = 1 - n$ (gas of holes), $\tilde{\epsilon}_c^{QP}(k) = -2t\cos(ka)$ [4.28]. The Drude weight is simply given by $\mathcal{D}_c(U = \infty) = (e^2/\hbar)2ta\sin(n_h\pi)/\hbar$. Hence, the effective number of carriers is indeed seen to vanish for $n \to 1$. Furthermore, the "diverging effective mass" can easily be understood in the hole picture since the one-dimensional density of states diverges at the band edges. Consequently, the viewpoint of a hole conduction with vanishing carrier density at $n \to 1$ appears to be the more natural one [4.43].

4.2 Hubbard Model with Long-Range Hopping

In this section we present exact results for the one-dimensional Hubbard model with long-range hopping. We start with the discussion of some basic properties of the $1/r$ Hubbard model. Next, we solve the two-electron problem on L lattice sites. We use this solution to construct an effective Hamilton operator for hard-core bosons, which we conjecture to reproduce the exact spectrum and thermodynamics of the $1/r$ Hubbard model for *all* electron numbers N. At half band-filling the model describes a metal–insulator transition at *finite* interaction strength, $U_c = W$, where $W = 2\pi t$ is the bandwidth. At the end of this section we briefly discuss the specific heat, magnetic susceptibility, and isothermal compressibility as functions of temperature and interaction strength.

4.2.1 Basic Properties

The applicability of the standard Bethe ansatz is restricted to the Hubbard model with nearest-neighbor hopping. It was only recently that modified Hubbard models were formulated for which exact results could be found. To identify exactly solvable Hubbard models with modified hopping matrix elements we start from the strong coupling limit at half band-filling; see Sect. 2.5. If a spin-1/2 Heisenberg model with its two degrees of freedom per lattice site is exactly solvable, we might be able to construct a corresponding Hubbard model with four degrees of freedom per lattice site which could also be integrable. Until recently, however, the class of exactly solvable Heisenberg models was also constrained to the Bethe-ansatz solvable cases, and a different path had to be found.

There is a whole class of *continuum* models with non-local interactions which were exactly solved many years ago. The ground-state wave function of those continuum models can be written as a determinant over powers of elliptic functions which depend on only the coordinate differences $(x_i - x_j)$ of pairs of two bosons; for a review, see [4.44]. Among this class is the Calogero–Sutherland–Moser model that describes bosons in one dimension with the two-particle interaction $V(x) = \lambda(\lambda - 1)/x^2$ [4.45]. For this model one can not only determine the ground-state wave function, ground-state energy, and thermodynamics, but even calculate the exact ground-state density–density correlation function for special λ values ($\lambda = 1, 2, 4$) [4.46].

In the past few years [4.47, 4.48] these continuum models have been used to construct successfully integrable *lattice* spin models. The starting point for this development was the observation that the spin–spin correlations in the Gutzwiller-projected half-filled paramagnetic Fermi sea [4.49], a wave function defined *on the lattice* (see Sect. 3.4), are described by the same function as the density–density correlations of the Calogero–Sutherland–Moser model for $\lambda = 4$. This was independently realized by Haldane [4.50] and Shastry [4.51] who successfully constructed the exactly solvable spin-1/2 Heisenberg model with $1/r^2$ exchange. The Hamiltonian of the $1/r^2$ Heisenberg or Haldane–Shastry model explicitly reads

$$\hat{H}_{\mathrm{HS}} = \sum_{l\neq m=1}^{L} J(l-m)\left[\hat{\boldsymbol{S}}_l \cdot \hat{\boldsymbol{S}}_m - \frac{1}{4}\right] , \quad J(l-m) = \frac{J}{d(l-m)^2} , \tag{4.40a}$$

where

$$d(l-m) = \frac{L}{\pi}\sin\left[\frac{\pi(l-m)}{L}\right] \tag{4.40b}$$

denotes the chord distance between the lattice sites l and m on a ring of circumference L. In fact, the Gutzwiller-projected half-filled Fermi sea is the exact ground state of this spin model [4.50, 4.51]. In the meanwhile the integrability structure has been studied in detail [4.52], all thermodynamic quantities [4.53], and even the dynamic correlation functions have been calculated analytically [4.54]; for a review, see [4.55].

According to (2.43), the following relation holds between the hopping matrix elements $t(l-m)$ of a Hubbard model and the exchange coupling $J(l-m)$ of its corresponding Heisenberg model:

$$J(l-m) = \frac{2\,|t(l-m)|^2}{U} . \tag{4.41}$$

Since a Hubbard model has twice as many degrees of freedom per lattice site as a Heisenberg model, there are infinitely many Hubbard models whose hopping matrix elements obey (4.41) for the exchange coupling in (4.40a). We will show below that the choice

$$t(l-m) = (-\mathrm{i}t)\frac{(-1)^{l-m}}{d(l-m)} \tag{4.42}$$

leads to an exactly solvable modified Hubbard model. Since $d(l-m)$ is antisymmetric under the permutation of l and m, the hopping matrix element has to be purely imaginary to guarantee $t(l-m) = t^*(m-l)$. For a fixed distance $r = |l-m|$, and in the thermodynamic limit ($L \to \infty$), the hopping decays proportionally to the inverse of the distance ("$1/r$ Hubbard model").

In the following we use the standard convention $\hat{c}_{l,\sigma}^{+} = \sqrt{1/L}\sum_k \mathrm{e}^{-\mathrm{i}kl}\hat{c}_{k,\sigma}^{+}$ for the Fourier transformation from position space ($\hat{c}_{l,\sigma}$) into momentum space ($\hat{c}_{k,\sigma}$). Unfortunately, the opposite convention was used in [4.56, 4.59]. The dispersion relation is given by

$$\epsilon(k) = \frac{1}{L}\sum_{l\neq m=1}^{L} t(l-m)\mathrm{e}^{-\mathrm{i}k(l-m)} , \tag{4.43a}$$

where we choose antiperiodic boundary conditions and an even number of lattice sites L in the following. Unlike in models with short-range hopping and/or interactions, the choice of L as even or odd enters the choice of boundary conditions. Periodic boundary conditions for even L may be used when we assign an additional factor $\cos[\pi(l-m)/L]$ to the hopping matrix elements. Apparently, this factor plays no role in the thermodynamic limit.

To perform the calculation of $\epsilon(k)$ we set $r = l - m$ and use $(-1)^L = 1$, $\sin(\pi r/L + \pi) = -\sin(\pi r/L)$, and $\mathrm{e}^{\mathrm{i}kL} = (-1)$ which allows us to carry out one lattice sum. The dispersion relation then becomes

$$\epsilon(k) = (-\mathrm{i}t)\sum_{r=1}^{L-1} \frac{(-1)^r \mathrm{e}^{-\mathrm{i}kr}}{(L/\pi)\sin(\pi r/L)}, \tag{4.43b}$$

with $k = (2m+1)\pi/L$ ($m = -(L/2), \ldots, (L/2) - 1$). We use the series representation of the cosecans function

$$\frac{1}{\sin(\pi r/L)} = (-1)^r \frac{\mathrm{i}}{L} \sum_{m=-(L/2)}^{(L/2)-1} (2m+1)\mathrm{e}^{\mathrm{i}\pi(2m+1)/L} \quad (r \neq 0) \tag{4.43c}$$

such that we can perform the sum over r, which gives $\delta_{k,\pi(2m+1)/L}$. The simple final result for the dispersion relation becomes

$$\epsilon(k) = tk \tag{4.43d}$$

for k in the first Brillouin zone. The dispersion relation is found to be *linear* within the Brillouin zone.

The unphysical properties of such a dispersion relation are immediately apparent. First, it has odd parity, $\epsilon(-k) = -\epsilon(k)$. Since the interaction part of the Hubbard model has even parity, parity is not a good quantum number for the $1/r$ Hubbard model, in general. Second, the dispersion relation develops a *jump discontinuity* at the Brillouin zone boundary in the thermodynamic limit. Since it is an isolated discontinuity we do not expect unphysical effects (e.g., divergences) in thermodynamic quantities. However, transport properties like the Drude weight will display some pathologies since the jump implies a (left-moving) electron with infinite velocity.

Now that the dispersion relation has odd parity one has to repeat some calculation for the symmetry properties of the Hubbard model; see Sect. 2.5. For an antisymmetric dispersion relation the transformation necessary to restrict the analysis to maximal half band-filling reads

$$\mathcal{T}_4: \quad \hat{c}_{k,\sigma}^+ \mapsto \hat{c}_{-k,\sigma}\,. \tag{4.44}$$

Electrons and holes are mapped into each other such that the kinetic energy operator stays invariant, whereas the potential energy operator is invariant up to the usual constants. In particular, we can again deduce that the band is half filled, $N = L$, at all temperatures when the chemical potential is given by $\mu(T) = U/2$.

In all Hubbard models the spectrum of the potential energy operator is highly degenerate since all states with D double occupancies have the same potential energy irrespective of their positions on the lattice. Eigenvalues of the potential energy operator differ by only an integer multiple of the interaction energy U. In the $1/r$ Hubbard model the spectrum of the kinetic energy operator is also highly degenerate. To verify this we imagine some (excited) eigenstate of the kinetic energy operator with total momentum Q. If we carry out a two-electron scattering with momentum transfer q, the two final momenta $k_1 + q$ and $k_2 - q$ may lie inside or outside the first Brillouin zone. The difference in the kinetic energy before and after the scattering is readily calculated as $\Delta T = \Delta T(q) + \Delta T(-q)$ with $\Delta T(q) = \epsilon(k+q) - \epsilon(k)$. Since the dispersion relation is linear we have $\Delta T(q) = tq$, if $|k+q| < \pi$, and $\Delta T(q) = t(q - 2\pi \mathrm{sgn}(q))$, if $k+q < -\pi$ or $k+q > \pi$, respectively ($\mathrm{sgn}(x)$ is the sign function). If the final momenta lie both inside or both outside the first Brillouin zone, we find $\Delta T = 0$. In the remaining two cases we have $\Delta T = \pm 2\pi t = \pm W$. For fixed Q all eigenstates of the kinetic energy operator thus differ by only an integer multiple of the bandwidth.

Since the spectra of both parts of the Hamilton operator are highly degenerate we may expect to find many simultaneous eigenstates of $\hat{T}$ and $\hat{D}$. The problem remains non-trivial, though, because the operators do not commute, $[\hat{T}, \hat{D}]_- \neq 0$.

4.2.2 General Two-Electron Problem in One Dimension

We address the L^2 states in the sector with $S^z = 0$. Since the total momentum is a good quantum number we classify the solutions according to Q.

$Q = 0$. The L eigenstates are ($n = 0, \ldots, (L-1)$)

$$|\psi(Q=0,n)\rangle = \frac{1}{L} \sum_{|p|<\pi} \mathrm{e}^{ipn} \hat{c}^+_{p,\uparrow} \hat{c}^+_{-p,\downarrow} |\mathrm{vacuum}\rangle \tag{4.45a}$$

$$= \frac{1}{L} \sum_{x=1}^{L} \hat{c}^+_{x+n,\uparrow} \hat{c}^+_{x,\downarrow} |\mathrm{vacuum}\rangle . \tag{4.45b}$$

The momentum representation (4.45a) shows that these states are eigenstates of the kinetic energy operator with eigenvalue $T = 0$ since $\epsilon(p) + \epsilon(-p) = 0$. The position space representation (4.45b) implies that they are also eigenstates of the potential energy operator with eigenvalues $D = 1$ for $n = 0$ and $D = 0$ for $n = 1, \ldots, (L-1)$, respectively.

$Q \neq 0$. For $Q = 2\pi m_Q / L$ ($m_Q = 1, \ldots, (L-1)$) we address the following orthonormal states:

$$|\psi_{\mathrm{i}}(Q, n_{\mathrm{i}})\rangle = \sqrt{\frac{1}{L - m_Q}} \sum_{-\pi<p<\pi-Q} \mathrm{e}^{\mathrm{i}n_{\mathrm{i}}pL/(L-m_Q)} \hat{c}^+_{Q+p,\uparrow} \hat{c}^+_{-p,\downarrow} |\mathrm{vacuum}\rangle , \tag{4.46a}$$

$$|\psi_{\mathrm{a}}(Q, n_{\mathrm{a}})\rangle = \sqrt{\frac{1}{m_Q}} \sum_{\pi-Q<p<\pi} \mathrm{e}^{\mathrm{i}n_{\mathrm{a}}pL/m_Q} \hat{c}^+_{Q+p-2\pi,\uparrow} \hat{c}^+_{-p,\downarrow} |\mathrm{vacuum}\rangle , \tag{4.46b}$$

with $n_\mathrm{i} = 0, \ldots, (L - m_Q - 1)$ and $n_\mathrm{a} = 0, \ldots, (m_Q - 1)$. For each $Q \neq 0$ there are L states; in sum there are $L(L-1)$ eigenstates with $Q \neq 0$. Together with those L states for $Q = 0$ the states after (4.46) form an orthonormal basis of the subspace for $S^z = 0$.

By construction we have

$$\hat{T}|\psi_\mathrm{i}(Q, n_\mathrm{i})\rangle = tQ|\psi_\mathrm{i}(Q, n_\mathrm{i})\rangle \ , \tag{4.47a}$$

$$\hat{T}|\psi_\mathrm{a}(Q, n_\mathrm{a})\rangle = t(Q - 2\pi)|\psi_\mathrm{a}(Q, n_\mathrm{a})\rangle \ , \tag{4.47b}$$

since the momenta of the electrons involved were chosen from the first Brillouin zone. In the position space representation of the wave functions (4.46) we are only interested at the moment in whether the wave functions do have a double occupancy or not. The amplitude for a double occupancy is proportional to

$$\sum_{-\pi<p<\pi-Q} \mathrm{e}^{\mathrm{i}n_\mathrm{i}pL/(L-m_Q)} = \delta_{n_\mathrm{i},0}(L - m_Q) \ , \tag{4.48a}$$

$$\sum_{\pi-Q<p<\pi} \mathrm{e}^{\mathrm{i}n_\mathrm{a}pL/m_Q} = \delta_{n_\mathrm{a},0} m_Q \ . \tag{4.48b}$$

All states with $n_\mathrm{i} \neq 0$, $n_\mathrm{a} \neq 0$ have no double occupancy, $D = 0$. Thus we have shown that

$$\hat{H}|\psi_\mathrm{i}(Q, n_\mathrm{i} \neq 0)\rangle = tQ|\psi_\mathrm{i}(Q, n_\mathrm{i} \neq 0)\rangle \ , \tag{4.49a}$$

$$\hat{H}|\psi_\mathrm{a}(Q, n_\mathrm{a} \neq 0)\rangle = t(Q - 2\pi)|\psi_\mathrm{a}(Q, n_\mathrm{a} \neq 0)\rangle \ . \tag{4.49b}$$

Of all L^2 states in the subspace $S^z = 0$ we have found $L^2 - 2(L-1)$ states which are *simultaneous* eigenstates of the kinetic and the potential energy operators.

For fixed $Q \neq 0$, two states, $n_\mathrm{i} = 0$ and $n_\mathrm{a} = 0$, mix. Explicitly,

$$\hat{D}|\psi_\mathrm{i}(Q, n_\mathrm{i} = 0)\rangle = \sqrt{L - m_Q}\frac{1}{L}\sum_x \mathrm{e}^{\mathrm{i}Qx}\hat{c}^+_{x,\uparrow}\hat{c}^+_{x,\downarrow}|\mathrm{vacuum}\rangle \ , \tag{4.50a}$$

$$\hat{D}|\psi_\mathrm{a}(Q, n_\mathrm{a} = 0)\rangle = \sqrt{m_Q}\frac{1}{L}\sum_x \mathrm{e}^{\mathrm{i}Qx}\hat{c}^+_{x,\uparrow}\hat{c}^+_{x,\downarrow}|\mathrm{vacuum}\rangle \ . \tag{4.50b}$$

The remaining $2(L-1)$ eigenvalues $E^\pm(Q)$ thus result from the diagonalization of the two-by-two matrices

$$\begin{pmatrix} tQ + U(1 - m_Q/L) & U\sqrt{(m_Q/L)(1 - m_Q/L)} \\ U\sqrt{(m_Q/L)(1 - m_Q/L)} & t(Q - 2\pi) + Um_Q/L \end{pmatrix} \tag{4.51a}$$

as

$$E^\pm(Q) = \frac{1}{2}\left[W(Q - \pi)/\pi + U \pm \sqrt{W^2 + U^2 - 2UW(Q - \pi)/\pi}\right] \ . \tag{4.51b}$$

This calculation for two electrons on L lattice sites strongly resembles the one we carried out for the Hubbard model for two electrons on two lattice sites; see Sect. 4.1.1. We have found two "coupling states" (called a "doublet" in the following) for fixed total momentum. All other states are simultaneous eigenstates of the kinetic and potential energy operators.

It is instructive to analyze the amplitudes $A(x_1, x_2)$ of the wave functions for $Q \neq 0$ in position space. A short calculation gives

$$A_{\mathrm{i}}(x_1, x_2; n_{\mathrm{i}}) = \phi_{\mathrm{i},n_{\mathrm{i}}}(Q) \frac{\mathrm{e}^{\mathrm{i}\pi x_1}\mathrm{e}^{\mathrm{i}(Q-\pi)x_2} - \mathrm{e}^{\mathrm{i}\pi x_2}\mathrm{e}^{\mathrm{i}(Q-\pi)x_1}}{\sin\left[(\pi/L)(x_1 - x_2 + n_{\mathrm{i}}L/(L - m_Q)\right]} , \tag{4.52a}$$

$$A_{\mathrm{a}}(x_1, x_2; n_{\mathrm{a}}) = \phi_{\mathrm{a},n_{\mathrm{a}}}(Q) \frac{\mathrm{e}^{\mathrm{i}(Q-\pi)x_1}\mathrm{e}^{-\mathrm{i}\pi x_2} - \mathrm{e}^{\mathrm{i}(Q-\pi)x_2}\mathrm{e}^{-\mathrm{i}\pi x_1}}{\sin\left[(\pi/L)(x_1 - x_2 + n_{\mathrm{a}}L/m_Q\right]} , \tag{4.52b}$$

where $\phi_{\mathrm{i},n_{\mathrm{i}}}(Q)$ and $\phi_{\mathrm{a},n_{\mathrm{a}}}(Q)$ are irrelevant phase factors. These equations cover all cases if the limits are properly taken for $n_{\mathrm{i}} = 0$ and/or $n_{\mathrm{a}} = 0$. The position space representation (4.52) shows that the wave functions are not plane waves, not even asymptotically for $|x_1 - x_2| \to \infty$. Hence, the $1/r$ Hubbard model is a very special problem that cannot be solved by standard methods.

4.2.3 Effective Hamiltonian for the N-Electron Problem

The solution of the two-electron problem showed that there are many simultaneous eigenstates of the kinetic and the potential energy operators. In effect, there were at most two states mixing with each other. This observation remains valid when the three-electron problem is analyzed. The discussion follows [4.56].

As already discussed in Sect. 4.2.1 the potential energy operator can change the energy of an eigenstate of $\hat{T}$ at most by $\pm W$. Analogously, the kinetic energy operator can change the energy of an eigenstate of $\hat{D}$ at most by $\pm U$. Thus, it is appears to be natural to split the operator for double occupancies into three parts which obey the following commutation relations with the kinetic energy operator:

$$\hat{D} = \hat{D}^0 + \hat{D}^+ + \hat{D}^- , \tag{4.53a}$$

$$[\hat{T}, \hat{D}^0]_- = 0 , \tag{4.53b}$$

$$[\hat{T}, \hat{D}^\pm]_- = \pm W \hat{D}^\pm . \tag{4.53c}$$

We set $\Delta = 2\pi/L$ and introduce the operators $\hat{\rho}_\sigma^{\mathrm{N}}(q)$ and $\hat{\rho}_\sigma^{\mathrm{U}}(q)$ $(0 < q < 2\pi)$ which scatter an electron within the Brillouin zone ("normal process") and out of the Brillouin zone ("umklapp process"), respectively:

$$\hat{\rho}^{\mathrm{N}}_{\sigma}(q) = \sum_{-\pi<k<\pi-q} \hat{c}^{+}_{k+q,\sigma}\hat{c}_{k,\sigma} = \left(\hat{\rho}^{\mathrm{N}}_{\sigma}(-q)\right)^{+} , \tag{4.54a}$$

$$\hat{\rho}^{\mathrm{U}}_{\sigma}(q) = \sum_{\pi-q<k<\pi} \hat{c}^{+}_{k+q,\sigma}\hat{c}_{k,\sigma} = \left(\hat{\rho}^{\mathrm{U}}_{\sigma}(-q)\right)^{+} . \tag{4.54b}$$

Unfortunately, the algebra of these operators is not closed. The construction of generalized operators with closed algebra is possible; their analysis, however, seems to be forbiddingly complicated.

With the help of (4.54) we can explicitly write

$$\hat{D}^{0} = \hat{N}_{\uparrow}\hat{N}_{\downarrow}/L + \sum_{0<q\leq 2\pi-\Delta} \hat{\rho}^{\mathrm{N}}_{\uparrow}(q)\hat{\rho}^{\mathrm{N}}_{\downarrow}(-q) + \hat{\rho}^{\mathrm{U}}_{\uparrow}(q)\hat{\rho}^{\mathrm{U}}_{\downarrow}(-q) , \tag{4.55a}$$

$$\hat{D}^{+} = \left(\hat{D}^{-}\right)^{+} = \sum_{0<q\leq 2\pi-\Delta} \hat{\rho}^{\mathrm{N}}_{\uparrow}(q)\hat{\rho}^{\mathrm{U}}_{\downarrow}(-q) . \tag{4.55b}$$

Equation (4.53b) shows that it is convenient to represent the Hamilton operator as

$$\hat{H} = \hat{H}_0 + \hat{H}_1 , \tag{4.56a}$$

$$\hat{H}_0 = \hat{T} + U\hat{D}^{0} , \tag{4.56b}$$

$$\hat{H}_1 = U(\hat{D}^{+} + \hat{D}^{-}) . \tag{4.56c}$$

For the first part of the solution we should diagonalize $\hat{H}_0$. According to (4.53b) we could always find simultaneous eigenstates of both the kinetic energy operator and $\hat{D}^0$. Unfortunately, this task is still too complicated to be carried out analytically. However, this first step in the analysis of the full $1/r$ Hubbard model is readily carried out numerically on small lattices with $L = 2, 4, 6$ sites and $N = 2, 4, 6$ electrons (half band-filling). The corresponding Hilbert spaces have dimension $H = 6, 70, 924$; altogether 1000 states. We first diagonalize $\hat{H}_0$ in the subspaces with fixed total spin S and fixed total momentum Q and calculate the matrix elements and eigenvalues of $\hat{H}_1$ in the new basis. In addition to many simultaneous eigenstates of $\hat{H}_0$ and $\hat{H}_1$ we thereby find that at most 2^r states ($r = 1, \ldots, (N/2)$) are coupled. The eigenvalues of the non-diagonal part $\hat{H}_1$ can be written as the sums of r contributions of the form $\sqrt{m(L-m)}/L$ ($m = 1, \ldots, (L-1)$).

We conclude that even the spectrum for more than two electrons is solely determined by the "doublets" of the two-electron problem. Our remaining task is to find a representation for the doublets and the non-mixing states which reproduces the numerically determined eigenenergies in detail.

For this purpose we introduce a new configuration space with "orbitals" $\mathcal{K} = (\pi/L)(2m_{\mathcal{K}} + 1)$ ($m_{\mathcal{K}} = -(L/2), \ldots, (L/2) - 1$). The values for $\mathcal{K}$ and k are identical. The *orbital index* thus reflects the properties of *momentum space*. Into each of these orbitals we place a hard-core boson which represents the charge degrees of freedom (double occupancy, $\hat{d}_{\mathcal{K}} \equiv \bullet$; hole, $\hat{e}_{\mathcal{K}} \equiv \circ$) and the spin degrees of freedom ($\uparrow$ spin, $\hat{s}_{\mathcal{K},\uparrow} \equiv \uparrow$; $\downarrow$ spin, $\hat{s}_{\mathcal{K},\downarrow} \equiv \downarrow$), respectively. For each orbital $\mathcal{K}$ the hard-core constraint implies

$$\hat{d}^+_{\mathcal{K}}\hat{d}_{\mathcal{K}} + \hat{e}^+_{\mathcal{K}}\hat{e}_{\mathcal{K}} + \hat{s}^+_{\mathcal{K},\uparrow}\hat{s}_{\mathcal{K},\uparrow} + \hat{s}^+_{\mathcal{K},\downarrow}\hat{s}_{\mathcal{K},\downarrow} = 1 \,. \tag{4.57}$$

The *orbital occupancy* thus reflects the properties of *position space*. The new configuration space thereby unifies properties of momentum and position space. Obviously, the number of states generated in the configuration space equals 4^L.

We now have to assign a value for energy and momentum to each configuration. To set up such an effective Hamilton operator we start from the two-electron problem on L lattice sites for which there are L^2 states for $S^z = 0$. L configurations have one double occupancy ($\bullet$) and $L-1$ holes ($\circ$). Of those L states only one is an eigenstate with $D=1, T=0$. The other $L-1$ states, e.g., $\circ \ldots \circ \bullet \circ \circ \ldots \circ$, mix with another configuration. We denote this by drawing a box around a neighboring pair of a double occupancy and a hole, $\boxed{\bullet \circ}_{\mathcal{K},\mathcal{K}+\Delta}$. Only the configuration with a double occupancy ($\bullet$) at the end of the chain does not mix. The remaining $L(L-1)$ configurations have a single $\uparrow$ and $\downarrow$ spin each as well as $L-2$ holes ($\circ$). We know from Sect. 4.2.1 that only $L-1$ configurations are *not* simultaneous eigenstates of the kinetic and potential energy operators. It is natural to also mark a neighboring pair of spins as a "mixing" configuration, $\boxed{\uparrow\downarrow}_{\mathcal{K},\mathcal{K}+\Delta}$.

The first part of the effective Hamiltonian is concerned with the unboxed parts of a configuration. We equally distribute momentum and kinetic energy onto spin and charge degrees of freedom and assign an interaction energy U to each double occupancy. Explicitly,

$$\begin{aligned}\hat{H}^{\text{eff}}_0 &= \sum_{-\pi<\mathcal{K}<\pi}{}' \left\{ \frac{t\mathcal{K}}{2}\left[\left(\hat{s}^+_{\mathcal{K},\uparrow}\hat{s}_{\mathcal{K},\uparrow} + \hat{s}^+_{\mathcal{K},\downarrow}\hat{s}_{\mathcal{K},\downarrow}\right) - \left(\hat{d}^+_{\mathcal{K}}\hat{d}_{\mathcal{K}} + \hat{e}^+_{\mathcal{K}}\hat{e}_{\mathcal{K}}\right)\right]\right.\\ &\qquad \left. + U\hat{d}^+_{\mathcal{K}}\hat{d}_{\mathcal{K}}\right\} ,\end{aligned} \tag{4.58}$$

where the prime on the sum indicates that we have not taken into account the boxed parts of a configuration yet.

We assign the momentum π and $-\pi$ to the configurations $\boxed{\uparrow\downarrow}_{\mathcal{K},\mathcal{K}+\Delta}$ and $\boxed{\bullet \circ}_{\mathcal{K},\mathcal{K}+\Delta}$, respectively. We further set $Q = \pi + (2\mathcal{K}+\Delta)/2$ [4.56] which allows us to deduce the second part of the effective Hamilton operator from the solution of the two-electron problem as

$$\begin{aligned}\hat{H}^{\text{eff}}_1 &= \sum_{-\pi<\mathcal{K}<\pi-\Delta} \left\{\left[\frac{U}{2} - \left(\pi t - \frac{U}{4\pi}(2\mathcal{K}+\Delta)\right)\right] \hat{d}^+_{\mathcal{K}}\hat{d}_{\mathcal{K}}\hat{e}^+_{\mathcal{K}+\Delta}\hat{e}_{\mathcal{K}+\Delta}\right.\\ &\quad + \left[\frac{U}{2} + \left(\pi t - \frac{U}{4\pi}(2\mathcal{K}+\Delta)\right)\right] \hat{s}^+_{\mathcal{K},\uparrow}\hat{s}_{\mathcal{K},\uparrow}\hat{s}^+_{\mathcal{K}+\Delta,\downarrow}\hat{s}_{\mathcal{K}+\Delta,\downarrow}\\ &\quad \left. + \frac{U}{4\pi}\sqrt{(2\pi)^2 - (2\mathcal{K}+\Delta)^2}\left[\hat{d}^+_{\mathcal{K}}\hat{e}^+_{\mathcal{K}+\Delta}\hat{s}_{\mathcal{K},\uparrow}\hat{s}_{\mathcal{K}+\Delta,\downarrow} + \text{h.c.}\right]\right\} .\end{aligned} \tag{4.59}$$

The total effective Hamilton operator is given by $\hat{H}^{\text{eff}} = \hat{H}^{\text{eff}}_0 + \hat{H}^{\text{eff}}_1$.

One can explicitly check that this effective Hamiltonian reproduces not only the eigenenergies for the two-electron problem for all L but also the spectrum of those 1000 states of the numerical diagonalization for $N = L = 2, 4, 6$. Further evidence for the validity of the effective Hamiltonian can be found by using perturbation theory in the interaction strength and the electron density at zero temperature [4.57], and the high-temperature expansion for all densities and interaction strengths. Furthermore, for $n = 1$, and to order t^2/U, we can reproduce the complete thermodynamics of the Haldane–Shastry spin model, (4.40a). In addition, the exact solution of the corresponding $\hat{T}$ model [see (2.40)] provides another confirmation for $n < 1$ in the limit $U = \infty$ and all temperatures [4.58]. Finally, we can also deduce the results for the $1/r$ Hubbard model from the analysis of the more general $1/\sinh r$ Hubbard model. The latter model is solved with the help of the asymptotic Bethe ansatz in the next section. Thus, we are confident that $\hat{H}^{\text{eff}}$ indeed gives the exact spectrum and thermodynamics of the $1/r$ Hubbard model.

We can easily write the effective Hamiltonian in a pure occupation number representation by a simple diagonalization of 2×2 matrices for each pair $\mathcal{K}, \mathcal{K}+\Delta$. We add an external magnetic field ($\mathcal{H}_0$) and the chemical potential $\mu_\sigma = \mu - \sigma(g\mu_B\mathcal{H}_0)/2$, and the diagonalized effective Hamiltonian finally becomes

$$\begin{aligned} \hat{H}^{\text{eff}} &= \sum_{-\pi<\mathcal{K}<\pi} \Big\{ \sum_\sigma h^{\text{s}}_{\mathcal{K},\sigma} \widetilde{n}^{\text{s}}_{\mathcal{K},\sigma} + h^{\text{d}}_{\mathcal{K}} \widetilde{n}^{\text{d}}_{\mathcal{K}} + h^{\text{e}}_{\mathcal{K}} \widetilde{n}^{\text{e}}_{\mathcal{K}} \\ &\qquad + J_{\mathcal{K}} \left[\widetilde{n}^{\text{d}}_{\mathcal{K}-\Delta} \widetilde{n}^{\text{e}}_{\mathcal{K}} - \widetilde{n}^{\text{s}}_{\mathcal{K}-\Delta,\uparrow} \widetilde{n}^{\text{s}}_{\mathcal{K},\downarrow} \right] \Big\} , \end{aligned} \tag{4.60}$$

with $h^{\text{s}}_{\mathcal{K},\sigma} = (t\mathcal{K})/2 - \mu_\sigma$, $h^{\text{d}}_{\mathcal{K}} = -(t\mathcal{K})/2 - 2\mu + U$, $h^{\text{e}}_{\mathcal{K}} = -(t\mathcal{K})/2$, and $J_{\mathcal{K}} = \left[t(2\mathcal{K} - \Delta) - U + \sqrt{(2\pi t)^2 + U^2 - 2tU(2\mathcal{K} - \Delta)} \right] /2 \geq 0$. Here we introduced the notation $\widetilde{n}^{\text{s}}_{\mathcal{K},\sigma} = \widetilde{s}^{+}_{\mathcal{K},\sigma} \widetilde{s}_{\mathcal{K},\sigma}$ etc., where the tilde denotes operators after the transformation. The restriction to $t \geq 0$ and $U \geq -2\pi t$ allows us to identify $\mathcal{K}_{\text{min}} - \Delta \equiv \mathcal{K}_{\text{max}}$ since the coupling vanishes at $\mathcal{K} = \mathcal{K}_{\text{min}} = -((L-1)/2)\Delta$, $J_{\mathcal{K}_{\text{min}}} = 0$.

4.2.4 Ground-State Energy and Metal–Insulator Transition

Since we were able to reduce the effective Hamiltonian to an occupation number representation it is very easy to calculate the ground-state energy. For even electron number $N \leq L$ the ground state consists of orbitals in the region $-\pi < \mathcal{K} < \mathcal{K}_{\text{F}} = \pi(2n - 1)$ filled with the spin and charge singlets of the *new* operators $\widetilde{s}^{+}_{\mathcal{K},\uparrow}, \widetilde{s}^{+}_{\mathcal{K},\downarrow}$ while all orbitals $\mathcal{K} > \mathcal{K}_{\text{F}}$ are occupied with holes $\widetilde{e}^{+}_{\mathcal{K}}$.

In the thermodynamic limit the ground-state energy density of the $1/r$ Hubbard model for at most half band-filling is readily calculated as

$$
\begin{aligned}
(E_0/L)(n \leq 1, U) &= \int_{\mathcal{K}_\mathrm{F}}^{\pi} \mathrm{d}\mathcal{K} \frac{(-t)\mathcal{K}}{2\pi} - \frac{1}{2}\int_{-\pi}^{\mathcal{K}_\mathrm{F}} \mathrm{d}\mathcal{K}\frac{J_\mathcal{K}}{2\pi} \\
&= \frac{Un - W(1-n)n}{4} \\
&\quad - \frac{1}{24WU}\left[(W+U)^3 - \left((W+U)^2 - 4WUn\right)^{3/2}\right] ,
\end{aligned} \tag{4.61}
$$

where we took into account that $\int_{-\pi}^{\pi} \mathcal{K}\mathrm{d}\mathcal{K} = 0$, and that only each *pair* gives a contribution to the interaction. The chemical potential at less than half-filling follows from its definition as

$$
\begin{aligned}
\mu(n \leq 1) &= \frac{\partial (E_0/L)(n \leq 1, U)}{\partial n} \\
&= \frac{1}{4}\left[U - W(1-2n)\right] - \frac{1}{4}\sqrt{(W+U)^2 - 4WUn} .
\end{aligned} \tag{4.62}
$$

Due to electron–hole symmetry we find $\mu(2-n) = U - \mu(n)$. Hence there is no gap in the excitation spectrum for all $n < 1$. For $n = 1$ the left-hand derivative of the ground-state energy is given by $\mu^- = [U + W - |U - W|]/4 = U - \mu^+$. The energy gap can thus be calculated from its definition $\Delta\mu = \mu^+ - \mu^-$ as

$$
\Delta\mu = \frac{1}{2}(U - W + |U - W|) = \begin{cases} 0 & \text{for} \quad U \leq W \\ U - W & \text{for} \quad U \geq W \end{cases} . \tag{4.63}
$$

Consequently, the $1/r$ Hubbard model describes a metal–insulator transition at half band-filling at a *finite* interaction strength, $U_\mathrm{c} = W$, and the gap opens continuously with exponent $\nu = 1$.

It remains to determine the "order" of the transition. As already mentioned in Sect. 3.1 we *define* a zero-temperature transition to be of nth order if there is a jump in the nth derivative of the ground-state energy density with respect to the interaction strength. For the $1/r$ Hubbard model at half band-filling the ground-state energy density is obtained from (4.61) as

$$
(E_0/L)(n = 1, U) = \begin{cases} \dfrac{-W + U}{4} - \dfrac{U^2}{12W} & \text{for} \quad U < W \\[2ex] -\dfrac{W^2}{12U} & \text{for} \quad U \geq W \end{cases} , \tag{4.64}
$$

and the mean double occupancy as its first derivative with respect to U follows as

$$
\overline{d}(n = 1, U) = \begin{cases} \dfrac{1}{4} - \dfrac{U}{6W} & \text{for} \quad U \leq W \\[2ex] \dfrac{W^2}{12U^2} & \text{for} \quad U \geq W \end{cases} , \tag{4.65}
$$

which is continuous at the transition. Actually, only the *third* derivative of the ground-state energy with respect to U is discontinuous so that the $1/r$ Hubbard model describes a "third-order" zero-temperature transition at $U_\mathrm{c} = W$.

It is seen that the "order" of the transition and the behavior of the gap are not directly related to each other. This is not too surprising because the two quantities probe the behavior of $(E_0/L)(U,n)$ as a function of the interaction strength ("order") and as a function of the density ("gap"), respectively. As a consequence, a "second-order" transition may still display a jump in the gap at the transition. This scenario could be realized in the infinite-dimensional Hubbard model; see Sect. 5.6.4.

Further ground-state properties can easily be derived from the occupation number representation of the effective Hamilton operator (4.60) [4.59]. A spin excitation of the half-filled ground state consists of a broken spin pair: a σ spin at $\mathcal{K}_1$ and a σ' spin at $\mathcal{K}_2$ ("spinons"). The energy of a single spinon is

$$\begin{aligned}\epsilon_{\rm s}(\mathcal{K}) &= \left(\sqrt{W^2+U^2-4WU\mathcal{K}/\pi}+2W\mathcal{K}/\pi-U\right)/4 \geq 0\,,\\ &\quad |\mathcal{K}|<\pi/2\,.\end{aligned} \tag{4.66}$$

The dispersion is gapless at $\mathcal{K}=-\pi/2$ for all U/W (right-moving spinons), and the gap for left-moving spinons closes at $U=W$ for half band-filling. The corresponding velocities for elementary spin excitations are

$$v_{\rm s}^{\rm R}(n) = \frac{v_{\rm F}}{1+U/W}\,,\quad v_{\rm F}=ta/\hbar\equiv t\,, \tag{4.67a}$$

$$v_{\rm s}^{\rm L}(n=1) = -\frac{v_{\rm F}}{U/W-1}\quad \text{for } U/W>1. \tag{4.67b}$$

In connection with the metal–insulator transition the charge excitations are more interesting since they describe the splitting of the upper and lower Hubbard bands at half band-filling. A charge excitation of the half-filled ground state consists of a broken spin pair, a hole at $\mathcal{K}_1$ ("holon"), and a double occupancy at $\mathcal{K}_2$ ("doublon"). In the thermodynamic limit these two objects move independently from each other with the dispersion relation [4.59]

$$\begin{aligned}\epsilon_{\rm c}(\mathcal{K}) &= \left(\sqrt{W^2+U^2-4WU\mathcal{K}/\pi}-2W\mathcal{K}/\pi+U\right)/4 \geq 0\,,\\ &\quad |\mathcal{K}|<\pi/2\,,\end{aligned} \tag{4.68}$$

where we equally distributed the potential energy of the "doublon" onto both objects. The charge dispersion relation is gapless in the metallic phase, and the velocity for (left-moving) charge excitations is given by

$$v_{\rm c}^{\rm L}(n=1) = -\frac{v_{\rm F}}{1-U/W}\quad \text{for } U/W<1. \tag{4.69}$$

The upper Hubbard band is given by the dispersion relation of double occupancies alone. The dispersion relation of the upper Hubbard band is thus given by

$$\begin{aligned}\epsilon_{\rm UHB}(k) &= \epsilon_{\rm c}(2k)+U/2\\ &= U-\left[Wk/\pi+U-\sqrt{W^2+U^2-2WUk/\pi}\right]/4\,, \\ &\quad |k|<\pi\,,\end{aligned} \tag{4.70}$$

where we rescaled the momenta to provide a better comparison with the results from the Bethe ansatz. The dispersion relation of the lower Hubbard band is given by $\epsilon_{\mathrm{LHB}}(k) = U - \epsilon_{\mathrm{UHB}}(k)$.

In the absence of an external magnetic field the Drude weight can be deduced from the velocity of the charge carriers and their interaction matrix element ξ^2 of the "dressed charge matrix"; see Sect. 4.1.7, (4.36). For the much simpler case of the $1/r$ Hubbard model the elementary charge and spin excitations do *not* interact with each other. Therefore, the S matrix of the excitations is unity. As a trivial consequence we have $\xi^2 = 1$, independent of the interaction strength. This observation also follows from the explicit calculation of the relation between the compressibility and the velocity of the charge excitations [4.59] according to (4.38). Since the $1/r$ Hubbard model with its linear dispersion relation describes right-moving electrons only, we have $K_\rho = \xi^2 = 1$ for the correlation exponent of the "$4k_{\mathrm{F}}$" oscillations; see, e.g., [4.38].

Now that the Drude weight depends on only the velocity of the charge excitations $\epsilon_{\mathrm{c}}(\mathcal{K})$ we may use (4.36) and (4.68) to obtain

$$\mathcal{D}_{\mathrm{c}} = \mathcal{D}_{\mathrm{c}}^0 \frac{|v_{\mathrm{c}}^{\mathrm{L}}|}{v_{\mathrm{F}}} . \tag{4.71}$$

The Drude weight for the non-interacting system is $\mathcal{D}_{\mathrm{c}}^0 = e^2 v_{\mathrm{F}}/\hbar$ (Fermi velocity $v_{\mathrm{F}} = ta/\hbar$), and the velocity of the left-moving holons follows from the derivative of (4.68) at $\mathcal{K} = \pi/2$ as

$$v_{\mathrm{c}}^{\mathrm{L}}(n) = -v_{\mathrm{F}} \left(1 + \frac{U}{\sqrt{(W+U)^2 - 4WUn}} \right) . \tag{4.72}$$

Hence, the Drude weight becomes

$$\mathcal{D}_{\mathrm{c}} = \frac{e^2 ta}{\hbar^2} \left[1 + \frac{U}{\sqrt{(W+U)^2 - 4WUn}} \right] . \tag{4.73}$$

Approaching the metal–insulator transition the Drude weight actually *diverges* proportionally to $1/(W-U)$ at half band-filling since the velocity of the charge carriers diverges. An increasing conductivity on the approach to a metal–insulator transition from the metallic side is unphysical. The reason for this behavior is the dispersion relation with its jump discontinuity at the Brillouin zone boundary. We will further comment on that in Sect. 4.4.

4.2.5 Thermodynamic Properties

Both the "standard" and the $1/r$ Hubbard model belong to the class of Luttinger liquids [4.43, 4.60]. In their metallic phase both the charge and spin degrees of freedom are gapless excitations with finite velocities, whereas in the insulating state the charge excitations are gaped. In a Luttinger liquid the velocities of the elementary excitations determine the compressibility κ and the magnetic susceptibility χ.

For example, for the $1/r$ Hubbard model one can calculate these quantities by taking the corresponding second derivatives of the ground-state energy density with respect to the electron density and the external magnetic field. One finds that the Luttinger liquid relations

$$\kappa(n=1, U<W, T=0) = \frac{1}{\pi |v_{\mathrm{c}}^{\mathrm{L}}|}, \tag{4.74a}$$

$$\kappa(n=1, U>W, T=0) = 0, \tag{4.74b}$$

$$\chi(n=1, U<W, T=0) = \frac{1}{\pi v_{\mathrm{s}}^{\mathrm{R}}}, \tag{4.74c}$$

$$\chi(n=1, U>W, T=0) = \frac{1}{\pi v_{\mathrm{s}}^{\mathrm{R}}} + \frac{1}{\pi |v_{\mathrm{s}}^{\mathrm{L}}|} \tag{4.74d}$$

are satisfied [4.59]. Similar equations also hold for the case of the Hubbard model with nearest-neighbor hopping [4.43, 4.60].

If we further assume that the system is conformal invariant at low enough temperatures (conformal charge $c = 1$ for Luttinger liquids), we may even deduce the Sommerfeld coefficient of the specific heat γ, $c_{\mathrm{V}}(T/W \to 0) = \gamma T$:

$$\gamma(n, U) = \frac{\pi^2}{3} \left[\frac{\kappa(n, U) + \chi(n, U)}{2} \right] . \tag{4.75}$$

Using the arguments of conformal field theory or the Luttinger liquid approach, we can also deduce the exponents for the long-range decay of ground-state correlation functions [4.59]. The same program could also be carried out for the Hubbard model with nearest-neighbor interaction [4.43].

In general, the analytical calculation of the magnetic susceptibility, the compressibility, and the specific heat as functions of the interaction strength and temperature poses a very difficult problem even for formally solved problems like the standard Hubbard model. Based on the string hypothesis on the quasi-particle quantum numbers for the excited states Takahashi [4.61] developed an infinite hierarchy of equations ("thermodynamic Bethe ansatz") and calculated the Sommerfeld coefficient for half band-filling as a function of U [4.62]. For the Hubbard model at and close to half band-filling a truncated subset was numerically solved for the magnetic susceptibility, compressibility, and specific heat for $U = W, 2W, 3W$ and temperatures $k_{\mathrm{B}}T \leq W$ in [4.63]. Although the thermodynamic Bethe ansatz does not take the complete spectrum into account, the free energy to leading order in the system size can still be obtained correctly from it because the missing "non-highest-weight states" give only a contribution of $\mathcal{O}(\ln(L)/L)$ to the free energy density.[1]

[1] This is essentially the same argument by which one can show that, in the thermodynamic limit, all states of a spin-1/2 model with $S^z = 0$ are spin singlets; see Sect. 5.2.2. I am very grateful to Fabian Eßler for pointing this out to me.

It was only very recently that the exact free energy density of the Hubbard model could be obtained in closed form. The approach by Klümper and Bariev [4.64] is based on the Suzuki–Trotter representation of the partition function [4.65] which defines a two-dimensional vertex model. The largest eigenvalue of the corresponding (quantum) transfer matrix then gives the free energy density of the Hubbard model in terms of four coupled non-linear integral equations [4.64]. As an important result it was proven that the Sommerfeld coefficient is indeed given by (4.75) so that the assumptions of conformal invariance are justified. At finite temperature it is very difficult to evaluate the integral equations for the free energy and its derivatives.

Fortunately, the situation is much simpler for the $1/r$ Hubbard model. Since the effective Hamiltonian (4.60) is equivalent to a *classical* one-dimensional Ashkin–Teller model [4.66], the free energy density can easily be found with the help of the transfer matrix technique [4.11]. We define the abbreviations $S_{\mathcal{K},\sigma} = \exp\left(-\beta h^{\mathrm{s}}_{\mathcal{K},\sigma}\right)$, $D_{\mathcal{K}} = \exp\left(-\beta h^{\mathrm{d}}_{\mathcal{K}}\right)$, $E_{\mathcal{K}} = \exp\left(-\beta h^{\mathrm{e}}_{\mathcal{K}}\right)$, $P_{\mathcal{K}} = \exp\left(-\beta J_{\mathcal{K}}\right)$, and $X_{\mathcal{K}} = S_{\mathcal{K},\uparrow} + S_{\mathcal{K},\downarrow} + D_{\mathcal{K}} + E_{\mathcal{K}}$ where $\beta = 1/k_{\mathrm{B}}T$ is the inverse temperature. The transfer matrices between sites $\mathcal{K} - \Delta$ and $\mathcal{K}$ has two vanishing eigenvalues. The larger of the two remaining eigenvalues is given by

$$\lambda^{+}_{\mathcal{K}} = \frac{1}{2}\left[X_{\mathcal{K}} + \sqrt{X^2_{\mathcal{K}} - 4\left[S_{\mathcal{K},\uparrow}S_{\mathcal{K},\downarrow}\left(1 - P^{-1}_{\mathcal{K}}\right) + D_{\mathcal{K}}E_{\mathcal{K}}\left(1 - P_{\mathcal{K}}\right)\right]}\,\right] . \tag{4.76}$$

In the thermodynamic limit the free energy density for all chemical potentials μ, magnetic fields $\mathcal{H}_0$, interaction strengths U, and temperatures T becomes[2]

$$f(\mu, \mathcal{H}_0, U, T) = -\frac{1}{\beta}\int_{-\pi}^{\pi} \frac{\mathrm{d}\mathcal{K}}{2\pi} \ln \lambda^{+}_{\mathcal{K}} . \tag{4.77}$$

This compact expression of the free energy density reflects the fact that the excitations of the $1/r$ Hubbard model are essentially independent of each other.

The expressions for the free energy density, the compressibility, and the magnetic susceptibility become particularly simple for half band-filling in the absence of an external field. For the free energy density we find [4.59]

$$\begin{aligned} f(n=1, U, T) &= -U/2 + e_0(n=1, U) \\ &\quad - \frac{2}{2\pi\beta}\int_{-\pi/2}^{\pi/2} \mathrm{d}\mathcal{K}\Big\{\ln\left[1 + \exp\left(-\beta\epsilon_{\mathrm{s}}(\mathcal{K})\right)\right] \\ &\qquad + \ln\left[1 + \exp\left(-\beta\epsilon_{\mathrm{c}}(\mathcal{K})\right)\right]\Big\} , \end{aligned} \tag{4.78}$$

[2] The transfer matrices between two sites only commute to order $1/L$. Hence, Equation (27) of [4.59] is wrong but the expression (28) of [4.59] for the free energy density remains correct in the thermodynamic limit. I am grateful to Holger Frahm for pointing this out to me.

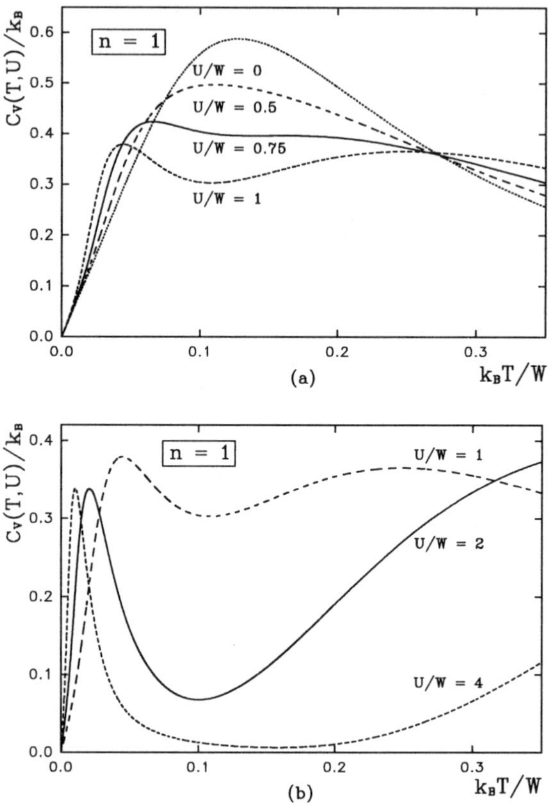

Fig. 4.5. Specific heat as a function of temperature for the $1/r$ Hubbard model at half band-filling for various values of U/W: (**a**) below ($U/W = 0, 0.5, 0.75, 1$), (**b**) above ($U/W = 1, 2, 4$), the metal–Mott–Hubbard insulator transition (from [4.59]).

where the dispersion relations for the spin and charge excitations were given in (4.66) and (4.68). The specific heat is found as

$$\begin{aligned} c_{\mathrm{V}}(n=1,U,T) &= c_{\mathrm{V}}^{\mathrm{c}}(n=1,U,T) + c_{\mathrm{V}}^{\mathrm{s}}(n=1,U,T) , \\ c_{\mathrm{V}}^{\mathrm{c,s}}(n=1,U,T) &= k_{\mathrm{B}} \int_{-\pi}^{\pi} \frac{\mathrm{d}\mathcal{K}}{\pi} \left[\frac{\beta\epsilon^{\mathrm{c,s}}(\mathcal{K})/2}{\cosh\bigl(\beta\epsilon^{\mathrm{c,s}}(\mathcal{K})/2\bigr)} \right]^2 . \end{aligned} \tag{4.79}$$

It is displayed as a function of temperature in Fig. 4.5.

As is seen from Fig. 4.5a, the Sommerfeld coefficient agrees with (4.75) which shows [4.67] that the low-energy physics of the model can indeed be described in terms of a conformal field theory with central charge $c = 1$. Even below the metal-to-insulator transition the specific heat displays a two-peak structure for $U/W \geq 1/2$ since the density of states for the spin excitations develops a van-Hove singularity above $U_{\mathrm{spin}} = W/2$. At this value the spinon dispersion changes its qualitative behavior. The dispersion at small interactions is monotonous, whereas it has a maximum within the Brillouin zone above U_{spin}. Hartree–Fock theory would predict long-range order to set in at U_{spin} [4.57]. Instead, the metal–Mott–Hubbard insulator transition occurs at $U_{\mathrm{c}} = W$, and is driven by charge correlations, not by spin correlations.

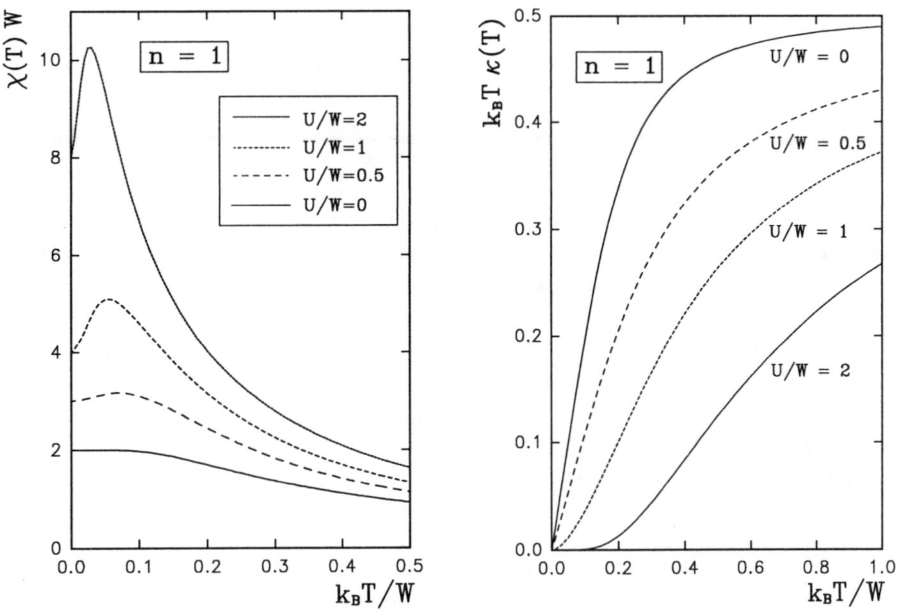

Fig. 4.6. (**a**) Magnetic susceptibility and (**b**) fluctuation of the electron density as a function of temperature for the $1/r$ Hubbard model at half band-filling for various values of U/W (from [4.59]).

Above the Mott transition (see Fig. 4.5b) the upper and lower Hubbard bands are separated. For $U \gg W$ we find a narrow peak in the specific heat around $T = J \sim t^2/U$ due to the spin excitations, and a broad maximum around $T = \mathcal{O}(U - W)$ due to the charge excitations. Despite the peculiar properties of the $1/r$ Hubbard model the overall shape of the specific heat agrees well with that for the standard Hubbard model for interacting strength $U \geq W$, as obtained in [4.63]. Hence, the specific heat as displayed in Fig. 4.5 should be rather generic for a correlated electron system.

The isothermal compressibility at half band-filling becomes

$$\kappa(n = 1, U, T) = \frac{\beta}{\pi} \int_{-\pi/2}^{\pi/2} \mathrm{d}\mathcal{K} \left[\exp\left(-\beta\epsilon_{\mathrm{s}}(\mathcal{K})\right) + \exp\left(\beta\epsilon_{\mathrm{c}}(\mathcal{K})\right)\right]^{-1} , \quad (4.80)$$

and the magnetic susceptibility, which can be obtained form electron–hole symmetry, reads

$$\chi(n = 1, U, T) = \frac{\beta}{\pi} \int_{-\pi/2}^{\pi/2} \mathrm{d}\mathcal{K} \left[\exp\left(\beta\epsilon_{\mathrm{s}}(\kappa)\right) + \exp\left(-\beta\epsilon_{\mathrm{c}}(\mathcal{K})\right)\right]^{-1} . \quad (4.81)$$

The magnetic susceptibility is depicted in Fig. 4.6 together with the fluctuation of the electron density $\langle(\Delta\hat{N})^2\rangle/L = T\kappa$ where $\Delta\hat{N} = \hat{N} - \langle\hat{N}\rangle$.

The magnetic susceptibility at low temperatures shows Pauli behavior which is strongly enhanced by the interaction, especially above the metal–insulator transition; see (4.74). For all $U > 0$ it develops a peak. For large U/W it is located around $k_{\mathrm{B}}T = \mathcal{O}(J \sim t^2/U)$, the energy scale for spin excitations in the corresponding Heisenberg model. At high temperatures, $k_{\mathrm{B}}T \gg J$, the magnetic susceptibility follows a Curie law. The Curie "constant" gives the size of the local magnetic moments. For large U/W, and temperatures above the spin excitation energy but below the charge gap, the system behaves as if it contained only individual spins on each lattice site, i.e., $\chi(n = 1, J \ll k_{\mathrm{B}}T \ll \Delta\mu = U - W) = 1/T$, in units of $n_{\mathrm{e}}(\mu_{\mathrm{B}})^2$; see [4.35, Chap. 31]). The correlated moments cease to exist for temperatures $k_{\mathrm{B}}T = \mathcal{O}(U)$, and a crossover to the atomic limit behavior can be observed, $\chi(n = 1, U \ll k_{\mathrm{B}}T) = 1/(2T)$, since now (non-magnetic) double occupancies and holes are equally allowed.

The fluctuation in the electron density shown in Fig. 4.6b is linear for small temperatures in the metallic phase, i.e., the compressibility is finite and given by (4.74). Above the transition the charge gap has opened, and electron number fluctuations are exponentially suppressed at temperatures below the gap. At high temperatures the fluctuations saturate at the atomic limit value $T\kappa(n = 1, U \ll k_{\mathrm{B}}T) = 1/2$ where each site is equally likely to be singly occupied, doubly occupied, or empty. Again, the overall behavior of the fluctuation of the electron density as shown in Fig. 4.6b is generic for correlated electron systems.

As is seen from (4.78), (4.80), and (4.81) the thermodynamics at half band-filling is wholly determined by the elementary excitations whose dispersion relations were given in (4.66) and (4.68). This is peculiar to the $1/r$ Hubbard model since the elementary excitations are essentially independent of each other. Nevertheless we can most easily infer from the free energy and the thermodynamic response functions that the gap is "robust" since the (gaped) charge dispersion relation enters all thermodynamic quantities. Hence, the metal–Mott–Hubbard insulator transition in the $1/r$ Hubbard model provides an explicit example for a quantum phase transition.

4.3 Hubbard Model with Variable-Range Hopping

In this section we present exact results for the one-dimensional Hubbard model with variable-range hopping. We start with the discussion of some basic properties of the $1/\sinh r$ Hubbard model in which the hopping is of variable range. Next, we solve the two-electron problem from which we derive the scattering phase shift and the two-electron S matrix. Under the *assumption* of integrability of the model we may apply Sutherland's asymptotic Bethe ansatz to obtain the generalized Lieb–Wu equations from which we calculate the ground-state energy. At half band-filling the model is an insulator for all finite interaction strengths for all finite hopping ranges.

4.3.1 Basic Properties

Our analysis in Sect. 4.2.1 shows how to construct lattice spin models from integrable continuum models with two-body interaction $V(x)$. Among those is the class with the interaction $V(x) \sim 1/\sinh^2(\kappa x)$. Here, κ^{-1} is the effective range of the interaction. On the lattice the corresponding $1/\sinh^2 r$ Heisenberg model,

$$\begin{aligned} \hat{H}_{\rm I} &= \sum_{l \neq m=-\infty}^{\infty} J(l-m) \left[\hat{\boldsymbol{S}}_l \cdot \hat{\boldsymbol{S}}_m - \frac{1}{4} \right] , \\ J(l-m) &= J \left(\frac{\sinh(\kappa)}{\sinh[\kappa(l-m)]} \right)^2 , \end{aligned} \tag{4.82}$$

is defined on an infinite lattice to avoid problems with boundary terms. Inozemtsev [4.68] explicitly constructed the wave functions for $2 \leq M < \infty$ magnons for the ferromagnetic model ($J < 0$) in the thermodynamic limit. Thus we will call this lattice model for spin-1/2 Heisenberg operators the "Inozemtsev model". The general mapping of continuum models onto lattice spin models can now be performed [4.69] such that the full spectrum and the thermodynamics of the Inozemtsev model is known [4.70]. The explicit form of the wave functions for *finite* electron densities, however, could not be determined at this stage.

In analogy to the approach in Sect. 4.2 we may conjecture that the Hubbard model that canonically belongs to the Inozemtsev model will also be integrable. The generalization of integrable spin models to Hubbard models is canonical but not trivial; see, e.g., [4.71]. In the following we will use the hypothesis that the $1/\sinh r$ Hubbard model with hopping amplitudes

$$t(l-m) = (-\mathrm{i}t)(-1)^{l-m} \frac{\sinh(\kappa)}{\sinh[\kappa(l-m)]} \tag{4.83}$$

is indeed integrable. As for the $1/r$ Hubbard model we have to choose purely imaginary hopping matrix elements to guarantee that $t(l-m) = t^*(m-l)$ because $\sinh[\kappa(l-m)]$ is antisymmetric under the exchange of l and m.

In contrast to the Hubbard models discussed thus far, the $1/\sinh r$ Hubbard model contains a parameter κ which controls the hopping range. Up to the distance of the order of $\mathcal{O}(\kappa^{-1})$ the amplitude for electron transfer decays linearly, whereas it decays exponentially for $|l-m| \gg \kappa^{-1}$. As a consequence, we obtain the Hubbard model with nearest-neighbor hopping in the limit $\kappa \to \infty$ when we perform the additional transformation $\hat{c}_{l,\sigma}^{+} \mapsto \mathrm{e}^{\mathrm{i}\pi l/2}\, \hat{c}_{l,\sigma}|_{\rm HM}$. This corresponds to a shift $k_{\rm HM} = k + \pi/2$ in momentum space. We recover the $1/r$ Hubbard model in the thermodynamic limit for $\kappa \to 0$ since then the hopping decays proportionally to the inverse of the distance at all length scales. The proper order of limits is to take the thermodynamic limit first, and let $\kappa \to 0$ afterwards.

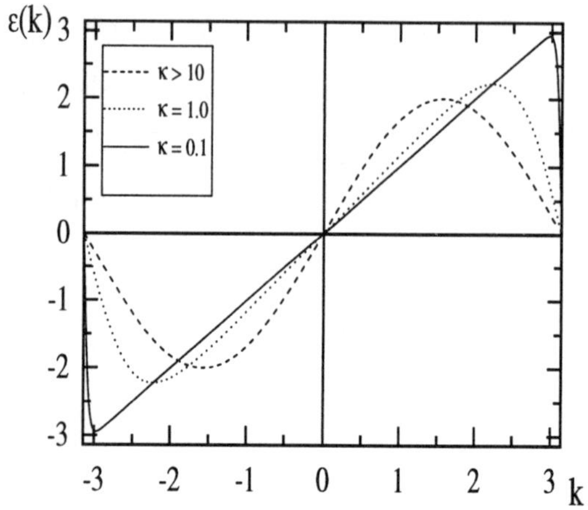

Fig. 4.7. Dispersion relation of the $1/\sinh r$ Hubbard model for $\kappa > 10$ (short-range hopping), $\kappa = 1.0$ and $\kappa = 0.1$ (medium-size hopping ranges). The dispersion is linear for $\kappa = 0$ (long-range hopping) (from [4.74]).

We can readily calculate the dispersion relation from

$$\epsilon(k) = \sum_{r\neq 0} t(r) \mathrm{e}^{-\mathrm{i}kr} \; . \tag{4.84a}$$

With the definitions $p = (k+\pi)/2$, $q = \mathrm{e}^{-\kappa}$ we rewrite this in the form

$$\epsilon(k) = (-4t)\sinh(\kappa) \sum_{r=1}^{\infty} \frac{q^r}{1-q^{2r}} \sin(2pr) \; . \tag{4.84b}$$

This expression is found to be a logarithmic derivative of a Jacobi theta function [4.72, 4.73],

$$\epsilon(k) = -t\sinh(\kappa) \left. \frac{\partial}{\partial x} \ln \theta_4(x, \mathrm{e}^{-\kappa}) \right|_{x=(k+\pi)/2} \; . \tag{4.84c}$$

The dispersion relation is shown in Fig. 4.7.

As in the $1/r$ Hubbard model, parity is, in general, not a good quantum number for $\kappa < \infty$ since the dispersion relation has odd parity, $\epsilon(-k) = -\epsilon(k)$, and the interaction has even parity. For dispersion relations with odd parity we may again use the electron–hole transformation $\mathcal{T}_4$ (4.44) to restrict our investigations to at most half band-filling.

4.3.2 General Two-Electron Problem in One Dimension

As in the case of the Hubbard model with nearest-neighbor hopping we analyze the two-electron state

$$|\psi\rangle = \sum_{x_1,x_2} \psi(x_1,x_2) \hat{c}^+_{x_1,\uparrow} \hat{c}^+_{x_2,\downarrow} |\text{vacuum}\rangle \tag{4.85}$$

in the spin singlet channel such that we again have $\psi(x_1,x_2) = \psi(x_2,x_1)$ for spin $S = S^z = 0$. For $x_1 \neq x_2$ the Schrödinger equation $\hat{H}|\psi\rangle = E|\psi\rangle$ gives

$$E\psi(x_1,x_2) = F(x_1,x_2) + F(x_2,x_1) + t(x_1-x_2)\left[\psi(x_2,x_2) - \psi(x_1,x_1)\right] , \tag{4.86a}$$

and for $x_1 = x_2$ we obtain

$$(E-U)\psi(x_1,x_1) = 2F(x_1,x_1) . \tag{4.86b}$$

Here we used the abbreviation

$$F(x_1,x_2) = \sum_{x\neq x_1,x_2} t(x_1-x)\psi(x,x_2) . \tag{4.86c}$$

Both equations (4.86a) and (4.86b) have to be solved simultaneously. The wave function must have the following properties.

(i) It asymptotically describes an incoming plane wave and an outgoing scattered wave,

$$\psi(x_1 \ll x_2) \sim \mathrm{e}^{\mathrm{i}(k_1x_1+k_2x_2)} - \mathrm{e}^{\mathrm{i}\theta(k_1,k_2)}\mathrm{e}^{\mathrm{i}(k_2x_1+k_1x_2)} , \tag{4.87}$$

where $\theta(k_1,k_2)$ is the scattering phase shift and k_1, k_2 are quasi-momenta.

(ii) For $\kappa=\infty$ it has to reduce to the Bethe-ansatz wave function (4.4). This implies that the sector condition $x_1<x_2$ has to be extended in a proper way.

(iii) The function $F(x_1,x_2)$ of (4.86c) must give terms of the form $[\epsilon(k_1)+\epsilon(k_2)]\psi(x_1,x_2)$; see (4.86a).

The form of the hopping matrix elements and the items (ii) and (iii) suggest that the following trigonometric identity will be helpful:

$$\frac{1}{\sinh(z_1)\sinh(z_2)} = \frac{\coth(z_2)-\coth(z_1)}{\sinh(z_1-z_2)} . \tag{4.88}$$

In (4.86c) it guarantees that there will be terms in $F(x_1,x_2)$ proportional to $1/\sinh(x_1-x_2)$ if we set $\psi(x_1,x_2) \sim 1/\sinh(x_1-x_2)$. Together with condition (i) our ansatz for the wave function becomes

$$\begin{aligned}\psi(x_1,x_2) &= \frac{1-\delta_{x_1,x_2}}{2\sinh[\kappa(x_2-x_1)]}\Psi(x_1,x_2) + \delta_{x_1,x_2}\lambda\mathrm{e}^{\mathrm{i}(k_1+k_2)x_1} ,\\ \Psi(x_1,x_2) &= \mathrm{e}^{\kappa(x_2-x_1)}\left(A\mathrm{e}^{\mathrm{i}(k_1x_1+k_2x_2)} - B\mathrm{e}^{\mathrm{i}(k_2x_1+k_1x_2)}\right)\\ &\quad -\mathrm{e}^{-\kappa(x_2-x_1)}\left(A\mathrm{e}^{\mathrm{i}(k_1x_2+k_2x_1)} - B\mathrm{e}^{\mathrm{i}(k_1x_1+k_2x_2)}\right) .\end{aligned} \tag{4.89}$$

Again, we may calculate the scattering matrix and phase shift for the spin singlet as $s=-B/A=-\mathrm{e}^{\mathrm{i}\theta(k_1,k_2)}$.

The wave function in (4.89) has the form of Inozemtsev's two-magnon state [4.68]. A generalization of the Bethe ansatz can thus be achieved by a "softening" of the sector condition $x_1<x_2$ into the Bose function

$$\frac{\mathrm{e}^{\kappa(x_2-x_1)}}{2\sinh[\kappa(x_2-x_1)]} = \frac{1}{1-\mathrm{e}^{-2\kappa(x_2-x_1)}} \xrightarrow[\kappa\to\infty]{} \begin{cases} 1 & \text{for} \quad x_1<x_2 \\ 0 & \text{for} \quad x_1>x_2 \end{cases} . \tag{4.90}$$

This extension of the Bethe ansatz had already been recognized by Inozemtsev [4.68] in the context of the $1/\sinh^2 r$ Heisenberg model.

We have to be careful when we evaluate the function $F(x_1, x_2)$ because infinite lattice sums appear. After we have applied the trigonometric identity in (4.88) we must not directly split the sum over x into two separate sums because we would obtain diverging expressions. Depending on the prefactors $\exp[-\kappa(x-x_2)]$ and $\exp[\kappa(x-x_2)]$, respectively, we have to write

$$\begin{aligned} \coth[\kappa(x_1-x)] - \coth[\kappa(x_2-x)] &= [\coth[\kappa(x_1-x)] \mp 1] \\ &\quad - [\coth[\kappa(x_2-x)] \mp 1] \end{aligned} \tag{4.91}$$

before we split the sum over x into two sums. The rest of the calculation is straightforward but lengthy. The result for $x_1 \neq x_2$ is

$$F(x_1,x_2) + F(x_2,x_1) = [\epsilon(k_1)+\epsilon(k_2)]\psi(x_1,x_2) + G(x_1,x_2) \tag{4.92a}$$

with the abbreviation

$$\begin{aligned} G(x_1,x_2) &= -t(x_1-x_2)\left[\mathrm{e}^{\mathrm{i}(k_1+k_2)x_2} - \mathrm{e}^{\mathrm{i}(k_1+k_2)x_1}\right] \\ &\quad \times\left[(A-B) + (A+B)\frac{\epsilon(k_1)-\epsilon(k_2)}{2\mathrm{i}t\sinh(\kappa)}\right] . \end{aligned} \tag{4.92b}$$

Some unwanted terms cancel because of the exchange symmetry of $F(x_1,x_2)+F(x_2,x_1)$ in $x_1 \leftrightarrow x_2$. Thus, (4.86a) is indeed solved by the ansatz (4.89) if

$$E = \epsilon(k_1)+\epsilon(k_2) , \tag{4.93a}$$

$$\lambda = (A-B) + \frac{(A+B)(\epsilon(k_1)-\epsilon(k_2))}{2\mathrm{i}t\sinh(\kappa)} . \tag{4.93b}$$

The calculation of $F(x_1,x_1)$ can be carried out without further problems and leads to

$$2F(x_1,x_1) = \mathrm{e}^{\mathrm{i}(k_1+k_2)x_1}\left[E(A-B) - \mathrm{i}(A+B)\left(\eta(k_1)-\eta(k_2)\right)\right] \tag{4.94a}$$

with the auxiliary function

$$\eta(k) = \eta(-k) = 2t\sinh(\kappa)\sum_{n=1}^{\infty}(-1)^n\cos(kn)\cosh(\kappa n)/\sinh^2(\kappa n) . \tag{4.94b}$$

This function obeys $(\partial\eta(k))/(\partial k) = (\partial\epsilon(k))/(\partial\kappa)$. With the help of these results and (4.86b) we can determine the scattering matrix in the singlet channel, $s=-B/A$, from

$$(A-B)E - \mathrm{i}(A+B)(\eta(k_1)-\eta(k_2)) = \lambda(E-U) . \tag{4.95}$$

Hence, we have formally solved the two-electron problem. The solution is valid in the thermodynamic limit only. Therefore, it is not possible to formulate a formal quantization condition for the quasi-momenta. A way out of this dilemma is provided by Sutherland's asymptotic Bethe ansatz which we will apply to obtain the generalized Lieb–Wu equations and thus the spectrum of our model in the thermodynamic limit.

4.3.3 *S* Matrix and Generalized Lieb–Wu Equations

Inozemtsev [4.68] succeeded in explicitly constructing many-electron wave functions in the thermodynamic limit. Unfortunately, the exponential tails of the wave functions prevented the construction of wave functions for *finite* electron densities. It seems likely that one has to construct a lattice-periodic version of the $1/\sinh$ exchange before further progress can be made.

Sutherland [4.12, 4.75] recognized that for integrable systems the two-electron scattering phase is enough to derive the spectrum as long as the wave functions asymptotically ($x_1 \ll x_2 \ll \ldots \ll x_N$) describe plane waves ("asymptotic Bethe ansatz"). It is by no means obvious that the $1/\sinh r$ Hubbard model is indeed integrable. In the following we will nevertheless investigate the consequences of this hypothesis; we are confident that the results are exact in the thermodynamic limit.

The solution of the two-electron problem gives the following result for the scattering matrix for $S = 0$:

$$s = -\frac{B}{A} = \frac{\mathrm{i}(H(k_1) - H(k_2)) + U/(2t)}{\mathrm{i}(H(k_1) - H(k_2)) - U/(2t)} , \tag{4.96a}$$

with the auxiliary function

$$H(k) = \frac{1}{2t}\left[-\eta(k) + \frac{\epsilon(k)(\epsilon(k) - U)}{2t\sinh(\kappa)}\right] . \tag{4.96b}$$

When we compare this with (4.8) we immediately see that the generalization of the Bethe-ansatz result is achieved by the replacement

$$\sin(k_{\mathrm{HM}}) \mapsto H(k) . \tag{4.97}$$

Since $\lim_{\kappa\to\infty} H(k) = \lim_{\kappa\to\infty}[-\eta(k)/(2t)] = \cos(k) = \sin(k_{\mathrm{HM}})$, we can already verify that all results for the Hubbard model with nearest-neighbor hopping will indeed be reproduced in the limit $\kappa \to \infty$.

In analogy to the results in Sect. 4.1 we find the total two-electron scattering matrix as

$$S^{12} = \frac{(H(k_1) - H(k_2))I + \mathrm{i}U/(2t)P^{12}}{H(k_1) - H(k_2) + \mathrm{i}U/(2t)} . \tag{4.98}$$

This two-electron S matrix does fulfill the Yang–Baxter equation (4.15a). However, this does *not* prove integrability because the sector condition has been softened into a Bose function.

The scattering phase shift is found as

$$\theta(k_1, k_2) = (-2)\arctan\left[\frac{H(k_2) - H(k_1)}{U/(2t)}\right] . \tag{4.99}$$

The assumption of the asymptotic Bethe ansatz then immediately yields the generalized Lieb–Wu equations [see (4.19) and (4.97)]

$$\begin{aligned} k_j L &= 2\pi I_j + \sum_{\alpha=1}^{N_\downarrow} \Theta\bigl(2H(k_j) - 2\Lambda_\alpha\bigr) , \\ & j = 1, \ldots, N , \end{aligned} \tag{4.100a}$$

$$\begin{aligned} -\sum_{\beta=1}^{N_\downarrow} \Theta(\Lambda_\alpha - \Lambda_\beta) &= -2\pi J_\alpha + \sum_{j=1}^{N} \Theta\bigl(2H(k_j) - 2\Lambda_\alpha\bigr) , \\ & \alpha = 1, \ldots, N_\downarrow . \end{aligned} \tag{4.100b}$$

As before we use $\Theta(x) = -2\arctan\bigl[x/(U/(2t))\bigr]$. The quantum numbers I_j are half odd-integer if $N_\downarrow$ is odd, and integer otherwise. The values of $2\pi I_j/L$ are restricted to the first Brillouin zone ($N \leq L$). The quantum numbers J_α are half odd-integer if $N - N_\downarrow - 1$ is odd, and integer otherwise. We further have $2\pi J_\alpha \leq \pi(N - N_\downarrow)$.

Strictly speaking, the equations (4.100) do not belong to the $1/\sinh r$ Hubbard model with hopping amplitudes (4.83) but to a model defined on a finite ring of circumference L. Both models agree only in the thermodynamic limit. Fortunately, the error we introduce on the quantization of the quasi-momenta k_j in (4.100) is only exponentially small, of the order of $\mathcal{O}(\exp(-\kappa L))$, and thus negligible for $L \gg \kappa^{-1}$. Obviously, the thermodynamic limit $L \to \infty$ has to be taken *before* the limit $\kappa \to 0$ to obtain the proper results for the $1/r$ Hubbard model. For this reason the two-electron wave function for the $1/r$ Hubbard model in Sect. 4.2, (4.52), does not agree with the wave functions of this section, (4.89), in the limit $\kappa \to 0$.

4.3.4 Ground-State Energy at Half Band-Filling

In complete analogy to our considerations in Sect. 4.1.4 we can reduce the generalized Lieb–Wu equations to integral equations in the thermodynamic limit. We formulate these equations and explicitly solve them for half band-filling in Appendix A.3. With the auxiliary functions

$$\widetilde{J}_0^{\binom{c}{s}}(\omega) = \begin{pmatrix}\mathrm{Re}\\ \mathrm{Im}\end{pmatrix} \int_{-\pi}^{\pi} \frac{\mathrm{d}k}{2\pi} \mathrm{e}^{\mathrm{i}\omega H(k)} , \tag{4.101a}$$

$$\widetilde{J}_1^{\binom{c}{s}}(\omega) = \begin{pmatrix}\mathrm{Re}\\ \mathrm{Im}\end{pmatrix} \int_{-\pi}^{\pi} \frac{\mathrm{d}k}{2\pi} \mathrm{e}^{\mathrm{i}\omega H(k)} \epsilon(k) \frac{\partial H(k)}{\partial k} \tag{4.101b}$$

the ground-state quasi-particle densities can be written as

$$\sigma_0(\Lambda) = \int_0^\infty \frac{\mathrm{d}\omega}{2\pi \cosh(\omega U/(4t))} \Bigl[\cos(\omega\Lambda)\widetilde{J}_0^c(\omega) + \sin(\omega\Lambda)\widetilde{J}_0^s(\omega)\Bigr] , \tag{4.102a}$$

$$\begin{aligned} \rho_0(k) = \frac{1}{2\pi} + \frac{\partial H(k)}{\partial k} \int_0^\infty \frac{\mathrm{d}\omega}{\pi(1 + \mathrm{e}^{\omega U/(2t)})} \Bigl[&\cos(\omega H(k))\widetilde{J}_0^c(\omega) \\ &+ \sin(\omega H(k))\widetilde{J}_0^s(\omega)\Bigr] . \end{aligned} \tag{4.102b}$$

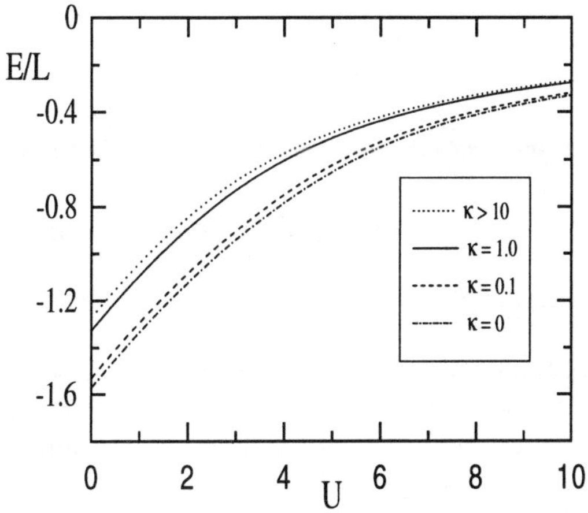

Fig. 4.8. Ground-state energy of the $1/\sinh r$ Hubbard model at half band-filling for $\kappa > 10$ (short-range hopping), $\kappa = 1.0$ and $\kappa = 0.1$ (medium hopping ranges), and $\kappa = 0$ (long-range hopping) ($t = 1$; from [4.74]).

For $\kappa \to \infty$ we have $\widetilde{J}_0^{\rm c}(\omega) \to J_0(\omega)$, $\widetilde{J}_0^{\rm s}(\omega) \to 0$, and $\widetilde{J}_1^{\rm c}(\omega) \to -2J_1(\omega)/\omega$, $\widetilde{J}_1^{\rm s}(\omega) \to 0$, where $J_0(\omega)$ and $J_1(\omega)$ are Bessel functions. Now that we have $\lim_{\kappa\to\infty} H(k) = \sin(k_{\rm HM})$ we recover the result obtained by Lieb and Wu [4.16] for the quasi-particle densities in the case of the standard Hubbard model, (4.23).

The ground-state energy density for half band-filling is found as

$$(E_0/L)(N = L, U) = \int_0^\infty \frac{2\,{\rm d}\omega}{1 + {\rm e}^{\omega U/(2t)}} \left[\widetilde{J}_0^{\rm c}(\omega)\widetilde{J}_1^{\rm c}(\omega) + \widetilde{J}_0^{\rm s}(\omega)\widetilde{J}_1^{\rm s}(\omega)\right] . \tag{4.103}$$

For $\kappa \to \infty$ we again recover the Lieb–Wu result for the ground-state energy (4.24) [4.16]. Furthermore, we obtain the correct result for the ground-state energy of the $1/r$ Hubbard model at half band-filling, (4.61), if we carefully carry out the limit $\kappa \to 0$. This is shown in Appendix A.3.

The ground-state energy as a function of the interaction strength U/t for various hopping ranges is displayed in Fig. 4.8. For small U/t all curves have the inclination $U/(4t)$. This corresponds to the Hartree energy for half band-filling. For large U/t the $1/\sinh r$ Hubbard model transforms into the Inozemtsev model, (4.82). We find $\lim_{U\to\infty}(UE/(Lt^2)) = -4\ln 2 \approx -2.773, -2.826, -3.197, -\pi^2/3 \approx -3.290$ for the values $\kappa = \infty, 1, 0.1, 0$. In [4.74] it is shown that all the results in [4.70] for the spin model are completely reproduced. As a by-product [4.74] one even obtains an explicit solution to the coupled integral equations for the Inozemtsev model [4.70].

4.3.5 Metal–Insulator Transition at Half Band-Filling

For the analysis of the metal–insulator transition at half band-filling we require the quasi-particle dispersions; see Sect. 4.1.5. As shown in Appendix A.3 we can obtain a closed solution for the corresponding integral equations in this case. We find

$$\tilde{\epsilon}_{\mathrm{s}}^{\mathrm{QP}}(\Lambda) = \int_0^\infty \frac{\mathrm{d}\omega}{\cosh(\omega U/(4t))}\left[\cos(\omega\Lambda)\widetilde{J}_1^{\mathrm{c}}(\omega) + \sin(\omega\Lambda)\widetilde{J}_1^{\mathrm{s}}(\omega)\right] , \tag{4.104a}$$

$$\tilde{\epsilon}_{\mathrm{c}}^{\mathrm{QP}}(k) = \epsilon_{\mathrm{c}}^{0}(k) + 2\int_0^\infty \frac{\mathrm{d}\omega}{1+\mathrm{e}^{\omega U/(2t)}}\Big[\cos(\omega H(k))\widetilde{J}_1^{\mathrm{c}}(\omega) + \sin(\omega H(k))\widetilde{J}_1^{\mathrm{s}}(\omega)\Big] , \tag{4.104b}$$

with $\epsilon_{\mathrm{c}}^{0}(k) = -\mu^{-} + \epsilon(k)$. The chemical potential μ^{-} follows from the condition $\max_{\{k\}}[\tilde{\epsilon}_{\mathrm{c}}^{\mathrm{QP}}(k)] = 0$.

This result can be further analyzed for weak and strong interactions. For $U = 0$ we have $\tilde{\epsilon}_{\mathrm{c}}^{\mathrm{QP}}(k) = 2\epsilon_{\mathrm{c}}^{0}(k)$ for $-\pi < k < 0$ and $\tilde{\epsilon}_{\mathrm{c}}^{\mathrm{QP}}(k) = 0$ for $0 < k < \pi$. Thus $\mu^{-}(U = 0) = \mu^{+}(U = 0) = 0$ holds, and the system is metallic. For large U/t we find instead that $\tilde{\epsilon}_{\mathrm{c}}^{\mathrm{QP}}(k) = -\mu^{-} + \epsilon(k)$. Hence, $\mu^{-}(U \gg W) = W/2$ and $\mu^{+}(U \gg W) = U - W/2$. The system has an excitation gap of the size $\Delta\mu(U \gg W) = U - W$. The upper and lower Hubbard bands have the full bandwidth $W_1 = W_2 = W$ for $U \gg W$, in complete agreement with the general ideas of Mott and Hubbard.

The calculation of the energy gap in the limit of small interactions is very difficult, as could already be seen for the case of the Hubbard model with nearest-neighbor interactions. For the $1/\sinh r$ Hubbard model we could not find a simple analytical expression for the energy gap. Nevertheless we can make some qualitative estimate on the behavior of the energy gap. For small U/t and all $\kappa > 0$ the function $H(k)$ is a symmetrical function, independent of U; see (4.96b). As in the standard Hubbard model the decisive U dependence of the quasi-particle dispersions, (4.104), stems from the factor $1/(1+\exp(\omega U/(2t)))$ in the ω integral. This exponential dependence is the reason for an exponentially small energy gap in the Hubbard model with nearest-neighbor hopping, (4.33). Consequently, we *qualitatively* find the same exponential dependence of the energy gap on U/t for all $\kappa > 0$ for small U/t. Hence the system is an insulator at half band-filling, $U_{\mathrm{c}}(\kappa > 0) = 0^{+}$. We need a new limit, $\kappa \to 0$, to change this analytical behavior. We show in Appendix A.3 that (4.104b) indeed gives $U_{\mathrm{c}}(\kappa = 0) = W$, in agreement with (4.63).

Now we have completed our analytical studies of modified Hubbard models with variable hopping ranges. We will discuss the results for the metal–insulator transition in more detail in the next section.

4.4 Mott–Hubbard Transition in One Dimension

In this section we first discuss peculiarities of our one-dimensional exactly solvable models. At small energies all our modified Hubbard models with finite hopping range ($\kappa > 0$) display the perfect nesting property which results in an insulating ground state at half band-filling for all finite interaction strengths. We discuss the effect of umklapp scattering within the field-theoretical "g-ology" Hamiltonian to which our models reduce at low energies. The nesting property is absent in the dispersion relation of the $1/r$ Hubbard model, but it displays a jump discontinuity at the Brillouin zone boundary. This results in some unphysical properties of the metal–Mott–Hubbard insulator transition in this model. Despite some peculiarities the generic features of the Mott–Hubbard transition are clearly borne out in one dimension.

4.4.1 g-ology Hamiltonian

With the exception of the $1/r$ Hubbard model all one-dimensional Hubbard models display an insulating behavior at half band-filling for all $U > 0$. Since the effect appears at arbitrarily small interaction strengths it should be possible to understand this transition within a more general model description for one-dimensional electrons with weak interactions.

In the absence of any interactions the system is metallic. In the ground state all states in momentum space in the region $|k| \leq k_{\mathrm{F}}$ are occupied, and the system possesses two Fermi points at $k = \pm k_{\mathrm{F}}$. If the Coulomb interaction is weak, the bandwidth W is large compared to all Coulomb parameters U, V, etc. In this case we may *linearize* the dispersion relation around the Fermi points since excitations are limited to small energy transfers. The band edges $\pm W/2$, and the Brillouin zone boundaries $\pm\pi/a$, appear only as cut-off parameters for otherwise formally divergent frequency and momentum integrals.

After the linearization of the dispersion relation around the two Fermi points we obtain the general effective low-energy Hamilton operator for one-dimensional electrons. This field-theoretical model, the "g-ology" Hamiltonian, is extensively discussed in [4.76]; see also [4.43, 4.60, 4.77] and references therein. In the following we will limit our discussion to those model aspects that are relevant for the metal–insulator transition.

According to its derivation the g-ology model describes "left-moving" electrons at $k = -k_{\mathrm{F}}$ ($\hat{\ell}_{k,\sigma}^{+}$) and "right-moving" electrons at $k = +k_{\mathrm{F}}$ ($\hat{r}_{k,\sigma}^{+}$). They interact with each other such that the corresponding energy transfer in a scattering process is small compared to the bandwidth. As in Fermi-liquid theory we thereby assume that energetically low-lying states are not drastically influenced by the interaction. In particular, there is always a filled (Dirac) sea of states with arbitrarily negative energy below the Fermi energy. This infinite sea of electrons allows for an exact solution of the g-ology model for some important generic cases, namely the Luttinger model and the Luther–Emery model; see below. However, the existence of the Dirac sea requires some care in practical calculations.

In the following we use the convention that all operators are normal ordered with respect to the non-interacting ground state (Dirac sea). In addition, momentum indices on the operators for right- and left-moving electrons always denote momenta in regions $|k - k_{\mathrm{F}}| < k_0 \ll \pi/a$ and

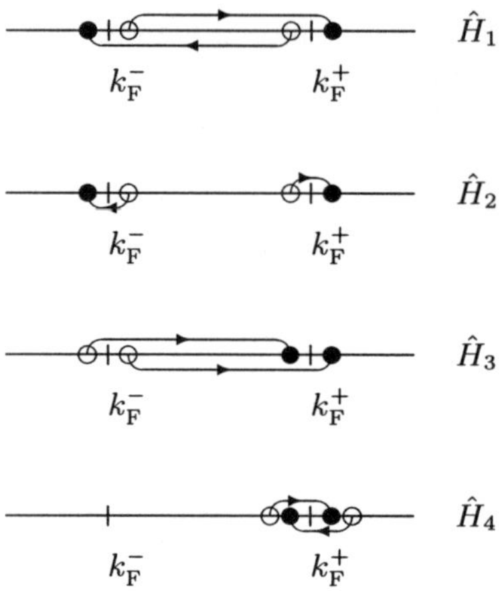

Fig. 4.9. Elementary scattering processes in the g-ology model. $\hat{H}_1$: Backscattering. $\hat{H}_3$: Umklapp scattering. $\hat{H}_2$, $\hat{H}_4$: Scattering terms of the Luttinger model.

$|k + k_F| < k_0 \ll \pi/a$ around the corresponding Fermi point (k_0 is a cut-off momentum). The Hamilton operator for the g-ology model can then be formulated as

$$\hat{H}_g = \hat{H}_0 + \sum_{l=1}^{4} \hat{H}_l \, , \tag{4.105a}$$

$$\hat{H}_0 = \sum_{k,\sigma} \hbar v_F^r (k - k_F) \hat{r}_{k,\sigma}^+ \hat{r}_{k,\sigma} - \sum_{k,\sigma} \hbar v_F^\ell (k + k_F) \hat{\ell}_{k,\sigma}^+ \hat{\ell}_{k,\sigma} \, , \tag{4.105b}$$

$$\hat{H}_1 = \frac{1}{L} \sum_{\substack{k_1,k_2,q \\ \sigma,\sigma'}} g_1^{\sigma,\sigma'} \hat{r}_{k_1,\sigma}^+ \hat{\ell}_{k_2,\sigma'}^+ \hat{r}_{k_2+2k_F+q,\sigma'} \hat{\ell}_{k_1-2k_F-q,\sigma} \, , \tag{4.105c}$$

$$\hat{H}_2 = \frac{1}{L} \sum_{\substack{k_1,k_2,q \\ \sigma,\sigma'}} g_2^{\sigma,\sigma'} \hat{r}_{k_1,\sigma}^+ \hat{\ell}_{k_2,\sigma'}^+ \hat{r}_{k_2+q,\sigma'} \hat{\ell}_{k_1-q,\sigma} \, , \tag{4.105d}$$

$$\hat{H}_3 = \frac{1}{2L} \sum_{\substack{k_1,k_2,q \\ \sigma,\sigma'}} g_3^{\sigma,\sigma'} \left[\hat{r}_{k_1,\sigma}^+ \hat{r}_{k_2,\sigma'}^+ \hat{\ell}_{k_2-2k_F+q,\sigma'} \hat{\ell}_{k_1+2k_F-q-G,\sigma} + \text{h.c.} \right] \, , \tag{4.105e}$$

$$\hat{H}_4 = \frac{1}{2L} \sum_{\substack{k_1,k_2,q \\ \sigma,\sigma'}} g_4^{\sigma,\sigma'} \left[\hat{r}_{k_1,\sigma}^+ \hat{r}_{k_2,\sigma'}^+ \hat{r}_{k_2+q,\sigma'} \hat{r}_{k_1-q,\sigma} + (l \leftrightarrow r) \right] \, . \tag{4.105f}$$

The Fermi velocities v_F^r and v_F^ℓ of the right- and left-moving electrons need not necessarily be the same. The coupling strengths $g_l^{\sigma,\sigma'}$ are functions of the parameters in the original lattice model under consideration. In general, an identification of the two parameter sets is only possible for small interaction strengths; see below.

The interaction contributions $\hat{H}_l$ ($l = 1, \ldots, 4$) define the four possible two-electron scattering processes, compatible with small energy transfers.

They are displayed in Fig. 4.9. $\hat{H}_1$ and $\hat{H}_3$ describe processes with large momentum transfers but only small momentum transfers are involved in $\hat{H}_2$ and $\hat{H}_4$. $\hat{H}_1$ is called the backscattering term since a right-moving electron is scattered into a left-moving electron, and vice versa. $\hat{H}_3$ is called the umklapp scattering term since two left-moving electrons are scattered into two right-moving electrons, and vice versa. Since the total momentum in a scattering process is conserved, this process is only possible on a lattice where the total momentum is defined modulo a reciprocal lattice vector ($G = 2\pi/a$). Even then the umklapp process is forbidden for general band-filling because such a process requires $Q = 4k_{\mathrm{F}} - G \equiv 0$ which is fulfilled only at half band-filling where $k_{\mathrm{F}} = n\pi/(2a) = G/4$ for $n = 1$.

In all cases the Hamiltonian can be decomposed into two commuting parts for the charge and spin degrees of freedom. This is a first indication that charge and spin propagate differently in one dimension (charge–spin separation). For the charge and the spin sector we define

$$g_l^{\mathrm{c}} = g_l^{\sigma,\sigma} + g_l^{\sigma,\sigma'} , \qquad l = 1, \ldots, 4 , \tag{4.106a}$$

$$g_l^{\mathrm{s}} = g_l^{\sigma,\sigma} - g_l^{\sigma,\sigma'} , \qquad l = 1, \ldots, 4 , \tag{4.106b}$$

as new coupling parameters.

When both backscattering and umklapp scattering are absent, $g_1^{\nu} = 0$ and $g_3^{\nu} = 0$ ($\nu = \mathrm{c}$ and/or $\nu = \mathrm{s}$), the general g-ology Hamiltonian reduces to the Luttinger model [4.78], which was first solved exactly by Lieb and Mattis [4.79]. One special variant is the "chiral" Luttinger model for $g_2^{\nu} = 0$ in which the left-moving and right-moving charges and/or spins are independent of each other.

In the presence of backscattering and/or umklapp scattering the g-ology model for the spin and/or charge degrees of freedom is exactly solvable only for a special choice of the coupling constants. The exactly solvable "Luther–Emery line" [4.80] is defined by the relations

$$\frac{g_2^{\nu}}{\pi\hbar v_{\mathrm{F}} + g_4^{\nu}} = \frac{3}{5} \tag{4.107}$$

in the charge sector ($\nu = \mathrm{c}$) and/or the spin sector ($\nu = \mathrm{s}$), respectively [4.43, 4.60, 4.76].

4.4.2 Perfect Nesting in One Dimension

For modified Hubbard models we have $g_l^{\sigma,-\sigma} = U + \mathcal{O}(U^2/W)$, $g_l^{\sigma,\sigma} = \mathcal{O}(U^2/W)$ for all $l = 1, \ldots, 4$, and we should address the full g-ology model, at least in principle. In this case we can employ renormalization group theory to analyze which of the g parameters are relevant or irrelevant for the physical properties described by the model. Such an analysis shows [4.60] that the umklapp scattering $\hat{H}_3$ and the backscattering term $\hat{H}_1$ are irrelevant for all modified Hubbard models away from half band-filling. Hence, the Luttinger model describes the (small-g) fix-point of the g-ology model. In particular, this implies that the Hubbard models with finite-range hopping are metallic away from half band-filling.

Since the Luttinger model can be solved exactly, one can calculate the velocities of the elementary excitations as functions of the parameters $g_2^{\mathrm{c,s}}$ and $g_4^{\mathrm{c,s}}$ as

$$v_\nu = \sqrt{\left[\hbar v_{\mathrm{F}} + g_4^\nu/\pi\right]^2 - \left[g_2^\nu/\pi\right]^2}\ , \quad \nu = \mathrm{c,s}\ . \tag{4.108}$$

Since these two velocities are generally unequal, charge and spin excitations propagate differently through the system. Because the g-ology model and the Luttinger model, as one of its important fix-points, are the generic low-energy Hamiltonians in one dimension, the phenomenon of charge–spin separation is common to all one-dimensional models at low energies [4.36]. Therefore, in one dimension the concept of a "Luttinger liquid" replaces Landau's Fermi-liquid picture of interacting electrons.

To make contact between the Luttinger model and exactly solved Hubbard models we may compare the velocities in (4.108) with those of the elementary excitations in the exact solutions. This allows us to identify the g parameter as a function of U/t for *all* interaction strengths. In this way it is possible to transfer all known results for the Luttinger model to the (modified) Hubbard models at small excitation energies [4.34]. The comparison of the results from a direct analysis of the Bethe ansatz equations for the Hubbard model and those from the analogy to the Luttinger model explicitly shows that the Hubbard model away from half band-filling belongs to the general class of Luttinger liquids [4.36]. The effective values for $g_{2,4}^\nu(U/t)$ are finite, and the system is a metal away from half band-filling. We have already made use of this fact in Sect. 4.1.

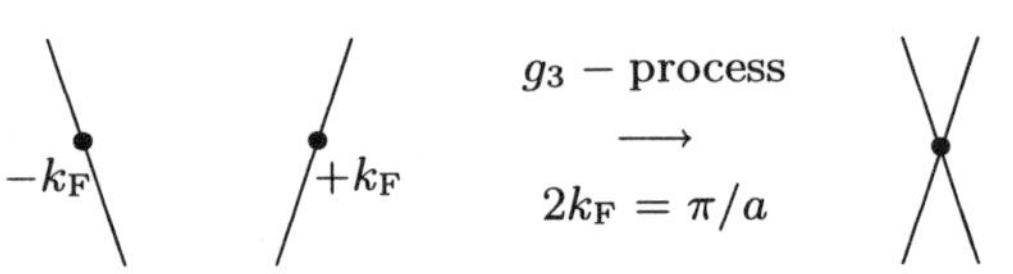

Fig. 4.10. umklapp scattering at half band-filling: both dispersion relations effectively overlap due to the umklapp shift of the momenta, and thus split for $U > 0$.

At half band-filling, however, the umklapp scattering becomes *relevant* in the sense of the renormalization group theory [4.81]. It is precisely this umklapp contribution that is responsible for the metal–insulator transition at $U_{\mathrm{c}}^{d=1} = 0^+$ because one-dimensional systems with two Fermi points always fulfill the perfect nesting condition $\epsilon(k_{\mathrm{F}} + \delta_k) = -\epsilon(k_{\mathrm{F}} + \delta_k - \pi/a)$ for $\delta_k \ll k_{\mathrm{F}}$. The effect of the umklapp term can easily be understood by a look at Fig. 4.10. In the presence of umklapp scattering one branch of the dispersion relation can be viewed as being shifted by half a reciprocal lattice vector $2k_{\mathrm{F}} = \pi/a$. Hence both branches effectively *cross*, and any finite interaction $g_3^{\uparrow,\downarrow} = U > 0$ will split the dispersion relation and an excitation gap will form. The system describes an insulator, $U_c^{d=1} = 0^+$. The property shown in Fig. 4.10 is exact for the g-ology model on the Luther–Emery line [4.43, 4.76] on which the gap is finite:

$$\Delta\mu = \frac{|g_3^c|}{2\pi\alpha} \tag{4.109}$$

(α is a cut-off parameter).

The “band crossing argument”, however, has to be used with care since it is based on the ideas of perturbation theory for non-interacting electrons (band structure theory). For example, the argument does *not* work for *spinless* fermions with nearest-neighbor interaction in one dimension for which the argument would formally be the same. The result, however, would be *wrong*. Spinless fermions at half band-filling show a transition from a metal into an insulating charge-density wave state at a finite value of the interaction strength, as seen from the exact solution [4.82]; see also Sect. 5.4. Therefore, it is necessary to scrutinize whether the umklapp term indeed destabilizes the Luttinger liquid fix-point. For spinless fermions this term is in fact only marginal [4.81], but it is relevant for the electrons in the Hubbard model which is a spinful system with momentum-independent interactions.

We can thus state that the perfect nesting property of all one-dimensional Hubbard models with left- and right-moving electrons is the reason for the insulating ground state at half band-filling and all $U > 0$.

4.4.3 Long-Range Hopping

The decisive advantage of the $1/r$ Hubbard model over the other modified Hubbard models is the absence of the perfect nesting property, because there are right-moving electrons only. Therefore, it corresponds to a chiral Luttinger model which is, however, defined and solved *on the lattice.* In its metallic phase at low energies the $1/r$ Hubbard model can be decomposed into two independent chiral Luttinger models for the charge and spin degrees with g-ology parameters $g_1^{c,s} = g_2^{c,s} = g_3^{c,s} = 0$. We identify the two parameters g_4^c and g_4^s as functions of U/t by a comparison of the velocities in the exact solution of the $1/r$ Hubbard model, (4.67) and (4.72), with those as functions of $g_4^{c,s}$ in the exact solution of the chiral Luttinger model, (4.108). The result is ($W = 2\pi t$, $v_F = \hbar t a \equiv t$) [4.59]

$$g_4^c = U\frac{W}{\sqrt{(W+U)^2 - 4WUn}} > 0\,, \tag{4.110a}$$

$$g_4^s = (-U)\frac{W}{U+W} < 0\,. \tag{4.110b}$$

For a chiral Luttinger model we further have $K_\rho = \xi^2 = 1$; see, e.g., [4.38]. According to (4.36) we may now determine the Drude weight. The resulting equation (4.73) shows that the Drude weight for half band-filling and for $U \to W^-$ *diverges* like

$$\mathcal{D}_{\rm c} = \frac{e^2 ta}{\hbar^2}\left[1 + \frac{U}{W-U}\right] . \tag{4.111}$$

A diverging Drude weight seems to be impossible on behalf of the usual f-sum rule for the conductivity. The integral over the real part of the optical conductivity obeys (see, e.g., [4.59])

$$\begin{aligned}\int_{-\infty}^{\infty} {\rm d}\omega {\rm Re}\left[\sigma(\omega)\right] &= \lim_{q\to 0}\frac{\pi e^2}{L}\sum_{k,\sigma}\langle\hat{n}_{k,\sigma}\rangle \\ &\quad\times\left\{\frac{1}{q}\left[\frac{\epsilon(k+q)-\epsilon(k)}{q}+\frac{\epsilon(k-q)-\epsilon(k)}{q}\right]\right\} ,\end{aligned} \tag{4.112a}$$

with $\sigma(\omega) \equiv \sigma(\omega, q = 0)$. For twice-differentiable dispersion relations $\epsilon(k)$ this reduces to

$$\int_{-\infty}^{\infty} {\rm d}\omega {\rm Re}\left[\sigma(\omega)\right] = \frac{\pi e^2}{L}\sum_{k,\sigma}\langle\hat{n}_{k,\sigma}\rangle\frac{\partial^2\epsilon(k)}{\partial k^2} . \tag{4.112b}$$

Hence the frequency integral over the conductivity has an upper bound. For the $1/r$ Hubbard model, however, we have to insert the minimum value for $q = \Delta = 2\pi/L$, which allows us to reduce (4.112a) to

$$\int_{-\infty}^{\infty} {\rm d}\omega {\rm Re}\left[\sigma(\omega)\right] = L\frac{e^2}{4\pi}\sum_{\sigma}\left[\langle\hat{n}_{-\pi+\Delta/2,\sigma}\rangle - \langle\hat{n}_{\pi-\Delta/2,\sigma}\rangle\right] . \tag{4.113}$$

The unphysical jump of the dispersion relation at the Brillouin zone boundary leads to a total conductivity that diverges proportionally to the system size. Correspondingly, the Drude weight can diverge at the transition proportionally to $\sqrt{L}$; see (4.73) and (4.110a).

At the metal-to-insulator transition the divergence of the charge velocity implies unphysical effects such as a diverging Drude weight and a vanishing effective mass of the charge carriers. Obviously, the details of the metal–Mott–Hubbard insulator transition are not properly described by the $1/r$ Hubbard model.

4.4.4 Comparison with Basic Theoretical Concepts

As the last point in this section we discuss which properties of our exactly solvable models are special to one dimension, and which can be viewed as generic, i.e., also valid in higher dimensions. This will allow us to compare the exact solutions of one-dimensional model systems to the more general views on the nature of the metal–insulator transition presented in Sect. 1.4.

Limits of Strong and Weak Interactions. All the exact solutions in this chapter show that two separate bands for charge excitations are formed for half band-filling and $U \gg W$. Their dispersion relation is given by $\epsilon_{\text{LHB}}(k) = \epsilon(k)$ and $\epsilon_{\text{UHB}}(k) = U - \epsilon(k)$, respectively. The lower and upper Hubbard bands describe the charge excitations of the half-filled ground state. The size of the energy gap therefore is of the order of $\Delta\mu^{\text{exact}} = U - W + \mathcal{O}(W^2/U)$. These exact results are in complete agreement with Mott's views (see Sect. 1.4) which were first quantified by Hubbard (Sect. 3.3) as well as Brinkman and Rice (Sect. 3.4). The charge excitations stay mobile for strong interactions.

This is in contrast to the predictions of Hartree–Fock mean-field theory (Sect. 3.1). Here, the electrons become localized onto their corresponding sublattice sites to optimize the exchange energy between their spins. Consequently, charge excitations are immobile for $U \gg W$ and the energy gap is given by $\Delta\mu^{\text{HF}} = U - \mathcal{O}(W^2/U)$. The fact that the energy gap is correctly reproduced to order U is often interpreted as a success for mean-field theory. Any theory, however, that localizes electrons on the lattice sites will reproduce this atomic limit contribution to the energy gap, e.g., the SIC-LSDA (Sect. 3.2). The conclusive information on the character of the insulating state is thus contained in the corrections to the leading order.

For weak interactions we have to leave out the special case of perfect nesting because there the system is always an insulator. The situation is typical for one-dimensional systems but rather an exception in higher dimensions. The generic case of the absence of perfect nesting can thus be discussed only for the $1/r$ Hubbard model for which the g_3 process is not present.

At half band-filling the $1/r$ Hubbard model describes a metal below the (finite) critical coupling, $U_{\text{c}} = W$. Although the perturbation series in U/t for the ground-state energy obviously converges for $U < W$, the system is not a Fermi liquid with quasi-particle excitations near the Fermi energy. Instead, as typical for one-dimensional systems, the metal is a Luttinger liquid in which (collective) charge and spin excitations propagate with different velocities.

In one dimension, however, the non-interacting Fermi gas should instead be viewed as a "Luttinger gas" such that there again is a one-to-one correspondence between the elementary excitations in both systems [4.83]. In this sense the physics described by the $1/r$ Hubbard model is qualitatively the same for all $0 \leq U < W$. In this respect the physics in one dimension parallels the expected situation in higher dimensions, where the metallic Fermi-liquid state exists for $0 \leq U < U_{\text{c}}$, and breaks down at the metal-to-insulator transition.

Metal–Insulator Transition. The metal–insulator transition in the $1/r$ Hubbard model is a consequence of the diverging effective *charge* coupling constant g_4^{c} for $U \to W^-$ at half band-filling; the spin channel of the corresponding Luttinger model does not display an anomaly in the metallic regime [see (4.110)]. The Mott–Hubbard gap at $U_{\text{c}} = W$ is due to charge correlations, in agreement with Mott's arguments in Sect. 1.4.

Furthermore, the Mott–Hubbard transition in the $1/r$ Hubbard model shows all the features of a generic quantum phase transition given in Mott's example in Sect. 1.4.6. It occurs at a finite value of the interaction strength, $U_c = W$, the gap opens continuously (exponent $\nu = 1$), and no ordering of the magnetic moments occurs. In addition, the gap is found to be robust since the (gaped) charge dispersion relation enters the thermodynamic quantities at all temperatures. This scenario should be generic for higher dimensions.

As already discussed in Sect. 1.4 the quantum phase transition is usually hidden since the magnetic moments are ordered at low enough temperature. One-dimensional Hubbard models do not show true long-range order in the ground state. Even in the presence of perfect nesting the spin–spin correlations decay with the inverse of the distance, and signal "quasi" long-range order only. Owing to perfect nesting, however, the point of the quantum phase transition is shifted to $U_{\rm c} = 0^+$. The energy scale for spin excitations is $J/t = \mathcal{O}(t/U)$, and the spin correlations can give rise to an exponentially small gap for small U/t, $\Delta\mu \sim \exp\left(-\mathcal{O}(J/t)\right)$. The insulating states for small and for large interactions are continuously connected for one-dimensional Hubbard models with the perfect nesting property, and no true long-range order of the moments occurs. For these reasons we characterize the insulating state as a Mott–Hubbard insulator.

The Hubbard model will display long-range magnetic order at half band-filling in higher dimensions. In three dimensions we thus expect a Mott–Heisenberg insulator which turns into a Mott–Hubbard insulator at some critical temperature where the long-range magnetic order is lost. Unfortunately, there are no systematic and controlled analytical methods to investigate the metal–insulator transition in lattice models in $d = 2, 3$ dimensions. The investigation of correlated electron systems in infinite dimensions ($d = \infty$) is one of the few available tools for studying this question in dimensions larger than one. We discuss the Hubbard model in infinite dimensions in the next chapter.

5. Hubbard Model in Infinite Dimensions

In this chapter we study the Hubbard model in infinite dimensions. The limit $d \to \infty$ is introduced in the first section. For models with localized spins it has been known for a long time that the results of mean-field approximations become exact in infinite dimensions. The intensive study of quantum-mechanical models with itinerant degrees of freedom in infinite dimensions has begun only recently. Hence, we expect further substantial progress in this direction in the future.

In infinite dimension the dispersion relation is a random variable over momentum space. In the second section we introduce the "random dispersion approximation" as a generalized construction principle for itinerant mean-field models. It is shown that Hubbard models with a random dispersion relation can be mapped onto an effective one-dimensional model.

In the third section we discuss the simplifications for Hubbard-type models that follow in infinite dimensions. It is shown that the proper self-energy is site-diagonal, i.e., independent of momentum for translationally invariant systems. This fact allows for the formally exact mapping of the infinite-dimensional Hubbard model onto a dynamic mean-field theory for itinerant electrons.

In the fourth section we present exact results that can be analytically obtained in infinite dimensions. For example, calculations in (self-consistent) perturbation theory are feasible, variational wave functions can be explicitly evaluated, and the Falicov–Kimball model as well as the problem of spinless fermions with nearest-neighbor interaction become exactly solvable.

In the fifth section we describe approximate methods for the solution of the Hubbard model in infinite dimensions. We address the possibilities and limitations of the quantum Monte-Carlo technique, exact diagonalization studies, iterated perturbation theory, the non-crossing approximation, and the local-moment approach.

In the last section we present and discuss results of the approximate methods for the metal–insulator transition in the infinite-dimensional Hubbard model. The phase diagram apparently supports Mott's general views on the metal–insulator transition in interacting electron systems as genuine correlation effect.

5.1 Limit of Infinite Dimensions

In the first subsection we introduce the limit of infinite dimensions for models of localized spins. In infinite dimensions the Curie–Weiß (or Bragg–Williams) mean-field theory for the Ising model becomes exact. In addition, the thermodynamics of the Heisenberg model is identical to that of the Ising model; quantum-mechanical fluctuations of the spins are suppressed and do not play any role in infinite dimensions. In the second subsection we define the large-d limit for itinerant electrons and analyze the properties of momentum space for infinite dimensions. The kinetic energy is found to be a randomly distributed variable over momentum space, and the notion of a "Fermi surface" looses its meaning.

Early introductions to infinite dimensions are given by Vollhardt [5.1] and Müller-Hartmann [5.2]; for recent reviews on the subject, see [5.3, 5.4].

5.1.1 Spin Models in Infinite Dimensions

The Hamilton operator of the Heisenberg model for quantum-mechanical spins reads

$$\hat{H}_{\text{Heis}} = \sum_{\boldsymbol{R}_1 \neq \boldsymbol{R}_2} J(\boldsymbol{R}_1 - \boldsymbol{R}_2) \hat{\boldsymbol{S}}_{\boldsymbol{R}_1} \cdot \hat{\boldsymbol{S}}_{\boldsymbol{R}_2} \tag{5.1}$$

(see Sect. 2.5). The spin operators $\hat{S}^z_{\boldsymbol{R}} = (\hat{n}_{\boldsymbol{R},\uparrow} - \hat{n}_{\boldsymbol{R},\downarrow})/2$, $\hat{S}^+_{\boldsymbol{R}} = \hat{S}^x_{\boldsymbol{R}} + \mathrm{i}\hat{S}^y_{\boldsymbol{R}} = \hat{c}^+_{\boldsymbol{R},\uparrow}\hat{c}_{\boldsymbol{R},\downarrow}$, and $\hat{S}^-_{\boldsymbol{R}} = \left(\hat{S}^+_{\boldsymbol{R}}\right)^+$ obey the spin algebra

$$\left[\hat{S}^z_{\boldsymbol{R}_i}, \hat{S}^\pm_{\boldsymbol{R}_j}\right]_- = \pm\delta_{\boldsymbol{R}_i,\boldsymbol{R}_j}\hat{S}^\pm_{\boldsymbol{R}_i} \, , \tag{5.2a}$$

$$\left[\hat{S}^+_{\boldsymbol{R}_i}, \hat{S}^-_{\boldsymbol{R}_j}\right]_- = 2\delta_{\boldsymbol{R}_i,\boldsymbol{R}_j}\hat{S}^z_{\boldsymbol{R}_i} \, . \tag{5.2b}$$

The Ising model as a classical model contains only the (commuting) z components of the Heisenberg model:

$$\hat{H}_{\text{Ising}} = \sum_{\boldsymbol{R}_1 \neq \boldsymbol{R}_2} J(\boldsymbol{R}_1 - \boldsymbol{R}_2) \hat{S}^z_{\boldsymbol{R}_1} \hat{S}^z_{\boldsymbol{R}_2} \, . \tag{5.3}$$

To keep the notation simple we further consider a nearest-neighbor interaction only,

$$J(\boldsymbol{R}_1 - \boldsymbol{R}_2) = J\delta_{\boldsymbol{R}_1,\boldsymbol{R}_2+\boldsymbol{\tau}} \, , \tag{5.4}$$

where $\boldsymbol{\tau}$ denotes the vectors to the Z nearest-neighbor sites.

When Z becomes large we have to scale properly the exchange coupling J such that, e.g., the energy of the fully ferromagnetically polarized state for $J < 0$, or of the Néel state for $J > 0$, remains finite in the limit $Z \to \infty$. Since each neighbor contributes an interaction energy J we obviously have to scale

$$J = \frac{\widetilde{J}}{Z} \tag{5.5}$$

with a new coupling $\widetilde{J}$ which is independent of Z. In the limit of infinite dimensions the number of nearest-neighbor sites on a d-dimensional cubic lattice diverges like $Z = 2d$ such that we should set $J = \widetilde{J}/(2d)$ in this case. Furthermore, on bipartite lattices we may redefine the spins as $\hat{F}^z(\boldsymbol{R}) = \hat{S}^z(\boldsymbol{R})\exp(\mathrm{i}\boldsymbol{QR})$ where $\boldsymbol{Q}$ is the nesting vector. Then, the ferromagnetic Ising model in terms of the spins $\hat{S}^z$ becomes the antiferromagnetic Ising model in terms of the spins $\hat{F}^z$. This shows that the ferromagnetic Ising model ($\widetilde{J} < 0$) and the antiferromagnetic Ising model ($\widetilde{J} > 0$) are equivalent to each other on bipartite lattices, and the results for thermodynamic quantities do not depend on the sign of $\widetilde{J}$.

Brout [5.5] and later Thompson [5.6] showed that the exact free energy of the Ising model, (5.3), with nearest-neighbor interaction (5.4) and scaling according to (5.5) is given by the mean-field expression of Curie–Weiß theory. For $Z \to \infty$ we may thus replace

$$\hat{H}_{\text{Ising}} \mapsto H^{\text{MF}} = \frac{\widetilde{J}}{Z}\sum_{\boldsymbol{R},\boldsymbol{T}}\left[\hat{S}^z_{\boldsymbol{R}}\langle\hat{S}^z_{\boldsymbol{R}+\boldsymbol{T}}\rangle + \langle\hat{S}^z_{\boldsymbol{R}}\rangle\hat{S}^z_{\boldsymbol{R}+\boldsymbol{T}} - \langle\hat{S}^z_{\boldsymbol{R}}\rangle\langle\hat{S}^z_{\boldsymbol{R}+\boldsymbol{T}}\rangle\right] \tag{5.6a}$$

$$= 2\widetilde{J}\sum_{\boldsymbol{R}} S^z_{\boldsymbol{R}} h_{\boldsymbol{R}} - \widetilde{J}\sum_{\boldsymbol{R}}\langle S^z_{\boldsymbol{R}}\rangle h_{\boldsymbol{R}} \tag{5.6b}$$

with the effective local Curie–Weiß field (spontaneous magnetization)

$$h_{\boldsymbol{R}} = \frac{1}{Z}\sum_{\boldsymbol{T}}\langle\hat{S}^z_{\boldsymbol{R}+\boldsymbol{T}}\rangle\,, \tag{5.6c}$$

where $\langle\ldots\rangle$ denotes the thermodynamic average. The summation over the large number of nearest neighbors thus corresponds to taking the thermodynamic average. For the Ising model the (Hartree) decoupling

$$\hat{A}\hat{B} \mapsto \hat{A}\langle\hat{B}\rangle + \langle\hat{A}\rangle\hat{B} - \langle\hat{A}\rangle\langle\hat{B}\rangle \tag{5.7}$$

becomes exact in the limit $Z \to \infty$. The mean-field model of (5.6) describes single spins in the Curie–Weiß field which is site-independent in the ferromagnetic case, $\langle\hat{S}^z_{\boldsymbol{R}}\rangle \equiv h_{\boldsymbol{R}} \equiv h$. In the antiferromagnetic case and for bipartite lattices the sign of the Curie–Weiß field alternates between the A and B lattice sites, $\langle\hat{S}^z_{\boldsymbol{R}}\rangle \equiv h_{\boldsymbol{R}} \equiv h\exp(\mathrm{i}\boldsymbol{QR})$.

In the ground state we obtain the completely ferromagnetically polarized state for $\widetilde{J} < 0$ or the Néel state with alternating ↑ and ↓ spins on the A and B lattice sites for $\widetilde{J} > 0$. The theory describes a second-order thermodynamic phase transition from the (anti-)ferromagnetic state into the paramagnetic state at the critical Curie–Weiß temperature, T_{C}, and Néel temperature, T_{N}, respectively. In natural units the spontaneous magnetization follows from the relation [5.6, 5.7][1]

$$2h(T) = \tanh\left[|\widetilde{J}|h(T)/T\right] . \tag{5.8a}$$

It is non-zero if $T < T_\mathrm{C} = |\widetilde{J}|/2$. The free energy density is given by [5.6, 5.7]

$$f(T, h(T)) = -\frac{T}{2}\ln\left(\frac{4}{1-4h^2}\right) + |\widetilde{J}|h^2 . \tag{5.8b}$$

In particular, we find $f(T > T_\mathrm{C,N}, h = 0) = -T\ln 2$ above the Curie and Néel temperature. The interaction between the spins becomes totally irrelevant above the transition temperature. In contrast to the exact solution in one dimension the mean-field model contains a phase transition which indeed occurs for the Ising model in all dimensions $d \geq 2$. Although the values of the critical exponents provided by the mean-field model are quantitatively unsatisfactory in finite dimensions [5.7], the qualitative behavior (prediction of a phase transition) is correctly reproduced by the large-d approach.

As already mentioned in [5.5] the free energy of the Heisenberg model reduces to that of the Ising model for $Z \to \infty$. Thus, its free energy density is also given by (5.8b). We can easily understand the irrelevance of the spin-flip terms in the Heisenberg model if we study the spin operators on nearest-neighbor shells,

$$\hat{h}_{\boldsymbol{R}}^{x,y,z} = \frac{1}{Z}\sum_{\boldsymbol{T}} \hat{S}_{\boldsymbol{R}+\boldsymbol{T}}^{x,y,z} . \tag{5.9}$$

When we calculate the commutators between two of these operators one of the nearest-neighbor sums involved cancels, and we find

$$[\hat{h}_{\boldsymbol{R}}^{\alpha}, \hat{h}_{\boldsymbol{R}'}^{\beta}]_- = \mathcal{O}(1/Z) . \tag{5.10}$$

This implies that the $\hat{h}_{\boldsymbol{R}}^{\alpha}$ commute with each other in the limit $Z \to \infty$. Since they also commute with the Hamilton operator to lowest order in $1/Z$ they are conserved quantities which may be replaced by their expectation values. After we have chosen the axis of quantization, only the z component remains and we again reach at (5.6).

The equivalence of the free energy for the Ising and the Heisenberg model in infinite dimensions already indicates that spatial quantum-mechanical fluctuations are suppressed in the limit $Z \to \infty$. In this way the Néel state becomes an exact ground state of the antiferromagnetic Heisenberg model in infinite dimensions and this state is considerably changed by spin fluctuations for all finite dimensions; see, e.g., [5.8].

---

[1] In statistical mechanics the length of the Ising spins is unity. The results for spin $S = 1/2$ are obtained by the scaling $h_\mathrm{SM} \to 2h$, $J_\mathrm{SM} \to J/4$. Note that often the interaction is counted per nearest-neighbor *pair*.

5.1.2 Non-Interacting Itinerant Electrons

For the Hubbard model we also have to scale our parameters to obtain a sensible model in the limit $Z \to \infty$. It is obvious that the purely local Hubbard interaction U remains unchanged and the nearest-neighbor tunnel amplitude t has to be rescaled. We again use the connection between the Heisenberg model and the Hubbard model for strong interactions and half band-filling. Since $J = 2t^2/U$ holds independent of dimension [see (2.43), Sect. 2.5] we immediately find from (5.5) that

$$t = \frac{\widetilde{t}}{\sqrt{Z}}, \tag{5.11}$$

with a Z-independent hopping amplitude $\widetilde{t}$. This scaling was first formulated for the Hubbard model by Zaitsev and Dushenat [5.9]. A more systematic investigation of the simplifications of itinerant electron models in infinite dimensions started with the work of Metzner and Vollhardt [5.10] and Müller-Hartmann [5.11].

To be specific we address non-interacting electrons on a d-dimensional cubic lattice ($Z = 2d$) to exemplify the consequences of the large-d limit on itinerant electrons. Their dispersion relation for nearest-neighbor hopping reads

$$\epsilon(\boldsymbol{k}) = -\frac{2\widetilde{t}}{\sqrt{2d}} \sum_{l=1}^{d} \cos(k_l) . \tag{5.12}$$

The dispersion relation fulfills the perfect nesting condition $\epsilon(\boldsymbol{k}+\boldsymbol{Q}) = -\epsilon(\boldsymbol{k})$ with the nesting vector $\boldsymbol{Q} = (\pi, \pi, \ldots, \pi)$. The probability density for the event that $\epsilon(\boldsymbol{k})$ takes a certain value ϵ in momentum space is given by the one-electron density of states for non-interacting electrons,

$$D_{\sigma,0}(\epsilon) = \frac{1}{L} \sum_{\boldsymbol{k}} \delta\left(\epsilon - \epsilon(\boldsymbol{k})\right) . \tag{5.13}$$

Each cosine term in (5.12) is bounded between the values -1 and $+1$ and the sum in the dispersion relation $\epsilon(\boldsymbol{k})$ runs over d independent terms. Consequently, $\epsilon(\boldsymbol{k})$ becomes a random variable in the limit $d \to \infty$. The density of states is Gauss distributed around zero with a standard deviation $\widetilde{t}$ [5.10],

$$D_{\sigma,0}(\epsilon) = \frac{1}{\sqrt{2\pi}\widetilde{t}} \exp\left(-\frac{(\epsilon/\widetilde{t})^2}{2}\right) . \tag{5.14}$$

The energy variable in the density of states is unbounded. The reason for this behavior lies in the kinetic energy of single states with special $\boldsymbol{k}$ vectors, e.g., $\boldsymbol{k} = \boldsymbol{0} = (0, 0, \ldots, 0)$, whose kinetic energy diverges proportionally to $\sqrt{d}$. In infinite dimensions, $d = \infty$, and for an exponentially small density, $n \to 0$, the kinetic energy per particle is infinite. Consequently, the limit of small densities, $n \to 0$, does not commute with the limit of infinite dimensions, $d \to \infty$.

We may define a bandwidth W even for a density of states with an unbounded energy variable. We use the second moment of the distribution and define

$$(W/4)^2 = \int \mathrm{d}\epsilon D_{\sigma,0}(\epsilon)\epsilon^2 \ . \tag{5.15}$$

For the Gauss distribution we obtain $W = 4\tilde{t}$. For the simple cubic lattice in infinite dimensions the bandwidth and the average kinetic energy of the Fermi gas at finite densities are always finite quantities, despite the unbounded density of states. The singular points in momentum space with diverging kinetic energy are of measure zero and thus have no influence on extensive quantities that are averages over momentum space. For the Bethe lattice with infinite coordination number $Z \to \infty$ the density of states is given by Hubbard's semi-ellipse (3.43). Its finite bandwidth W agrees with the definition in (5.15).

To avoid pathological effects of the exponential tails in the density of states it is desirable to perform calculations with a bounded density of states, e.g., Hubbard's semi-ellipse. In addition, van-Hove singularities, which are typical and important for finite dimensions, are absent in the large-d density of states. To include the effects of van-Hove singularities one should work with the proper d-dimensional density of states. This has the further advantage that corrections, e.g., to the ground-state energy density, are of algebraic order in $1/d$.

We may directly calculate the density of states and further properties of momentum space with the help of Fourier transformation [5.11, 5.12]. A particularly interesting quantity is the correlation function between the kinetic energies of two momenta $\boldsymbol{k}$ and $\boldsymbol{k}+\boldsymbol{q}$,

$$D_{\boldsymbol{q}}(\epsilon_1, \epsilon_2) = \frac{1}{L}\sum_{\boldsymbol{k}} \delta\left(\epsilon_1 - \epsilon(\boldsymbol{k}+\boldsymbol{q})\right) \delta\left(\epsilon_2 - \epsilon(\boldsymbol{k})\right) \ . \tag{5.16}$$

It gives the probability for the event that the kinetic energy changes on average from ϵ_1 to ϵ_2 due to a scattering with momentum transfer $\boldsymbol{q}$. In all finite dimensions we expect that small momentum transfers result in small energy changes. In contrast, in the limit $d \to \infty$ we obtain [5.11, 5.12]

$$D_{\boldsymbol{q}}(\epsilon_1, \epsilon_2) = \frac{\exp\left(-\dfrac{(\epsilon_1+\epsilon_2)^2}{4\tilde{t}^2(1+\eta(\boldsymbol{q}))}\right)}{\tilde{t}\sqrt{2\pi(1+\eta(\boldsymbol{q}))}} \frac{\exp\left(-\dfrac{(\epsilon_1-\epsilon_2)^2}{4\tilde{t}^2(1-\eta(\boldsymbol{q}))}\right)}{\tilde{t}\sqrt{2\pi(1-\eta(\boldsymbol{q}))}} \tag{5.17a}$$

with the function

$$\eta(\boldsymbol{q}) = \frac{1}{d}\sum_{l=1}^{d} \cos(q_l) = -\frac{\epsilon(\boldsymbol{q})}{\tilde{t}\sqrt{2d}} \ . \tag{5.17b}$$

For $\boldsymbol{q} = \boldsymbol{0}$ and $\boldsymbol{q} = \boldsymbol{Q}$ there is a perfect correlation between the energies ϵ_1 and ϵ_2, $D_{\boldsymbol{0}}(\epsilon_1, \epsilon_2) = D_{\sigma,0}(\epsilon_1)\delta(\epsilon_1 - \epsilon_2)$ and $D_{\boldsymbol{Q}}(\epsilon_1, \epsilon_2) = D_{\sigma,0}(\epsilon_1)\delta(\epsilon_1 + \epsilon_2)$, respectively. Furthermore, on the simple cubic lattice, some correlations remain in the nesting direction, $\boldsymbol{q} = \lambda \boldsymbol{Q}$ ($|\lambda| \leq 1$). For all other $\boldsymbol{q}$ with $|\boldsymbol{q} - \lambda \boldsymbol{Q}| > \mathcal{O}(\sqrt{1/d})$, however, the two energies are completely *uncorrelated*,

$$D_{\boldsymbol{q}}(\epsilon_1, \epsilon_2) = D_{\sigma,0}(\epsilon_1) D_{\sigma,0}(\epsilon_2) \qquad \text{for "almost all" } \boldsymbol{q}, \tag{5.18}$$

since for those $\boldsymbol{q}$ we have $\eta(\boldsymbol{q}) = \mathcal{O}(\sqrt{1/d}) \to 0$ in $d \to \infty$ dimensions.

In all finite dimensions small momentum transfers result in small changes in the kinetic energy since the dispersion relation is continuous. In infinite dimensions, however, the kinetic energies are essentially *randomly distributed* up to few constraints like the perfect nesting property. Since the dispersion relation $\epsilon(\boldsymbol{k})$ is a random function of the momenta $\boldsymbol{k}$, notions like "the Fermi surface" as a connected object that has an "inside" and an "outside" become totally meaningless in infinite dimensions. In this limit a $(d-1)$-dimensional "surface" in momentum space cannot have anything but a "random" topology. Only the Fermi energy E_{F} can still be used to separate occupied from unoccupied states for non-interacting electrons.

In the next section we discuss a broader class of models that all display the same mean-field properties as the Ising, Heisenberg, and Hubbard model in infinite dimensions. We identify the randomness of the dispersion relation as essential ingredient for the formulation of the mean-field approach to itinerant electron systems.

5.2 Generalized Mean-Field Approaches

In this section we present an alternative approach to mean-field theories for Ising and Heisenberg spin models. All models reduce to the Curie–Weiß theory if the Fourier transform of the exchange interaction is non-zero only for special momenta. This behavior can be obtained from Hubbard models at half band-filling and large interactions if we assume that the dispersion relation is a random quantity over momentum space. This "random dispersion approximation" generalizes the concept of infinite dimensions and provides new insights into the large-d limit for itinerant electron systems. In addition, it allows for a mapping of the problem onto a one-dimensional effective Hamiltonian.

5.2.1 Mean-Field Spin Models

Mean-field theories for spin systems, like the Ising model, (5.3), become exact not just in the limit of a large coordination number. For example [5.8], we may study a bipartite lattice and choose the exchange coupling to be the same between any pair of an A and a B site. The antiferromagnetic Ising model with infinite-range exchange thus reads

$$J(\boldsymbol{R}_1-\boldsymbol{R}_2)=\frac{J}{L}\left(1-\mathrm{e}^{\mathrm{i}\boldsymbol{Q}(\boldsymbol{R}_1-\boldsymbol{R}_2)}\right)\quad (J>0)\;. \tag{5.19}$$

For this case we also find that the Curie–Weiß mean-field theory becomes exact. The antiferromagnetic Ising model with infinite-range interactions may be written as

$$\hat{H}^{\mathrm{IR}}_{\mathrm{Ising}} = J\left(\hat{S}^z\hat{S}^z-\hat{S}^z(\boldsymbol{Q})\hat{S}^z(-\boldsymbol{Q})\right)\;, \tag{5.20a}$$

$$\hat{S}^z = \sqrt{\frac{1}{L}}\sum_{\boldsymbol{R}}\hat{S}^z_{\boldsymbol{R}}\;, \tag{5.20b}$$

$$\hat{S}^z(\boldsymbol{Q}) = \sqrt{\frac{1}{L}}\sum_{\boldsymbol{R}}\mathrm{e}^{-\mathrm{i}\boldsymbol{Q}\boldsymbol{R}}\hat{S}^z_{\boldsymbol{R}}\;. \tag{5.20c}$$

The method of steepest descent [5.7, Chap. 3] applied to the partition function to $\hat{H}^{\mathrm{IR}}_{\mathrm{Ising}}$ gives the mean-field equation (5.8a) for the spontaneous magnetization, and Stirling's formula allows us to re-derive the mean-field free energy density (5.8b). Hence, the Ising model with nearest-neighbor exchange in infinite dimensions gives the same result as the Ising model with infinite-range interactions.

The reason for this behavior is readily understood when we consider the Ising model in momentum space. The Hamiltonian reads

$$\hat{H}_{\mathrm{Ising}}=\sum_{\boldsymbol{q}}J(\boldsymbol{q})\hat{S}^z(\boldsymbol{q})\hat{S}^z(-\boldsymbol{q})\;. \tag{5.21a}$$

The exchange coupling between the spin operators,

$$\hat{S}^z(\boldsymbol{q})=\sqrt{\frac{1}{L}}\sum_{\boldsymbol{R}}\mathrm{e}^{-\mathrm{i}\boldsymbol{q}\boldsymbol{R}}\hat{S}^z(\boldsymbol{R})\;, \tag{5.21b}$$

is given by the Fourier transform of $J(\boldsymbol{R})$ as

$$J(\boldsymbol{q})=\sum_{\boldsymbol{R}}J(\boldsymbol{R})\mathrm{e}^{\mathrm{i}\boldsymbol{q}\boldsymbol{R}}\;. \tag{5.21c}$$

For a nearest-neighbor interaction on a d-dimensional simple cubic lattice we find

$$J(\boldsymbol{q})=\frac{2\widetilde{J}}{Z}\sum_{l=1}^{d}\cos(q_l)=\widetilde{J}\eta(\boldsymbol{q}) \tag{5.22}$$

[see (5.17b)]. In infinite dimensions, $J(\boldsymbol{q})$ vanishes for almost all $\boldsymbol{q}$, i.e., the spin coupling is non-zero only for a few vectors in momentum space. In addition, $J(\boldsymbol{0})$ and $J(\boldsymbol{Q})$ are the only relevant couplings for the thermodynamics because the free energy density was found to be the same for the infinite-range and the infinite-dimensional Ising model. The reason for the irrelevance of the couplings in the nesting direction can be understood within linear spin-wave theory for the infinite-dimensional Heisenberg model. This approximation becomes exact to order $\mathcal{O}(1/d)$ [5.13] and gives the dispersion relation

$$\omega(\boldsymbol{q}) = |\widetilde{J}|\sqrt{1-\eta^2(\boldsymbol{q})} \, . \tag{5.23}$$

This reduces to $\omega(\boldsymbol{q}) = |\widetilde{J}|$ for almost all $\boldsymbol{q}$. The spin waves along the $\boldsymbol{Q}$ direction display a finite dispersion, $\omega(\lambda\boldsymbol{Q}) = |\widetilde{J}\sin(\lambda\pi)|$, but a single direction is irrelevant for the thermodynamics as long as there is no instability along that line. The question of incommensurate order becomes relevant if we study possible symmetry breakings away from half band-filling; see Sect. 5.4.5.

The Ising model in infinite dimensions is seen to reduce to the mean-field model in (5.20). Hence, the main criterion for the applicability of mean-field theory is the reduction of the interaction in momentum space to a few relevant spin couplings such that the dynamics is governed by a few modes only.

The above infinite-range spin model is readily generalized to the Hubbard model with infinite-range hopping. This model is trivial in the sense that the kinetic and potential energy operators commute in the thermodynamic limit; see Sect. 6.3. Consequently, we have to find a more generic construction principle that guarantees that the coupling $J(\boldsymbol{q})$ is non-zero only for a few values of $\boldsymbol{q}$ if we want to extend the approach to itinerant electrons.

A close look at (5.21c) shows that we do not have to demand that the coupling $J(\boldsymbol{R})$ is the same for all lattice distances. Instead, we can recover the antiferromagnetic mean-field solution if we demand that $J(\boldsymbol{R})$ is a *random variable*. If we assume that there is no coupling between sites on the same sublattice, $J(\boldsymbol{R}) = J(\boldsymbol{R})(1-\exp(\mathrm{i}\boldsymbol{Q}\boldsymbol{R}))/2$, and the values $J(\boldsymbol{R}) = \mathcal{O}(1/L)$ are randomly distributed around their mean $J/L = (1/L)\sum_{\boldsymbol{R}} J(\boldsymbol{R})$, we also recover the mean-field model (5.20) in *the thermodynamic limit*. The couplings $J(\boldsymbol{q})$ for $\boldsymbol{q} \neq \boldsymbol{0}, \boldsymbol{Q}$ are small, of the order of $\mathcal{O}(\sqrt{1/L})$, since in (5.21c) we sum over L independent random variables of size $\mathcal{O}(1/L)$.

Before we carry on with our considerations we emphasize that we do *not* address spin systems with random couplings $J(\boldsymbol{R}_i, \boldsymbol{R}_j)$, or itinerant electron systems with hopping amplitudes $t(\boldsymbol{R}_i, \boldsymbol{R}_j)$, that individually depend on both lattice sites. The corresponding spin models are used for the description of spin glasses [5.8], and the infinite-dimensional Hubbard model with random hopping amplitudes is studied in [5.14]. The random dispersion models we will specify below are still translationally invariant.

5.2.2 Random Dispersion Approximation

Again, we use the link between the antiferromagnetic Heisenberg model and the Hubbard model at half band-filling and strong coupling, (2.43). The Fourier transformation of the Heisenberg coupling (2.44) into momentum space gives

$$\frac{U}{2}J(\boldsymbol{q}) = \frac{1}{L}\sum_{\boldsymbol{k}} \epsilon(\boldsymbol{k})\epsilon(\boldsymbol{k}+\boldsymbol{q}) = \int_{-\infty}^{\infty} \mathrm{d}\epsilon_1\, \epsilon_1 \int_{-\infty}^{\infty} \mathrm{d}\epsilon_2\, \epsilon_2\, D_{\boldsymbol{q}}(\epsilon_1, \epsilon_2) \, , \tag{5.24}$$

where we used (5.16) for the definition of the energy correlation function between the kinetic energies at the momenta $\boldsymbol{k}$ and $\boldsymbol{k}+\boldsymbol{q}$. In principle, itinerant electron models should be reducible to mean-field theories if their dispersion relation is uncorrelated in the sense of (5.18). For these dispersion relations we have

$$D_{\boldsymbol{q}}(\epsilon_1,\epsilon_2) = D_{\sigma,0}(\epsilon_1)D_{\sigma,0}(\epsilon_2) \qquad \text{for “almost all” } \boldsymbol{q}. \tag{5.25}$$

Thus, $J(\boldsymbol{q}) = 0$ for almost all $\boldsymbol{q}$ since $\int d\epsilon\, \epsilon\, D_{\sigma,0}(\epsilon) = 0$. Consequently, if (5.25) holds, the corresponding spin model is solved by its mean-field free energy density. As we already saw in Sect. 5.1.2, models for itinerant electrons in infinite dimensions obey this criterion. Hence, equation (5.25) can be viewed the basic condition which allows for the exact mapping of an itinerant electron model onto a mean-field theory.

These considerations show that the limit of infinite dimensions is a special example for systems with a random dispersion relation. They are defined as follows.

We randomly assign the kinetic energy $\epsilon(\boldsymbol{k})$ to each $\boldsymbol{k}$ point in the Brillouin zone where the probability $P(\epsilon)$ of finding the energy $\epsilon(\boldsymbol{k})$ in the small but finite interval $\Gamma(\epsilon) = [\epsilon - (\delta\epsilon)/2, \epsilon + (\delta\epsilon)/2[$ is given by the density of states $D_{\sigma,0}(\epsilon)$ as $P(\epsilon) = D_{\sigma,0}(\epsilon)(\delta\epsilon)$. The energy correlation function $D_{\boldsymbol{q}}(\epsilon_1,\epsilon_2)$ vanishes for all but a finite number of momenta $\boldsymbol{q}$.

Since we are interested in the half-filled case where the spin order is commensurate, we do not include correlations in certain directions as occurred on the infinite-dimensional simple cubic lattice. Later in this chapter we will again distinguish between Mott–Hubbard and Mott–Heisenberg insulators. Hence, we separately consider dispersion relations with and without the perfect nesting property.

If the dispersion relation fulfills the perfect nesting property, we set

$$\begin{aligned} D_{\boldsymbol{q}}(\epsilon_1,\epsilon_2) &= D_{\sigma,0}(\epsilon_1)\delta(\epsilon_1-\epsilon_2)\delta_{\boldsymbol{q},\boldsymbol{0}} + D_{\sigma,0}(\epsilon_1)\delta(\epsilon_1+\epsilon_2)\delta_{\boldsymbol{q},\boldsymbol{Q}} \\ &\quad +(1-\delta_{\boldsymbol{q},\boldsymbol{0}})(1-\delta_{\boldsymbol{q},\boldsymbol{Q}})D_{\sigma,0}(\epsilon_1)D_{\sigma,0}(\epsilon_2)\,. \end{aligned} \tag{5.26}$$

In this case the half-filled Hubbard model at large interactions reduces to the antiferromagnetic mean-field Heisenberg model,

$$\hat{H}_{\text{Heis}}^{\text{AF-MF}} = J\left(\hat{\boldsymbol{S}}\cdot\hat{\boldsymbol{S}} - \hat{\boldsymbol{S}}(\boldsymbol{Q})\cdot\hat{\boldsymbol{S}}(-\boldsymbol{Q}) - \frac{L}{4}\right), \tag{5.27a}$$

$$\hat{\boldsymbol{S}} = \sqrt{\frac{1}{L}}\sum_{\boldsymbol{R}}\hat{\boldsymbol{S}}_{\boldsymbol{R}}, \tag{5.27b}$$

$$\hat{\boldsymbol{S}}(\boldsymbol{Q}) = \sqrt{\frac{1}{L}}\sum_{\boldsymbol{R}}\mathrm{e}^{-\mathrm{i}\boldsymbol{Q}\boldsymbol{R}}\hat{\boldsymbol{S}}_{\boldsymbol{R}}, \tag{5.27c}$$

$$J = J_0 = \frac{2}{U}\int_{-\infty}^{\infty}\mathrm{d}\epsilon\,\epsilon^2 D_{\sigma,0}(\epsilon) = \frac{W^2}{8U} > 0\,. \tag{5.27d}$$

Again, this mean-field model gives the exact results (5.8) for the antiferromagnetic Heisenberg model in the limit of infinite dimensions [5.8], and the Néel state is its ground state.

If the dispersion relation does not have any particular symmetries, the energy correlation functions becomes

$$D_{\boldsymbol{q}}(\epsilon_1, \epsilon_2) = D_{\sigma,0}(\epsilon_1)\delta(\epsilon_1 - \epsilon_2)\delta_{\boldsymbol{q},\boldsymbol{0}} + (1 - \delta_{\boldsymbol{q},0})D_{\sigma,0}(\epsilon_1)D_{\sigma,0}(\epsilon_2) \, . \quad (5.28)$$

These models are sometimes called "fully-frustrated" Hubbard models [5.4]. For these types of dispersion relations the half-filled Hubbard model at large interactions reduces to the *ferromagnetic* mean-field Heisenberg model with an *antiferromagnetic* exchange interaction,

$$\hat{H}_{\text{Heis}}^{\text{F-MF}} = J\left(\hat{\boldsymbol{S}} \cdot \hat{\boldsymbol{S}} - \frac{L}{4}\right) \, . \quad (5.29)$$

The ground state of the fully-frustrated Hubbard models is vastly degenerate since all overall singlets have the same energy, $E_0 = -JL/4$. In the thermodynamic limit, the ground-state entropy density is $S_0/(k_{\text{B}}L) = \ln(2) - \mathcal{O}(\ln(L)/L)$ since almost all states with $S^z = 0$ also have total spin zero. This degeneracy would be lifted if we kept the interactions $J(\boldsymbol{q})$ for $\boldsymbol{q} \neq 0$. The ground state would be unique, and the third law of thermodynamics would hold. The example of the half-filled "fully-frustrated" Hubbard model at strong coupling shows that the limits of infinite dimensions and zero temperature do not commute, in general.

Within the random dispersion approximation the properties of momentum space are all contained in the dispersion relation $\epsilon(\boldsymbol{k})$ through its density of states and its energy correlation function (5.16). Thus, the Hubbard model with a random dispersion in any dimension can be mapped onto the *one-dimensional* model

$$\hat{H}_{\text{RDA}} = \sum_{|k| \leq \pi, \sigma} \epsilon_{\text{RDA}}(k)\hat{n}_{k,\sigma} + U\sum_{l} \hat{n}_{l,\uparrow}\hat{n}_{l,\downarrow} \, , \quad (5.30)$$

where $\epsilon_{\text{RDA}}(k)$ is randomly distributed over the momenta in the first Brillouin zone with probability density $D_{\sigma,0}(\epsilon)$.

We may also work with a continuous dispersion relation since we may choose a permutation $\mathcal{Q}$ of the L momenta in the first Brillouin zone such that

$$\epsilon_{\text{RDA}}(k) = \epsilon(\mathcal{Q}^{-1}(k)) \, , \quad (5.31)$$

where, e.g., $\epsilon(k) = tk$ for a constant density of states, or $\epsilon(k) = -2t\cos(k)$ for the one-dimensional density of states. Now we define new fermion operators via

$$\hat{c}_{k,\sigma} = \hat{d}_{\mathcal{Q}^{-1}(k),\sigma} \, , \quad (5.32a)$$

$$\hat{c}_{l,\sigma} = \sqrt{\frac{1}{L}}\sum_{k} \mathrm{e}^{\mathrm{i}\mathcal{Q}(k)l}\hat{d}_{k,\sigma} \, , \quad (5.32b)$$

which allows us to formulate the problem as

$$\hat{H}^{\mathcal{Q}} = \sum_{|k|\leq\pi,\sigma} \epsilon(k)\hat{d}^{+}_{k,\sigma}\hat{d}_{k,\sigma} + \frac{1}{L^2}\sum_{k_1,k_2,k_3,k_4} U^{\mathcal{Q}}(k_1,k_2,k_3,k_4)\hat{d}^{+}_{k_1,\uparrow}\hat{d}_{k_2,\uparrow}\hat{d}^{+}_{k_3,\downarrow}\hat{d}_{k_4,\downarrow} ,$$

$$U^{\mathcal{Q}}(k_1,k_2,k_3,k_4) = U\sum_{l}\exp\left[\mathrm{i}l\left(\mathcal{Q}(k_1)-\mathcal{Q}(k_2)+\mathcal{Q}(k_3)-\mathcal{Q}(k_4)\right)\right] . \tag{5.33}$$

In this way we can work with a continuous dispersion relation at the expense of the scattering potential $U^{\mathcal{Q}}(k_1,k_2,k_3,k_4)$.

In principle, we should calculate the Green function for each permutation $\mathcal{Q}$, and then take the average over all permutations. However, for large enough systems, and permutations $\mathcal{Q}$ that are "random enough", the Green function should be self-averaging. Nevertheless, it is obvious that we must *not* calculate with the *average* scattering potential $\langle U^{\mathcal{Q}}(k_1,k_2,k_3,k_4)\rangle_{\mathcal{Q}} = U$ because the average has to be done for the Green function. With this additional approximation, the model (5.30) would reduce to the Wolff model in which the Hubbard interaction is restricted to a single site. However, the Green function for the Wolff model and the Hubbard model with random dispersion relation are different from each other.

Unfortunately, an analytical solution of the model (5.33) has not been found yet but this representation might be useful for numerical approaches. Furthermore, the random dispersion approximation to itinerant electron systems provides some new insights into the physical meaning of the large-d limit. First, the lack of correlations of the kinetic energy distribution in momentum space is the decisive criterion for a mean-field theory. It also allows us to choose freely the density of states; this generalizes the concept of infinite dimensions to arbitrary lattice types. Second, there are real physical situations in which the momentum $\boldsymbol{k}$ of an electron relaxes on a much shorter time scale than its kinetic energy $\epsilon(\boldsymbol{k})$, e.g., in the presence of strong elastic scattering by impurities. This is the physical essence of the scattering potential $U^{\mathcal{Q}}(k_1,k_2,k_3,k_4)$ which smears out the Hubbard interaction in position space. Hence, there are real physical situations for which the random dispersion approximation will work fairly well.

For non-interacting electrons the structure of momentum space is irrelevant for the thermodynamics of the system. The dispersion relation $\epsilon(\boldsymbol{k})$ influences the free energy only via the density of states $D_{\sigma,0}(\epsilon)$, a quantity that is averaged over all $\boldsymbol{k}$ states. We thus expect that also the thermodynamics of interacting electrons is qualitatively correctly reproduced even if we approximate the true dispersion relation by a random one. Transport properties such as the electrical conductivity, however, might not be described properly by such a formalism. There, correlations in momentum space are of utmost importance. For example, the coherent effect of Anderson localization is completely absent in the random dispersion approximation to itinerant electron systems.

In the next section we will derive the by now standard reformulation of models in infinite dimensions or, more generally, with random dispersion relation, in terms of dynamic single-site models. Again, such models are analytically unsolvable in general, and approximate analytical and numerical methods have to be employed. In some special cases, however, exact results can be found, which we will present in Sect. 5.4. Based on the approximate techniques discussed in Sect. 5.5 we present results on the Mott transition in the infinite-dimensional Hubbard model in Sect. 5.6.

5.3 Simplifications in Infinite Dimensions

In this section we discuss the essential simplifications for interacting electron systems that occur in the limit of infinite dimensions. In the first subsection we show that the n-electron irreducible n-vertices are diagonal in position space. In translationally invariant systems this implies that the one-electron irreducible ("proper") self-energy and the two-electron irreducible ("proper") two-electron vertex are independent of momentum. The fact that the proper self-energy is purely local permits the construction of dynamic single-site models that are equivalent to the Hubbard model in infinite dimensions.

5.3.1 Position-Space Collapse of Diagrams

The discussion follows [5.1]; see also [5.10]–[5.12], [5.15]. We may calculate the average kinetic energy of interacting electrons with hopping amplitudes $t(\boldsymbol{R}_i - \boldsymbol{R}_j)$ from the zero-temperature Green-function formalism as [5.16]

$$E_{\text{kin}} = \sum_{\boldsymbol{R}_i,\boldsymbol{R}_j,\sigma} t(\boldsymbol{R}_i - \boldsymbol{R}_j) \int_{-\infty}^{\infty} \frac{\mathrm{d}\omega}{2\pi \mathrm{i}} G_\sigma(\boldsymbol{R}_j, \boldsymbol{R}_i; \omega) \mathrm{e}^{\mathrm{i}\omega\eta} , \tag{5.34a}$$

where $G_\sigma(\boldsymbol{R}_i, \boldsymbol{R}_j; \omega)$ denotes the one-electron Green function of the interacting system, and $\eta = 0^+$ is a positive infinitesimal. On an d-dimensional cubic lattice the distance of a lattice site from the origin of our coordinate system is given by the number of nearest-neighbor steps from the origin to that site ("New-York metric"), $||\boldsymbol{R}|| = \sum_{l=1}^{d} |R_l|$. In general, we have to scale the hopping amplitudes at distance $||\boldsymbol{R}_i - \boldsymbol{R}_j||$ as

$$t(\boldsymbol{R}_i - \boldsymbol{R}_j) = \tilde{t}(\boldsymbol{R}_i - \boldsymbol{R}_j) d^{-||\boldsymbol{R}_i - \boldsymbol{R}_j||/2} \tag{5.34b}$$

since the number of neighboring sites at this distance is of the order of $\mathcal{O}(d^{||\boldsymbol{R}_i - \boldsymbol{R}_j||})$. Because the average kinetic energy is finite in the limit $d \to \infty$ the one-electron Green function must decay as

$$G_\sigma(\boldsymbol{R}_i, \boldsymbol{R}_j; \omega) \sim d^{-||\boldsymbol{R}_i - \boldsymbol{R}_j||/2} \,. \tag{5.34c}$$

The Green function obviously rapidly decays with distance. In the framework of a diagrammatic expansion, however, we have to sum over all inner vertices such that the decay of the Green function is partly compensated. The theory does not become trivial in infinite dimensions but a complicated many-body problem remains.

In the following we assume that we may treat the Hubbard interaction perturbatively, and analyze the corresponding diagrammatic perturbation theory; see, e.g., [5.16], for technical details. This will allow us to identify the simplifications of the theory in infinite dimensions. Let us consider two inner vertices $\boldsymbol{R}_1$ and $\boldsymbol{R}_2$ ($\boldsymbol{R} = \boldsymbol{R}_1 - \boldsymbol{R}_2$) of a diagram which are connected by (at least) three *independent* paths. Paths between vertices are defined as consecutive Green-function lines. They are independent if they do not share a common Green-function line. The summation over the $R = ||\boldsymbol{R}||$th-neighbor shell then provides a summation factor of the order of $\mathcal{O}(d^R)$. However, the (at least) three Green functions suppress the total contribution of the Rth shell by a factor of the order of $\mathcal{O}(d^{-3R/2})$. In total, the contribution for $R \geq 1$ vanishes in $d \to \infty$ proportionally to $d^{-R/2}$ and only the term for $R = 0$ remains. Consequently, the two vertices $\boldsymbol{R}_1$ and $\boldsymbol{R}_2$ can be identified ("position space collapse of diagrams" [5.10]). This result becomes exact in infinite dimensions but has earlier been used as *ad hoc* assumption for an analytical evaluation of diagrams; see, e.g., [5.17].

The random dispersion approximation leads to the same simplifications. To see this we analyze the Green function of non-interacting electrons at zero temperature. If $\Theta(t)$ denotes the step function, the definition of the Green function directly leads to

$$\begin{aligned} \mathrm{i}G_{\sigma,0}(\boldsymbol{R}_1 - \boldsymbol{R}_2, t) &= \Theta(t) \int_{-\infty}^{E_\mathrm{F}} \mathrm{d}\epsilon \mathrm{e}^{-\mathrm{i}\epsilon t} g_\epsilon^0(\boldsymbol{R}_1 - \boldsymbol{R}_2) \\ &\quad - \Theta(-t) \int_{E_\mathrm{F}}^{\infty} \mathrm{d}\epsilon \mathrm{e}^{-\mathrm{i}\epsilon t} g_\epsilon^0(\boldsymbol{R}_1 - \boldsymbol{R}_2) \end{aligned} \tag{5.35a}$$

with the abbreviation

$$g_\epsilon^0(\boldsymbol{R}) = \frac{1}{L} \sum_{\boldsymbol{k}} \delta(\epsilon - \epsilon(\boldsymbol{k})) \mathrm{e}^{\mathrm{i}\boldsymbol{k}\boldsymbol{R}} \,. \tag{5.35b}$$

In high dimensions the function $g_\epsilon^0(\boldsymbol{R})$ is of the order of unity only for $\boldsymbol{R} = \boldsymbol{0}$ and else vanishes proportionally to $\mathcal{O}\left(d^{-R/2}\right)$. For a random dispersion relation we may write

$$g_\epsilon^0(\boldsymbol{R}) = \frac{1}{L} \frac{1}{(\delta\epsilon)} \sum_{\epsilon(\boldsymbol{k}) \in \Gamma(\epsilon)} \left[\cos(\boldsymbol{k}\boldsymbol{R}) + \mathrm{i} \sin(\boldsymbol{k}\boldsymbol{R})\right] \,, \tag{5.36}$$

where $\Gamma(\epsilon)$ is the energy shell $|\epsilon - \epsilon(\boldsymbol{k})| \leq (\delta\epsilon)/2$. For $\boldsymbol{R} \neq \boldsymbol{0}$ the momentum sums over the (bounded) sine and cosine functions in (5.36) have zero average and a standard deviation of the order of $\mathcal{O}\left(\sqrt{||\Gamma(\epsilon)||}\right)$ where $||\Gamma(\epsilon)|| = LD_{\sigma,0}(\epsilon)(\delta\epsilon)$ denotes the number of $\boldsymbol{k}$ points in $\Gamma(\epsilon)$. In sum we find

$$g_\epsilon^0(\boldsymbol{R}) = \begin{cases} \mathcal{O}(1) & \text{for} \quad \boldsymbol{R} = \boldsymbol{0} \\ \mathcal{O}\left(\sqrt{1/L}\right) & \text{for} \quad \boldsymbol{R} \neq \boldsymbol{0} \end{cases} \tag{5.37}$$

since $LD_{\sigma,0}(\epsilon)(\delta\epsilon) = \mathcal{O}(L)$ is an extensive quantity. Obviously, the thermodynamic limit, $L \to \infty$, now takes over the role of the limit of infinite dimensions, $d \to \infty$.

We emphasize that the Hubbard interaction does not become trivial in infinite dimensions since not all inner vertices in a diagrammatic theory are connected by three independent paths and, furthermore, there are no simplifications in the time/frequency domain, i.e., the problem remains a fully dynamic one. The Hartree decoupling of the on-site interaction does *not* become exact in the limit of infinite dimensions. This is different for nearest-neighbor density–density interactions. In the same way as for the Heisenberg model they have to be scaled as

$$\hat{V} = \frac{1}{Z} \sum_{\boldsymbol{R},\boldsymbol{T}} \hat{n}_{\boldsymbol{R}} \hat{n}_{\boldsymbol{R}+\boldsymbol{T}} \; . \tag{5.38a}$$

Due to the necessary scaling factor $1/Z$ all diagrams in which two vertices are connected by Green-function lines are reduced by an additional factor $1/Z$ [5.11]. Thus, only the Hartree diagram remains in $d = \infty$,

$$\hat{V} \equiv \hat{V}^{\mathrm{H}} = \frac{1}{Z} \sum_{\boldsymbol{R},\boldsymbol{T}} \left[\langle \hat{n}_{\boldsymbol{R}} \rangle \hat{n}_{\boldsymbol{R}+\boldsymbol{T}} + \hat{n}_{\boldsymbol{R}} \langle \hat{n}_{\boldsymbol{R}+\boldsymbol{T}} \rangle - \langle \hat{n}_{\boldsymbol{R}} \rangle \langle \hat{n}_{\boldsymbol{R}+\boldsymbol{T}} \rangle \right] \; . \tag{5.38b}$$

The Hartree decomposition becomes exact in infinite dimensions for all density–density interactions with the exception of the purely local Hubbard interaction. The Hubbard model thus appears as the "generic" model of interacting electrons in the limit of infinite dimensions.

The remaining task is the identification of those objects in a diagrammatic theory for the Hubbard model in which two vertices are always connected by at least three independent Green-function lines. This is the case for the vertices $\boldsymbol{R}_1$ and $\boldsymbol{R}_2$ for the graphs which belong to the one-electron irreducible ("proper") self-energy $\Sigma_\sigma(\boldsymbol{R}_1, \boldsymbol{R}_2; \omega)$. By definition [5.16], the proper self-energy is one-electron irreducible, i.e., a graph does not separate into two distinct pieces by cutting one Green-function line. Hence, there are three independent paths from $\boldsymbol{R}_1$ to $\boldsymbol{R}_2$ in $\Sigma_\sigma(\boldsymbol{R}_1, \boldsymbol{R}_2; \omega)$. Consequently,

$$\Sigma_\sigma(\boldsymbol{R}_1, \boldsymbol{R}_2; \omega) = \Sigma_\sigma(\boldsymbol{R}_1, \boldsymbol{R}_1; \omega) \delta_{\boldsymbol{R}_1, \boldsymbol{R}_2} + \mathcal{O}(\sqrt{1/d}) \quad \text{for} \quad d \to \infty \, . \tag{5.39}$$

When we calculate physical quantities – like the average kinetic energy – not the self-energy but rather the product $\Sigma_\sigma(\boldsymbol{R}_1, \boldsymbol{R}_2; \omega) \; G_\sigma(\boldsymbol{R}_1, \boldsymbol{R}_2; \omega)$ enters. Hence, corrections to physical quantities are of the order of $\mathcal{O}(1/d)$ instead of $\mathcal{O}(\sqrt{1/d})$ as in (5.39).

Although the proper self-energy is diagonal in the position-space indices this does *not* imply that the position-space indices of all vertices in the diagrams for $\Sigma_\sigma(\boldsymbol{R},\boldsymbol{R};\omega)$ are identical. In general, two inner vertices $\boldsymbol{R}_a$, $\boldsymbol{R}_b$ to this quantity are not linked by a Green-function line of the non-interacting system ("bare Green function", $G_{\sigma,0}(\boldsymbol{R}_a,\boldsymbol{R}_b;\omega)$ [5.16]) but rather by a Green-function line of the interacting system ("dressed Green function", $G_\sigma(\boldsymbol{R}_a,\boldsymbol{R}_b;\omega)$ [5.16]). The latter depends on the proper self-energy on all other lattice sites via the Dyson equation,

$$\begin{aligned} G_\sigma(\boldsymbol{R}_a,\boldsymbol{R}_b;\omega) &= G_{\sigma,0}(\boldsymbol{R}_a,\boldsymbol{R}_b;\omega) \\ &\quad + \sum_{\boldsymbol{R}} G_{\sigma,0}(\boldsymbol{R}_a,\boldsymbol{R};\omega)\Sigma_\sigma(\boldsymbol{R},\boldsymbol{R};\omega)G_\sigma(\boldsymbol{R},\boldsymbol{R}_b;\omega)\,. \end{aligned} \tag{5.40}$$

The "skeleton diagrams" [5.16] are those for which, by definition, all self-energy insertions on the Green-function lines have been removed. It is only this diagram class for which all lattice sites can be identified in infinite dimensions.

These considerations show that the infinite-dimensional Hubbard model does not simply reduce to the Wolff model [5.18], as we have already pointed out in the context of the random dispersion approximation in Sect. 5.2.2. In the Wolff model *all* vertices in the proper self-energy are identical, but this only holds true for the vertices of the skeleton diagrams in the infinite-dimensional Hubbard model. The obvious similarity between the Wolff model and the Hubbard model in infinite dimensions already indicates, however, that we may construct *effective* models with an interaction on a single site, which become equivalent to the Hubbard model in infinite dimensions. We discuss such a mapping in Sect. 5.3.3.

When we calculate n-electron Green functions we have to identify the corresponding n-electron irreducible vertex insertions which become site-diagonal in infinite dimensions. For example, for the two-electron correlation functions the two-electron irreducible ("proper") two-electron vertex also becomes purely local:

$$\begin{aligned} &\Gamma_{\sigma_1,\sigma_2|\sigma_3,\sigma_4}(\boldsymbol{R}_1,\omega_1;\boldsymbol{R}_2,\omega_2|\boldsymbol{R}_3,\omega_3;\boldsymbol{R}_4,\omega_4) \equiv \\ &\quad \Gamma_{\sigma_1,\sigma_2|\sigma_3,\sigma_4}(\boldsymbol{R}_1,\omega_1;\boldsymbol{R}_1,\omega_2|\boldsymbol{R}_1,\omega_3;\boldsymbol{R}_1,\omega_4)\delta_{\boldsymbol{R}_1,\boldsymbol{R}_2}\,\delta_{\boldsymbol{R}_1,\boldsymbol{R}_3}\,\delta_{\boldsymbol{R}_1,\boldsymbol{R}_4}\,. \end{aligned} \tag{5.41}$$

This fact considerably simplifies the calculation of the conductivity in infinite dimensions; see below.

5.3.2 Irrelevance of Momentum Conservation

If our system is translationally invariant, we may also interpret the simplifications in infinite dimensions as "irrelevance of momentum conservation" at (some) inner vertices. If four Green-function lines with momenta $\boldsymbol{k}_i$ $(i=1,\ldots,4)$ meet at a vertex both energy and momentum conservation hold (see, e.g., [5.16]). The momentum conservation on the lattice is expressed in terms of the Laue function,

$$\mathcal{L}(\boldsymbol{k}) = \sum_{\boldsymbol{R}} \exp(\mathrm{i}\boldsymbol{R}\boldsymbol{k}) \ . \tag{5.42}$$

Since the propagator in the non-interacting system depends on momentum only via the dispersion relation we have to consider the vertex function

$$v(\epsilon_1, \epsilon_2, \epsilon_3, \epsilon_4) = \sum_{\boldsymbol{k}_1,\boldsymbol{k}_2,\boldsymbol{k}_3,\boldsymbol{k}_4} \mathcal{L}(\boldsymbol{k}_1 - \boldsymbol{k}_2 + \boldsymbol{k}_3 - \boldsymbol{k}_4) \prod_{i=1}^{4} \delta\left(\epsilon_i - \epsilon(\boldsymbol{k}_i)\right) . \tag{5.43}$$

A direct calculation shows [5.11] that this vertex function factorizes into independent functions,

$$v(\epsilon_1, \epsilon_2, \epsilon_3, \epsilon_4) = \prod_{i=1}^{4} D_{\sigma,0}(\epsilon_i) \quad \text{in } d = \infty \ . \tag{5.44}$$

Only the term of $\boldsymbol{R} = \boldsymbol{0}$ contributes to the Laue function, $\mathcal{L}(\boldsymbol{k}) \equiv 1$. We emphasize that momentum conservation can be ignored only for those diagram parts that are connected by three different Green-function lines. If two vertices are connected by a single Green-function line, the Green function carries the same momentum at both vertices. Equation (5.44) does not hold at all vertices since we may not always independently sum over all four vertex momenta.

In translationally invariant systems we have

$$G_{\sigma,0}(\boldsymbol{k}, \omega) = (\omega - \epsilon(\boldsymbol{k}) + \mu)^{-1} \ , \tag{5.45a}$$

where the frequency variable is always to be understood as $\omega + i\eta\mathrm{sgn}(\omega)$. According to (5.39) the proper self-energy is *independent* of momentum. With the help of the Dyson equation (5.40) the Green function thus becomes

$$\begin{aligned} G_\sigma(\boldsymbol{k}, \omega) &= \frac{1}{[G_{\sigma,0}(\boldsymbol{k}, \omega)]^{-1} - \Sigma_\sigma(\omega)} \\ &= \frac{1}{\omega - \epsilon(\boldsymbol{k}) + \mu - \Sigma_\sigma(\omega)} \\ &= G_{\sigma,0}(\boldsymbol{k}, \omega - \Sigma_\sigma(\omega)) \ . \end{aligned} \tag{5.45b}$$

The momentum dependence of the Green function is determined solely by the dispersion relation $\epsilon(\boldsymbol{k})$.

The proper two-electron vertex, (5.42), is also momentum independent and thus has even parity under the reflection $\boldsymbol{k} \mapsto -\boldsymbol{k}$. On the other hand the electron velocity as the gradient of the dispersion relation has odd parity. When we have to perform a momentum sum over the product of the two-electron vertex and the electron current the resulting contribution will vanish. Hence, all vertex corrections to the conductivity *vanish* in infinite dimensions [5.19]–[5.22]. As a consequence, the knowledge of the one-electron

Green function suffices for the explicit calculation of the conductivity in infinite dimensions. In natural units the functional dependence of the conductivity on the one-electron Green function is given by [5.21]

$$\sigma(\omega,T) = \pi\sum_{\sigma}\int_{-\infty}^{\infty} d\omega' \int_{-\infty}^{\infty} d\epsilon D_{\sigma,0}(\epsilon)A_{\sigma}(\epsilon,\omega')A_{\sigma}(\epsilon,\omega+\omega') \times \frac{f_{\text{FD}}(\omega') - f_{\text{FD}}(\omega+\omega')}{\omega}, \tag{5.46}$$

where $f_{\text{FD}}(x)$ is the Fermi–Dirac distribution, and $A_\sigma(\epsilon,\omega)$ is the one-electron spectral function. The calculation of the conductivity has enormously simplified since it now depends only on one-electron properties. For example, a gap in the one-electron spectral density (gap for charge excitations) immediately implies an insulating behavior. This plausible argument presented in Sect. 1.4 is seen to become exact for the Hubbard model in infinite dimensions. The irrelevance of vertex correction also implies, however, that Anderson localization by lattice disorder cannot be described in infinite dimensions [5.23].

5.3.3 Effective Models with Single-Site Interaction

According to (5.39) the proper self-energy is diagonal in position space in infinite dimensions, $\Sigma_\sigma(\boldsymbol{R}_1,\boldsymbol{R}_2;\omega) = \Sigma_\sigma(\boldsymbol{R}_1,\boldsymbol{R}_1;\omega)\delta_{\boldsymbol{R}_1,\boldsymbol{R}_2}$. Models in which the interaction between electrons is limited to a *single* site also display a self-energy that depends only on this site. Hence, it must be possible to construct an "effective" single-site model that is equivalent to the infinite-dimensional Hubbard model. In this sense mean-field theories that become exact in infinite dimensions not only exist for spin systems (see Sect. 5.1.1) but also for itinerant electron systems [5.1, 5.4], [5.24]–[5.27].

The static Weiß molecular field $h_{\boldsymbol{R}}$ in the spin models has to be replaced by a dynamic mean-field in the case of the Hubbard model. To implement this idea we view the dynamic Weiß field as the Green function $G_{\sigma,0}^{\text{bath}}(\boldsymbol{R},\boldsymbol{R};\omega)$ at site $\boldsymbol{R}$ of non-interacting electrons of an "electron bath". It is precisely the frequency dependence of the molecular field that prevents a complete analytical solution of the *dynamic* mean-field theory as compared to the static problem for the spin models. We emphasize that the bath Green function is not identical to the Green function of the Hubbard model at $U = 0$. If this was so, the infinite-dimensional Hubbard model would reduce to the Wolff model [5.18]. We have already excluded this possibility in the last subsection.

There are several ways to formulate the mean-field equations which provide the exact solution of the infinite-dimensional Hubbard model; for a review of methods, see [5.4]. Edwards [5.28] gives an explicit derivation of the self-consistency equations starting from the graph-theoretical approach in perturbation theory around the non-interacting Fermi gas. Janiš and Vollhardt formulate the equations with the help of generating functionals in strict analogy to the coherent potential approximation (CPA) for disordered systems. This derivation most directly illustrates the concept of a dynamic mean-field theory [5.1, 5.25]. In the following we concentrate on the way pursued in [5.26, 5.27, 5.29] in which the single-site problem is formulated as an effective single-impurity Anderson model; see Sect. 2.4. This representation allows us to discuss the results in close analogy to this thoroughly studied model; see, e.g., [5.30, Chap. 11] and [5.31]. To simplify the discussion we assume a translationally invariant system such that the proper self-energy depends on frequency only.

In terms of Fermi operators $\hat{\psi}_{\epsilon,\sigma}$ and their canonical commutation relations $\left[\hat{\psi}^+_{\epsilon,\sigma}, \hat{\psi}_{\epsilon',\sigma'}\right]_+ = \delta_{\sigma,\sigma'}\delta(\epsilon-\epsilon')$ the Hamilton operator of the single-impurity Anderson model reads

$$\begin{aligned} \hat{H}_{\text{And}} &= \sum_\sigma \int \mathrm{d}\epsilon \epsilon \hat{\psi}^+_{\epsilon,\sigma}\hat{\psi}_{\epsilon,\sigma} + E_d \sum_\sigma \hat{n}^d_\sigma + U\hat{n}_\uparrow \hat{n}_\downarrow \\ &\quad + \sum_\sigma \left[\int \mathrm{d}\epsilon \sqrt{D_{\sigma,0}(\epsilon)} V(\epsilon) \hat{\psi}^+_{\epsilon,\sigma}\hat{d}_\sigma + \text{h.c.}\right] \end{aligned} \tag{5.47}$$

with $\hat{n}^d_\sigma = \hat{d}^+_\sigma \hat{d}_\sigma$. The bath electrons $\hat{\psi}_{\epsilon,\sigma}$ hybridize with the impurity electrons $\hat{d}_\sigma$ with strength $\sqrt{D_{\sigma,0}}V(\epsilon)$. The energy E_d determines the position of the impurity level with respect to the Fermi energy of the bath and thus the average occupation number of the impurity level. In turn, this determines the electron density in the original (Hubbard) model. In particular, we find $E_d = -U/2$ for half band-filling in the Hubbard model such that we have to investigate the symmetric Anderson model to study the metal–insulator transition in the infinite-dimensional Hubbard model.

The non-interacting ($U = 0$) Green function of the impurity reads

$$G^{\text{And}}_{\sigma,0}(\omega) = \left[\omega - E_d - \int \frac{\mathrm{d}\epsilon}{\pi} \frac{\Delta(\epsilon)}{\omega - \epsilon}\right]^{-1}, \tag{5.48a}$$

$$\Delta(\epsilon) = \pi D_{\sigma,0}(\epsilon) V(\epsilon)^2 . \tag{5.48b}$$

The effective strength of the hybridization between the bath electrons and the impurity electrons is determined by $\Delta(\epsilon)$. If it is finite, the density of states at the Fermi energy acquires a ("Kondo") resonance whose width depends on the interaction strength U. A finite density of states at the Fermi energy signals the existence of Landau quasi-particles such that the state in the original Hubbard model is metallic. If the effective hybridization vanishes, the bath and the impurity electrons will decouple. This obviously corresponds to an insulating state in the Hubbard model.

Let us assume that we know the exact solution of the impurity problem in (5.47) for a given effective hybridization $\Delta(\epsilon)$ such that we know the *interacting* Green function of the impurity problem, $G^{\text{And}}_\sigma(\omega)$. We may express this quantity by a perturbation expansion in U with the help of a graph formalism [5.32, 5.33]. We thereby obtain the same diagrams as for the local Green function for the infinite-dimensional Hubbard model, $G_\sigma(\boldsymbol{R},\boldsymbol{R};\omega) \equiv G_\sigma(\omega)$.

Since we want to solve the infinite-d Hubbard model we have to demand that the effective single-site model of (5.47) gives the (local) Green function of the infinite-dimensional Hubbard model, (5.48). Hence, our first consistency condition reads

$$G_\sigma^{\mathrm{And}}(\omega) = G_\sigma(\omega) \,. \tag{5.49a}$$

On the other hand we learn from the diagrammatic formalism that the skeleton graphs for the proper self-energy are identical in both descriptions. If (5.49a) is indeed satisfied, both proper self-energies must also agree,

$$\Sigma_\sigma^{\mathrm{And}}(\omega) = \Sigma_\sigma(\omega) \,, \tag{5.49b}$$

which is the second of our consistency equations. The two conditions in (5.49a) and (5.49b) are our self-consistency equations for the mean-field theory, which determine the non-interacting bath Green function, $G_{\sigma,0}^{\mathrm{And}}(\omega)$.

To really see this let us choose a bath Green function $G_{\sigma,0}^{\mathrm{And,B}}(\omega)$ to begin with (upper index "B"). This is the same as choosing $\Delta(\epsilon)$ or $V(\epsilon)$ in (5.47). Now assume we could exactly solve the impurity problem. Then we know the full impurity Green function $G_\sigma^{\mathrm{And,B}}(\omega;\{G_{\sigma,0}^{\mathrm{And,B}}\})$. The argument in $\{\ldots\}$ indicates that the full impurity Green function is functionally dependent on the bath Green function. As next step we can derive the self-energy of the impurity problem with the help of the Dyson equation as

$$\Sigma_\sigma^{\mathrm{And,B}}(\omega;\{G_{\sigma,0}^{\mathrm{And,B}}\}) = \left[G_{\sigma,0}^{\mathrm{And,B}}(\omega)\right]^{-1} - \left[G_\sigma^{\mathrm{And,B}}(\omega;\{G_{\sigma,0}^{\mathrm{And,B}}\})\right]^{-1} \,. \tag{5.50a}$$

According to (5.49b) this self-energy must be identical to the self-energy of the Hubbard model,

$$\Sigma_\sigma(\omega;\{G_{\sigma,0}^{\mathrm{And,B}}\}) = \Sigma_\sigma^{\mathrm{And,B}}(\omega;\{G_{\sigma,0}^{\mathrm{And,B}}\}) \,. \tag{5.50b}$$

The Dyson equation of the Hubbard model (5.40) then gives

$$G_\sigma(\omega;\{G_{\sigma,0}^{\mathrm{And,B}}\}) = \int_{-\infty}^{\infty} \frac{D_{\sigma,0}(\epsilon)\mathrm{d}\epsilon}{\omega-\epsilon+\mu-\Sigma_\sigma(\omega;\{G_{\sigma,0}^{\mathrm{And,B}}\})} \,. \tag{5.50c}$$

According to (5.49a) this Green function has to be identical to the Green function of the impurity problem,

$$G_\sigma^{\mathrm{And,E}}(\omega;\{G_{\sigma,0}^{\mathrm{And,B}}\}) = G_\sigma(\omega;\{G_{\sigma,0}^{\mathrm{And,B}}\}) \,. \tag{5.50d}$$

With the help of the Dyson equation of the impurity problem,

$$G_{\sigma,0}^{\mathrm{And,E}}(\omega;\{G_{\sigma,0}^{\mathrm{And,B}}\}) = \left\{\left[G_\sigma^{\mathrm{And,E}}(\omega;\{G_{\sigma,0}^{\mathrm{And,B}}\})\right]^{-1} + \Sigma_\sigma^{\mathrm{And,B}}(\omega;\{G_{\sigma,0}^{\mathrm{And,B}}\})\right\}^{-1} \,, \tag{5.50e}$$

we may test whether self-consistency has been reached, i.e., whether the bath Green function at the beginning and the end (upper index "E") indeed coincide,

$$G_{\sigma,0}^{\mathrm{And,E}}(\omega;\{G_{\sigma,0}^{\mathrm{And,B}}\}) \stackrel{?}{=} G_{\sigma,0}^{\mathrm{And,B}}(\omega) . \tag{5.50f}$$

The equations (5.49) indeed define the self-consistency conditions for the effective single-impurity Anderson model. The actual iteration scheme follows the circle of equations (5.50). When convergence has been reached, the Green function for the infinite-dimensional Hubbard model becomes

$$G_\sigma(\boldsymbol{k},\omega) = \frac{1}{\omega - \epsilon(\boldsymbol{k}) + \mu - \Sigma_\sigma(\omega)} \tag{5.51}$$

with the self-energy from (5.50b).

In the above derivation we used translational invariance to relate the self-energy to the local Green function (5.50c). A more general analysis of a single site at $\boldsymbol{R}_0$ embedded in a host shows [5.4, 5.34] that the local Green function can be cast into the form

$$G_\sigma(\boldsymbol{R}_0,\omega) = \frac{1}{\omega + \mu - S_\sigma(\boldsymbol{R}_0,\omega) - \Sigma_\sigma(\boldsymbol{R}_0,\omega)} , \tag{5.52a}$$

where the so-called Feenberg self-energy $S_\sigma(\boldsymbol{R}_0,\omega)$ is determined by the host Green function $G_\sigma^{\boldsymbol{R}_0}(\boldsymbol{R}_1,\boldsymbol{R}_2;\omega)$ of a system in which the impurity site $\boldsymbol{R}_0$ has been removed,

$$S_\sigma(\boldsymbol{R}_0,\omega) = \sum_{\boldsymbol{R}_1,\boldsymbol{R}_2} t(\boldsymbol{R}_1,\boldsymbol{R}_0)t(\boldsymbol{R}_0,\boldsymbol{R}_2)G_\sigma^{\boldsymbol{R}_0}(\boldsymbol{R}_1,\boldsymbol{R}_2;\omega) . \tag{5.52b}$$

Since the "impurity site" is not really a special site but has been chosen for the sake of the analysis, the cavity Green function can be expressed by the interacting Green function as

$$G_\sigma^{\boldsymbol{R}_0}(\boldsymbol{R}_1,\boldsymbol{R}_2;\omega) = G_\sigma(\boldsymbol{R}_1,\boldsymbol{R}_2;\omega) - \frac{G_\sigma(\boldsymbol{R}_1,\boldsymbol{R}_0;\omega)G_\sigma(\boldsymbol{R}_0,\boldsymbol{R}_2;\omega)}{G_\sigma(\boldsymbol{R}_0,\boldsymbol{R}_0;\omega)} . \tag{5.52c}$$

Naturally, these relations are only exact in infinite dimensions.

A particularly simple case is the Bethe lattice since the removal of the impurity site disconnects all sites $\boldsymbol{R}_1 \neq \boldsymbol{R}_2$. A Bethe lattice is built from a root site with bonds to Z neighbors, which serve as root sites for the next shell. It is self-similar, bipartite, but obviously not translationally invariant. For a finite number of shells one obtains a Cayley tree. In the limit of infinite connectivity, $Z \to \infty$, (5.52b) gives

$$S_\sigma(\boldsymbol{R}_0,\omega) = \sum_{\boldsymbol{R}_1} |t(\boldsymbol{R}_1,\boldsymbol{R}_0)|^2 G_\sigma(\boldsymbol{R}_0,\boldsymbol{R}_1;\omega) \quad : \quad \text{Bethe lattice} . \tag{5.53}$$

If all lattice sites were equal, we would find the self-consistency relation

$$G_\sigma(\omega) = \frac{1}{\omega + \mu - (W/4)^2 G_\sigma(\omega) - \Sigma_\sigma(\omega)} \tag{5.54}$$

for the Bethe lattice with infinite coordination number; this result can also be obtained from (5.50c). In the case of antiferromagnetic order on two Bethe sublattices A and B we have ($Z = \infty$)

$$G_{A,\sigma}(\omega) = \frac{1}{\omega + \mu - (W/4)^2 G_{B,\sigma}(\omega) - \Sigma_{A,\sigma}(\omega)}, \quad (5.55a)$$

$$G_{B,\sigma}(\omega) = \frac{1}{\omega + \mu - (W/4)^2 G_{A,\sigma}(\omega) - \Sigma_{B,\sigma}(\omega)}. \quad (5.55b)$$

On the Bethe lattice with infinite coordination number the self-consistency equations are simple algebraic equations. However, the Bethe lattice is not translationally invariant, and the notion of momentum is meaningless on this lattice.

The principle obstacle against the analytical solution of the infinite-dimensional Hubbard model lies in the fact that the single-impurity Anderson model can be solved analytically only for the special case $\Delta(\epsilon) = \Delta$ and a constant density of states with the help of the Bethe ansatz [5.35]. We are far from a general solution of the problem with an arbitrary $\Delta(\epsilon)$ and a general density of states $D_{\sigma,0}(\epsilon)$, and approximate analytical and numerical techniques have to be employed to find the self-energies $\Sigma_\sigma(\omega)$ from the exact impurity Green function. Before we come to those in Sects. 5.5 and 5.6 we discuss some cases for which analytical results can be derived.

5.4 Analytical Results in Infinite Dimensions

In this section we present results that can be obtained analytically in infinite dimensions. In the first two subsections we discuss achievements within the perturbative approach to the Hubbard model, for both small and large interaction strengths. The limit allows us to calculate diagrams explicitly, and even to perform self-consistent diagram summations with a comparably small numerical effort. In the third subsection we briefly return to Gutzwiller-correlated variational wave functions, which we introduced in Sect. 3.4. Their evaluation leads to the same graph formalism as for the Hubbard model but lines and vertices are independent of time or frequency indices. The self-consistent but static single-site problem thus leads to an explicit closed set of self-consistency equations. In the fourth subsection we present the solution of the simplified Hubbard or Falicov–Kimball model. In this model one of the two spin species is localized. The dynamics of the mean-field theory thereby so drastically simplifies that we can formulate an explicit closed set of self-consistency equations. In the fifth subsection we treat spinless fermions with nearest-neighbor interactions in infinite dimensions. As shown in Sect. 5.3 the Hartree decoupling becomes exact for this interaction, which allows us to determine analytically the phase diagram in infinite dimensions, including $1/d$ corrections and disorder effects.

5.4.1 Perturbation Theory for Small Interactions

Perturbation theory in the interaction parameter provides the microscopic justification for Fermi-liquid theory [5.36]; see Sect. 1.4.1. The limit of infinite dimensions allows for the explicit calculation of diagrams in the perturbation expansion. The discussion follows [5.11]; see also [5.1].

In a translationally invariant system the zero-temperature Green function can be written in terms of the self-energy $\Sigma_\sigma(\omega)$ as

$$G_\sigma(\boldsymbol{k},\omega) = \frac{1}{\omega - \epsilon(\boldsymbol{k}) + E_{\mathrm{F}} - \Sigma_\sigma(\boldsymbol{k},\omega)} \tag{5.56}$$

in all dimensions d. If the perturbation series converges, the self-energy is bounded. In particular, the imaginary part of the self-energy vanishes at the Fermi energy ($\omega = 0$) with a power law, $\mathrm{Im}\{\Sigma_\sigma(\omega)\} \sim \omega^2$. The Fermi surface of the interacting system thus is defined by the relation $E_{\mathrm{F},\sigma} = \epsilon(\boldsymbol{k}) + \Sigma_\sigma(\boldsymbol{k},0)$. Although the shape of the Fermi surface changes because of the electron–electron interaction, the volume enclosed by it remains preserved (Luttinger theorem [5.36]), $n_\sigma = \sum_{\boldsymbol{k},\sigma} \Theta\left(E_{\mathrm{F}} - \epsilon(\boldsymbol{k}) - \Sigma_\sigma(\boldsymbol{k},0)\right)$ ($\Theta(x)$ is the step function). The arguments of Luttinger and Ward [5.36] are strictly valid only in infinite dimensions since they neglect momentum conservation at inner vertices in their derivation for dimensions $d > 1$ [5.17].

In the limit of infinite dimensions the self-energy becomes momentum independent; see Sect. 5.3. Hence, the Fermi energy is only shifted, $E_{\mathrm{F},\sigma} = E_{\mathrm{F}}^0 + \Sigma_\sigma(0)$, and the "Fermi surface" retains its shape. In the last section we have shown that only the Fermi energy, not the Fermi surface, is a reasonable concept in infinite dimensions because of the random structure of momentum space. In the sense of a $1/d$ expansion we may deduce, however, that the Fermi surface in $d = 3$ dimensions is not strongly affected by interaction effects, even if there are no additional symmetry arguments for a conserved shape of the Fermi surface.

With the help of the definition of the density of states (see Appendix A.2) it is not difficult to show [5.11] that the density of states at the Fermi energy in the metallic phase is not changed by the interaction in infinite dimensions. This is a direct consequence of the fact that the interacting Green function obeys

$$G_\sigma(\boldsymbol{k},\omega) = G_{\sigma,0}(\boldsymbol{k},\omega - \Sigma_\sigma(\omega)) \tag{5.57}$$

for a translationally and spin-rotationally invariant metallic system in infinite dimensions. Since both the real and the imaginary part of $\Sigma_\sigma(\omega = 0)$ vanish for a Fermi liquid, and the imaginary part of the one-electron Green function gives the one-electron density of states, the result

$$D_\sigma(\epsilon = 0) = D_{\sigma,0}(\epsilon = 0) \tag{5.58}$$

immediately follows. It is typical for a single-impurity Anderson model [5.32] and does not come as a surprise in view of the connection between the two models; see Sect. 5.3. If the density of states at the Fermi energy does not fulfill (5.58) the theory does not describe a regular Fermi liquid. This happens, for example, for Edward's improved alloy-analogy approximation [5.28]; compare Sect. 3.3.8.

We can derive further results for the infinite-dimensional Hubbard model in its paramagnetic metallic phase that are characteristic for a Fermi liquid [5.11, 5.37]. The momentum distribution $\langle \hat{n}_{\boldsymbol{k},\sigma} \rangle$ (see Appendix A.2) only indirectly depends on $\boldsymbol{k}$ via the dispersion relation $\epsilon(\boldsymbol{k})$. It has a discontinuity at the Fermi energy of size $Z_{\mathrm{F}} = m/m^* = (1 - \Sigma'_\sigma(0))^{-1}$, where $\Sigma'_\sigma(0)$ is the derivative of the self-energy at $\omega = 0$ and m^* is the effective electron mass. The slope of the density of states $D_\sigma(\epsilon)$ near the Fermi energy is renormalized accordingly:

$$\frac{1}{Z_{\mathrm{F}}} = \frac{m^*}{m} = \frac{D'_\sigma(\epsilon = 0)}{D'_{\sigma,0}(\epsilon = 0)} \; . \tag{5.59}$$

In infinite dimensions the one-electron density of states $D_\sigma(\epsilon)$ near the Fermi energy determines the effective mass and thereby the factor $Z_{\mathrm{F}} \leq 1$, which gives the weight of the quasi-particle resonance in the spectral function $A_\sigma(\boldsymbol{k}, \omega)$ [5.36].

We can formulate the standard perturbation expansion for the Hubbard interaction in all dimensions with the help of the usual diagrammatic theory. However, the actual *calculation* of diagrams becomes *technically* rather complicated already in second order. As shown by Metzner and Vollhardt [5.10] for the example of the correlation energy to second order in U^2/t, the evaluation of diagrams is drastically simplified in infinite dimensions. Furthermore, the self-energy can be calculated analytically in second-order self-consistent perturbation theory [5.11]. In addition, electron–hole and electron–electron RPA diagrams can be summed up exactly in infinite dimensions [5.38]. This allows us to study the influence of the respective diagram classes on the density of states and the effective mass. A detailed comparison between the summation of various diagram classes for the Hubbard model at half band-filling is given by Freericks [5.39].

Within perturbation theory the results of these calculations become exact for small interaction strength in infinite dimensions. To aim at *finite* dimensions we may start from the proper self-energy, which is purely local in infinite dimensions, and successively improve on it by including the contributions from neighboring shells [5.17, 5.40]. Since this series expansion rapidly converges in $d \geq 3$ and is numerically well behaved, we may *numerically exactly* evaluate diagrams even in finite dimensions.

For half band-filling we may use a trick that allows us to apply formally results of the perturbation theory in U/t also for strong interactions. We calculate the self-energy *relative* to the paramagnetic Hartree–Fock theory, i.e., we include the chemical potential terms with $\mu = U/2$. In this way we also obtain the signatures of the upper and lower Hubbard bands in the one-electron density of states for $U \gtrsim W$. The reason for this is the special structure of

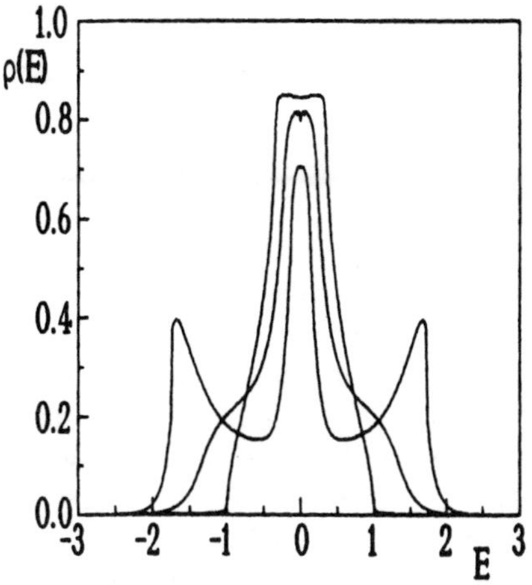

Fig. 5.1. Perturbative result for the one-electron density of states in the three-dimensional Hubbard model at half band-filling for $U/\widetilde{t} = 0, 1, 3$, calculated from second-order perturbation theory relative to the Hartree–Fock solution (from [5.17]).

the Green function at half band-filling in the *atomic* limit. From (3.30) we find that the atomic self-energy at half band-filling is given by

$$\Sigma_{\sigma}^{\text{atom}}(\omega, n=1) = \frac{U}{2} + \frac{U^2}{4\omega} \, . \tag{5.60}$$

Again, the frequency variable is to be understood as $\omega + \mathrm{i}\eta\mathrm{sgn}(\omega)$, where $\eta = 0^+$ is a positive infinitesimal and sgn is the sign function. We emphasize that the $1/\omega$ singularity is merely the consequence of the *definition* of $\Sigma_{\sigma}^{\text{atom}}(\omega)$ and bears no relation to the physics of the atomic limit; see Sect. 3.3.1.

Since the self-energy at half band-filling is only quadratic in U in the atomic limit, second-order perturbation theory around the non-interacting limit gives a *qualitatively* correct structure of the density of states even for large interactions as long as the system stays metallic. It not only gives the quasi-particle peak around $\omega = 0$, but also displays the features of the upper and lower Hubbard bands centered around $\omega = \pm U/2$ (for $U \gg W$). We emphasize that the theory properly locates the *position* of the Hubbard bands but not necessarily their respective widths since the theory becomes exact in the atomic limit (no electron hopping), and not necessarily in the limit of strong interactions ($U \gg W$).

The result of second-order perturbation theory relative to Hartree–Fock theory at half band-filling in three dimensions is shown in Fig. 5.1. Since the self-energy is momentum dependent in finite dimensions, (5.58) is fulfilled only approximately. The formation of the upper and lower Hubbard bands is clearly seen for intermediate values of the interaction strength. Examples for the $\boldsymbol{k}$-dependent self-energy can be found in [5.17].

The particularly symmetric situation at half band-filling also permits a qualitatively correct description of the periodic Anderson model with the help of second-order perturbation theory relative to Hartree–Fock theory; for details, see [5.20]. Naturally, this kind of perturbation theory is better known for the symmetric single-impurity Anderson model [5.32, 5.33]. Since the infinite-dimensional Hubbard model can be mapped onto a self-consistent single-impurity Anderson model, we may also try the perturbative approach on the iteration of the mean-field equations ("iterated perturbation theory" [5.4, 5.27, 5.29]); see Sect. 5.5.5.

5.4.2 Perturbation Theory for Large Interactions

Despite the considerable simplifications in infinite dimensions the Hubbard model for strong interactions cannot be solved exactly in general. For example, the Hubbard model at $U = \infty$ and less than half band-filling ($\hat{T}$ model, see Sect. 2.5.4) cannot be solved for general $n < 1$. Since we are mostly interested in the Mott transition we focus on the infinite-dimensional Hubbard model at half band-filling for which various exact results can be obtained in the limit $U/W \gg 1$. Unfortunately, an analytical solution of the infinite-dimensional Hubbard model for all U/W cannot be deduced from the large-U approach [5.9, 5.41].

Ground State. At half band-filling and large interactions the Hubbard model can be mapped onto the Heisenberg model; see Sect. 2.5.4. As we have already derived in Sect. 5.1.1, the Heisenberg Hamiltonian reduces to the Ising model in infinite dimensions, and the ground state is readily identified.

If the dispersion relation obeys the perfect nesting property, the Néel state is the ground state. The Néel state provides a convenient starting point for a perturbative analysis of the large-U Hubbard in infinite dimensions [5.9, 5.41, 5.42]. Since we expect a finite radius of convergence for the expansion around the limit of large interactions the antiferromagnetic order should persists in a finite region of interactions strengths. In dimensions $d \geq 3$ we also know that the antiferromagnetic order prevails up to the Néel temperature. Hence, the antiferromagnetic Mott–Heisenberg insulator covers a substantial part of the phase diagram of the half-filled Hubbard model in infinite dimension.

The Hubbard model whose dispersion relation does not have the perfect nesting property is more challenging. If the hopping is perfectly random (see Sect. 5.2.2) the half-filled Hubbard model at large interactions reduces to the ferromagnetic Heisenberg model with an antiferromagnetic coupling. As shown in Sect. 5.2.2 this implies that all states with zero total spin are ground states. Consequently, the Mott–Hubbard insulator is highly degenerate in infinite dimensions, its entropy density is $(S/L) = k_{\mathrm{B}} \ln(2)$ in the thermodynamic limit. In the singular limit of infinite dimensions this degeneracy is not lifted in higher orders in the perturbation theory in W/U since those terms that would generate a mixing of the states are of order $1/d$, or smaller. Thus, the Mott–Hubbard insulator in infinite dimensions has a finite ground-state degeneracy. It is very demanding to formulate an (approximate) theory that properly describes a transition from the unique metallic Fermi-liquid ground state into the highly degenerate Mott–Hubbard insulator.

Upper and Lower Hubbard Bands. Electron–hole symmetry means that it is sufficient at half band-filling to address the lower Hubbard band, which describes the motion of an extra hole in the background of 2^{L-1} spins ($\hat{T}$ model for $N = L - 1$). The contribution of the upper Hubbard band to the full density of states is found by shifting the lower Hubbard band by U. The problem of a single hole in a spin background can be solved exactly in various cases: (i) for a ferromagnetic spin background; (ii) if all hole paths are retraceable [5.43]; (iii) in infinite dimensions for the Néel and the random spin background [5.44].

For general lattices and/or dimensions the spin background strongly influences the motion of the hole. In fact, as shown by Nagaoka [5.45], the ferromagnetic background is the most favorable for the hole motion; see Sect. 2.5.4. In this case the hole motion is completely free, and the lower Hubbard band displays the full noninteracting bandwidth. For all other spin backgrounds, the Green function for the creation of a hole at site $\boldsymbol{R}_1$ and its annihilation at site $\boldsymbol{R}_2$ at some later time can be calculated exactly only in the other two cases (ii) and (iii).

Case (ii) covers one dimension and the Bethe lattice with an arbitrary coordination number. By definition, there are no paths on these lattices, which do not retrace each other when the local hole Green function ($\boldsymbol{R}_1 = \boldsymbol{R}_2$) is calculated. Furthermore, any given spin configuration gives the same result since the motion of the hole from $\boldsymbol{R}_1$ into the system and back to $\boldsymbol{R}_1$ does not change it. Consequently, the hole motion is independent of the spin configuration and we obtain the non-interacting density of states for the lower (and upper) Hubbard band, just as in case (i). In particular, the bands have the full width of the non-interacting case, $W_{1,2} = W$.

In infinite dimensions, case (iii), the Néel state traps the hole, i.e., the Green function is site diagonal because the hole *must* return to the point of its generation, $\boldsymbol{R}_2 = \boldsymbol{R}_1$, to recover the original spin configuration. In addition, only self-retraceable paths are allowed, and the density of states reduces to that of the Bethe lattice with infinite coordination number. Formally, the width of the Hubbard bands is given by $W_{1,2} = W$ such that they again display the full bandwidth. The situation is more complicated for a random spin background, a situation equivalent to temperatures large compared to the spin exchange energy J, i.e., above the Néel temperature. There, loops contribute [5.41] and the hole self-energy for the hypercubic lattice in the paramagnetic phase obeys

$$2\Sigma(\omega) = \omega - \frac{3}{G(\omega)}, \tag{5.61}$$

where $G(\omega)$ is the local hole Green function. This equation and the Dyson equation (5.45b) solve the problem of a single hole on an infinite-dimensional cubic lattice. Again, the width of the lower Hubbard band is found to be close to the full bandwidth, $W_1 \approx W$ [5.41].

The results in the last two subsection set bounds on approximate theories for the infinite-dimensional Hubbard model but they do not cover the whole range of interaction strengths. Explicit analytical results for all U/W can be achieved only if we further simplify the dynamics of the mean-field theory. In the next two subsections we present two problems in infinite dimensions whose mean-field theory can be solved exactly. Our first example are variational wave functions where the evaluation of expectation values is a time-independent problem. In the second example, the simplified Hubbard model, one spin species cannot move such that the dynamic problem reduces to a self-consistent solution of the CPA equations.

5.4.3 Variational Wave Functions

As shown in Sect. 3.4, we can exactly evaluate the class of Gutzwiller-correlated variational wave functions in infinite dimensions. They are defined as

$$|\Psi\rangle_g = g^{\hat{D}}|\Psi_0\rangle \, . \tag{5.62}$$

As in standard perturbation theory for the Hubbard model, a diagrammatic expansion can be formulated for the one-electron density matrix,

$$P_\sigma(\boldsymbol{f},\boldsymbol{h}) = \frac{{}_g\langle\Psi|\hat{c}^+_{\boldsymbol{f},\sigma}\hat{c}_{\boldsymbol{h},\sigma}|\Psi\rangle_g}{{}_g\langle\Psi|\Psi\rangle_g} \, , \tag{5.63}$$

which allows us to determine the variational ground-state energy. In the formulation that is most suitable for the one-dimensional problem the vertex (g^2-1) is purely local and lines correspond to the one-electron density matrix $P_{\sigma,0}(\boldsymbol{f},\boldsymbol{h}) = \langle\hat{c}^+_{\boldsymbol{f},\sigma}\hat{c}_{\boldsymbol{h},\sigma}\rangle_0$.

As for the Hubbard model the calculation of $P_\sigma(\boldsymbol{f},\boldsymbol{h})$ reduces to an effective single-site problem in infinite dimensions [5.10, 5.15]. Since the diagrams are time/frequency independent the self-consistent mean-field theory can be formulated explicitly. We denote the dressed lines by

$$\overline{P}_\sigma(\boldsymbol{f},\boldsymbol{h}) = P_{\sigma,0}(\boldsymbol{f},\boldsymbol{h}) + \sum_{\boldsymbol{g}_1,\boldsymbol{g}_2} P_{\sigma,0}(\boldsymbol{f},\boldsymbol{g}_1)S_\sigma(\boldsymbol{g}_1,\boldsymbol{g}_2)P_{\sigma,0}(\boldsymbol{g}_2,\boldsymbol{h}) \, , \tag{5.64a}$$

where $S_\sigma(\boldsymbol{f},\boldsymbol{h})$ is the "self-energy" in the variational theory. It is connected to the (purely local) "proper self-energy" $S^*_\sigma(\boldsymbol{f})$ via the Dyson equation

$$S_\sigma(\boldsymbol{f},\boldsymbol{h}) = S^*_\sigma(\boldsymbol{f})\left[\delta_{\boldsymbol{f},\boldsymbol{h}} + \sum_{\boldsymbol{g}} P_{\sigma,0}(\boldsymbol{f},\boldsymbol{g})S_\sigma(\boldsymbol{g},\boldsymbol{h})\right] \, , \tag{5.64b}$$

where we have used the fact that the proper self-energy is diagonal in position space in infinite dimensions.

The effective single-site problem can be exactly solved for Gutzwiller-correlated variational wave functions [5.10, 5.15] and provides the following closing relation for the self-consistency scheme

$$\overline{P}_\sigma(\boldsymbol{f},\boldsymbol{f}) = \frac{S^*_{-\sigma}(\boldsymbol{f})}{1-g^2-S^*_\sigma(\boldsymbol{f})S^*_{-\sigma}(\boldsymbol{f})} \; . \tag{5.64c}$$

The equations (5.64) are exact for Gutzwiller-correlated wave functions in infinite dimensions and constitute the variational mean-field theory. Since the variational theory is a time-independent problem the explicit closing relation for the single-site problem (5.64c) could be found. For the dynamic mean-field theory for the Hubbard model this could not be achieved.

Because of the lattice sums in (5.64a) and (5.64b) the variational mean-field equations can actually be evaluated as a function of g and further variational parameters contained in $|\Psi_0\rangle$ only for special cases [5.10, 5.15]. We have already encountered the same situation for Hartree–Fock theory; see Sect. 3.1. There, we were also limited to the discussion of those solutions of the equations which were particularly symmetric (translationally invariant or Néel-type antiferromagnetism on bipartite lattice structures). Fortunately, in the case of a Gutzwiller-correlated wave function we may apply a modified graph formalism in which the proper self-energy not only becomes local but completely vanishes in infinite dimensions [5.46]; see Sect. 3.4. Expectation values for general Gutzwiller-correlated wave functions can be explicitly calculated in this modified formalism, a $1/d$ expansion can be carried out, and its generalization to N_f-fold degenerate bands is feasible. In particular, the results in $d=\infty$ ($N_f=\infty$) become identical to those of the Kotliar–Ruckenstein (Barnes–Coleman) mean-field theory [5.47]; see Sect. 3.5.

5.4.4 Falicov–Kimball Model

The evaluation of Gutzwiller-correlated wave functions was possible in infinite dimensions since no time/frequency dependence had to be taken into account. The dynamics of the "simplified Hubbard" [5.48, 5.49] or "Falicov–Kimball" model [5.50] is also considerably reduced compared to that of the Hubbard model since the electrons of one spin species cannot hop. The Hamilton operator of the model is given by

$$\hat{H}_{\text{FK}} = \sum_{\boldsymbol{R}_1,\boldsymbol{R}_2} t(\boldsymbol{R}_1-\boldsymbol{R}_2)\hat{d}^+_{\boldsymbol{R}_1}\hat{d}_{\boldsymbol{R}_2} + U\sum_{\boldsymbol{R}} \hat{n}^d_{\boldsymbol{R}}\hat{n}^f_{\boldsymbol{R}} - \sum_{\boldsymbol{R}}\left(\mu^d\hat{n}^d_{\boldsymbol{R}} - E^f\hat{n}^f_{\boldsymbol{R}}\right) \; . \tag{5.65}$$

The notation of the mobile ($\hat{d}_{\boldsymbol{R}}$) and immobile electrons ($\hat{f}_{\boldsymbol{R}}$) results from the application of the model to systems of strong correlations between conduction and f electrons [5.50]. The model is also used for the description of crystallization [5.51]. In this case we describe a system of mobile electrons and immobile ions. We restrict ourselves to nearest-neighbor hopping, $t(\boldsymbol{R}_1-\boldsymbol{R}_2)=(-t)\delta_{\boldsymbol{R}_1-\boldsymbol{R}_2,\boldsymbol{\tau}}$. For $U \gg t$ the model reduces to the Ising model and thus provides a canonical generalization of this spin model to itinerant electron systems.

Although the f electrons have a zero hopping matrix element they are not localized because their distribution on the lattice depends on *temperature*. Hence, the model is still very complicated. If they were fixed, the model could be solved exactly in one dimension [5.52]. On bipartite lattices the electrons order in a charge-density wave state for finite U/t and half band-filling ($N^f = N^d = L/2$) below some finite critical temperature $T_c(d \geq 2, U > 0) > 0$ [5.51]. For $T < T_c(U)$ the f electrons (d electrons) preferably arrange on the A (B) lattice sites.

In infinite dimensions the model again reduces to an effective single-site problem. Brandt and Mielsch [5.53] were the first to formulate and solve the model in this way. Another formulation in terms of path integrals is given by Janiš [5.54]. On the Bethe lattice with infinite coordination number van Dongen and Vollhardt [5.55] found a representation of the mean-field theory in terms of a tight-binding Hamilton operator with interaction only on the root site of the Bethe lattice; for details, see [5.56].

The Hubbard model reduces to a single-impurity Anderson model. The occupation of the single site dynamically changes with time. In the Falicov–Kimball model, however, the f electrons cannot move away from the site. Hence, the single-site problem can exactly be solved in the presence of an *arbitrary* bath Green function that mimics the dynamic change in the d electron occupancy on the site. We denote the Matsubara Green function of the bath as

$$\mathcal{G}_{d,0}^{\text{atom}}(\mathrm{i}\omega_n) = \frac{1}{\mathrm{i}\omega_n + \mu^d - \Delta(\mathrm{i}\omega_n)}\,, \tag{5.66}$$

where $\omega_n = \pi(2n+1)T$ are the Matsubara frequencies [5.16], and $\Delta^d(\mathrm{i}\omega_n)$ is a general function of frequency that has to be determined self-consistently. These quantities are Hubbard's "scattering corrections" of Sect. 3.3.5.

The exact atomic partition function in the presence of the bath Green function becomes

$$\begin{aligned}\mathcal{Z} = 2\Biggl\{&\exp\left[\sum_n \left[\ln\left(\left(\mathcal{G}_{d,0}^{\text{atom}}(\mathrm{i}\omega_n)\right)^{-1}\right) - \ln(\mathrm{i}\omega_n)\right] \mathrm{e}^{\mathrm{i}\omega_n\eta}\right] \\ &+ \exp\left[-\beta E^f + \sum_n \left[\ln\left(\left(\mathcal{G}_{d,0}^{\text{atom}}(\mathrm{i}\omega_n)\right)^{-1} - U\right) - \ln(\mathrm{i}\omega_n)\right] \mathrm{e}^{\mathrm{i}\omega_n\eta}\right]\Biggr\}\,.\end{aligned} \tag{5.67a}$$

The partition function $\mathcal{Z}$ is given by two contributions since the lattice site is either unoccupied (free case) or occupied by an f electron such that the f-level energy and the interaction U contribute.

We differentiate the atomic free energy $f^{\text{atom}} = -T \ln \mathcal{Z}$ with respect to E^f which gives us the f-electron density,

$$n^f = \frac{2}{\mathcal{Z}} \exp\left[-\beta E^f + \sum_n \left[\ln\left(\left(\mathcal{G}_{d,0}^{\text{atom}}(\mathrm{i}\omega_n)\right)^{-1} - U\right) - \ln(\mathrm{i}\omega_n)\right] \mathrm{e}^{\mathrm{i}\omega_n\eta}\right]\,. \tag{5.67b}$$

In addition, we obtain the exact atomic Matsubara Green function of the d electrons by the differentiation of $-\beta f^{\text{atom}}$ with respect to $\Delta^d(\mathrm{i}\omega_n) \sim [\mathcal{G}_{d,0}^{\text{atom}}(\mathrm{i}\omega_n)]^{-1}$ since $\ln(\mathcal{Z})$ is the generating function for d-electron fluctuations on the site. The expected result is

$$\mathcal{G}_d^{\text{atom}}(\mathrm{i}\omega_n) = \frac{1-n^f}{[\mathcal{G}_{d,0}^{\text{atom}}(\mathrm{i}\omega_n)]^{-1}} + \frac{n^f}{[\mathcal{G}_{d,0}^{\text{atom}}(\mathrm{i}\omega_n)]^{-1} - U} . \tag{5.67c}$$

As discussed in Sect. 3.3.5 this result is identical to the alloy-analogy approximation for the Hubbard model.

Now that we have solved the atomic problem in the presence of an arbitrary bath, we are left with the implementation of (5.49) to solve exactly the Falicov–Kimball model in infinite dimensions:

$$\mathcal{G}_d^{\text{atom}}(\mathrm{i}\omega_n) = \mathcal{G}_d(\mathrm{i}\omega_n) \equiv \mathcal{G}_d(\boldsymbol{R},\boldsymbol{R};\mathrm{i}\omega_n) , \tag{5.68a}$$

$$\Sigma_d^{\text{atom}}(\mathrm{i}\omega_n) = \Sigma_d(\mathrm{i}\omega_n) . \tag{5.68b}$$

From (5.67c) and (5.68a) we find that

$$\begin{aligned} &1-(1-n^f)U\mathcal{G}_{d,0}^{\text{atom}}(\mathrm{i}\omega_n) \\ &= \frac{1}{2}\left[1-U\mathcal{G}_d(\mathrm{i}\omega_n) + \sqrt{(1-U\mathcal{G}_d(\mathrm{i}\omega_n))^2 + 4Un^f\mathcal{G}_d(\mathrm{i}\omega_n)}\right] \end{aligned} \tag{5.68c}$$

holds such that we may eliminate the bath Green function in favor of the local Green function of the d electrons. The Dyson equation for the atomic problem,

$$\begin{aligned} \Sigma_d^{\text{atom}}(\mathrm{i}\omega_n) &= \left[\mathcal{G}_{d,0}^{\text{atom}}(\mathrm{i}\omega_n)\right]^{-1} - \left[\mathcal{G}_d^{\text{atom}}(\mathrm{i}\omega_n)\right]^{-1} \\ &= Un^f + \frac{U^2 n^f(1-n^f)}{\left[\mathcal{G}_{d,0}^{\text{atom}}(\mathrm{i}\omega_n)\right]^{-1} - U(1-n^f)} \\ &= \frac{Un^f}{1-(1-n^f)U\mathcal{G}_{d,0}^{\text{atom}}(\mathrm{i}\omega_n)} , \end{aligned} \tag{5.68d}$$

(5.68b), and (5.68c) then provide the proper self-energy of the d electrons in the Falicov–Kimball model in infinite dimensions:

$$\Sigma_d(\mathrm{i}\omega_n) = \frac{U}{2} - \frac{1}{2\mathcal{G}_d(\mathrm{i}\omega_n)} + \sqrt{\left(\frac{1}{2\mathcal{G}_d(\mathrm{i}\omega_n)} - \frac{U}{2}\right)^2 + \frac{Un^f}{\mathcal{G}_d(\mathrm{i}\omega_n)}} . \tag{5.68e}$$

Here, the proper self-energy is still a functional of the local Green function at all frequencies since the occupation number of the f electrons still contains this dependence. Furthermore, the f-electron occupancy is site-dependent, $n^f \equiv n^f(\boldsymbol{R})$, such that $\Sigma_d(\mathrm{i}\omega_n) \equiv \Sigma_d(\boldsymbol{R};\mathrm{i}\omega_n)$. The Dyson equation for the lattice problem,

$$\begin{aligned} &(\mathrm{i}\omega_n + \mu^d - \Sigma_d(\boldsymbol{R}_1,\mathrm{i}\omega_n))\mathcal{G}_d(\boldsymbol{R}_1,\boldsymbol{R}_2;\mathrm{i}\omega_n) \\ &= \delta_{\boldsymbol{R}_1,\boldsymbol{R}_2} + \sum_{\boldsymbol{R}} t(\boldsymbol{R}_1-\boldsymbol{R})\mathcal{G}_d(\boldsymbol{R},\boldsymbol{R}_2;\mathrm{i}\omega_n) , \end{aligned} \tag{5.68f}$$

links the Green function and self-energy and thereby completes the system of equations that have to be solved self-consistently, namely (5.68e) and (5.68f).

The equations (5.68) are equivalent to the CPA equations for impurity scattering in metals [5.53, 5.54]. This does not come as a surprise since the coherent potential approximation becomes exact in infinite dimensions [5.23]. The decisive difference lies in the fact that the configuration of the "impurities" (f electrons) may change as a function of temperature. At half band-filling on bipartite lattices the electrons order on alternating sublattices in the low-temperature phase. Besides half band-filling, however, it is not clear yet how the translational symmetry is actually broken. This problem occurs in all self-consistent theories, e.g., in Hartree–Fock theory; see Sect. 3.1. The phase diagram at half band-filling is extensively discussed in [5.53, 5.55]. Since the nearest-neighbor interaction factorizes into its Hartree contributions in infinite dimensions we may even exactly solve the extended simplified Hubbard model in infinite dimensions. A thorough discussion of the thermodynamics for the Bethe lattice can be found with van Dongen [5.55].

The Falicov–Kimball model does *not* describe a Fermi liquid, not even in its translationally invariant metallic phase. In the presence of impurity scattering the imaginary part of the self-energy at the Fermi energy is always finite. After analytical continuation [5.16] we immediately see from (5.68e) that $\mathrm{Im}\{\Sigma(\omega)\} = 0$ at $T = 0$ only if $U = 0$ or $n^f = 0$. Hence, the Falicov–Kimball model is not a suitable starting point for the description of the dynamic properties of the Hubbard model. Nevertheless we can use it to construct a thermodynamically consistent approximation for the free energy of the Hubbard model [5.25, 5.57]; see Sect. 3.3.8. Furthermore, we can study the metal–insulator transition contained in the model, as we now discuss.

On the Bethe lattice the local part of the Green function for non-interacting electrons reads [5.58]

$$\begin{aligned} G_d(\boldsymbol{R},\boldsymbol{R};\omega,U=0) &\equiv G_d(\omega,U=0) \\ &= \frac{4}{W}\left[\frac{2\omega}{W} - \sqrt{\left(\frac{2\omega}{W}\right)^2 - 1}\right] , \end{aligned} \tag{5.69a}$$

where $W = 4\widetilde{t}$ is the bandwidth. The density of states is the Hubbard semi-ellipse

$$D_{d,0}(\epsilon) = \frac{4}{\pi W}\sqrt{1-\left(\frac{2\epsilon}{W}\right)^2} \qquad (|\epsilon| \leq W/2) . \tag{5.69b}$$

At half band-filling ($\mu^d(n=1) = U/2$) and for the special density of states of the Bethe lattice the Dyson equation (5.68f) for the Falicov–Kimball model gives

$$G_d(\omega) = G_d(\omega + U/2 - \Sigma_d(\omega), U=0) \tag{5.70}$$

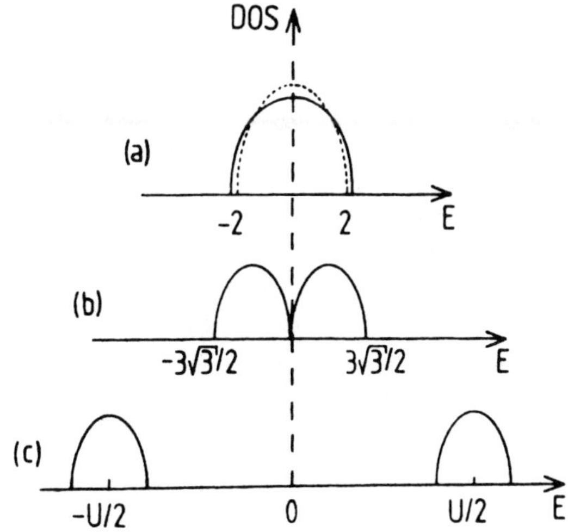

Fig. 5.2. Density of states of the Falicov–Kimball model at half band-filling on the Bethe lattice with infinite coordination number in the translationally invariant phase for (**a**) $U/W \ll 1$ (the *dashed line* is the result for $U = 0$), (**b**) $U/W = (U/W)_c = 1/2$, and (**c**) $U/W \gg 1$ ($W = 4\tilde{t}$, from [5.55]).

for the *local* interacting Green function; see Sect. 3.3.5. The self-energy is obtained from (5.68e), and the self-consistency problem for the Falicov–Kimball model on the half-filled Bethe lattice reduces to the following third-order equation for the Green function

$$[\widetilde{G}_d(\omega)]^3 - 2\widetilde{\omega}[\widetilde{G}_d(\omega)]^2 + \widetilde{G}_d(\omega)\Big(1 + \widetilde{\omega}^2 - (\widetilde{U}/2)^2\Big) - \widetilde{\omega} = 0\,, \tag{5.71}$$

where $\widetilde{G}_d(\omega) = WG_d(\omega)/4$, $\widetilde{U} = 4U/W$, and $\widetilde{\omega} = 4\omega/W$; see Sect. 3.3.5. Hence, we obtain a closed but lengthy expression for the exact Green function on the Bethe lattice with infinite coordination number; for a thorough discussion, see [5.55].

The exact result for the Falicov–Kimball model for $Z \to \infty$ is shown in Fig. 5.2. For weak interactions the density of states broadens and flattens out; in particular, its value at $\epsilon = 0$ ("at the Fermi energy") is reduced. Luttinger's theorem is not fulfilled, so the model does not describe a metallic Fermi-liquid state [5.59]. With the help of (5.71) the density of states at $\epsilon = 0$ is readily calculated from $D_d(\epsilon = 0) = -\mathrm{Im}\,\{G_d(\epsilon = 0)\}\,/\pi$ as

$$D_d(\epsilon = 0) = \begin{cases} \dfrac{4}{\pi W}\sqrt{1 - \left(\dfrac{2U}{W}\right)^2} & \text{for} \quad U \le U_c = W/2 \\ 0 & \text{for} \quad U \ge U_c = W/2 \end{cases} \,. \tag{5.72}$$

It continuously vanishes at the critical coupling $U_c = W/2$ and the single band splits into the upper and lower "Hubbard" bands. For $U \gg W$ the "center of gravity" of the two bands lies at $\pm U/2$. In this limit the density of states for both bands is semi-elliptic again but their width is only $W_1 = W_2 = W/\sqrt{2}$. The gap $\Delta\mu$ continuously opens at the transition, $\Delta\mu(U \to U_c^+) = [W/(3\sqrt{3})](U - U_c)^{3/2}$; see Sect. 3.3.5. This continuous band splitting persists in infinite dimensions even if long-range interactions that Hartree–Fock factorize are taken into account [5.55]. In the presence of long-range Coulomb interactions, however, Mott predicts a discontinuous transition [5.60, Chap. 4.2]. A "first-order" transition with a discontinuous gap as a function of U at $U = U_\mathrm{c}$ either has more subtle than purely energetic reasons, or important effects are neglected in the limit $d \to \infty$ or in the Falicov–Kimball model. Nevertheless, the present example shows that long-range interactions alone do not necessarily lead to a first-order transition or to a jump of the gap as a function of the interaction strength.

The transition at $U = U_\mathrm{c} = W/2$ is not a "true" metal–Mott insulator transition since the state for $U < U_\mathrm{c}$ is not a metallic Fermi-liquid state. Thus it appears to be more natural to analyze a model of spinless fermions to study the metal–insulator transition where the metallic state is a Fermi liquid. In the following we present some results on this model in infinite dimensions.

5.4.5 Spinless Fermions

The Hamilton operator for spinless fermions with nearest-neighbor interaction reads

$$\hat{H}_\mathrm{SF} = \sum_{\boldsymbol{R}_1,\boldsymbol{R}_2} t(\boldsymbol{R}_1 - \boldsymbol{R}_2)\hat{d}^+_{\boldsymbol{R}_1}\hat{d}_{\boldsymbol{R}_2} + \frac{V}{Z}\sum_{\boldsymbol{R},\boldsymbol{\tau}} \hat{n}_{\boldsymbol{R}}\hat{n}_{\boldsymbol{R}+\boldsymbol{\tau}} - \mu\sum_{\boldsymbol{R}} \hat{n}_{\boldsymbol{R}} \,, \tag{5.73}$$

where usually the hopping is restricted to nearest neighbors. For $V = 0$ the model describes a Fermi-liquid metal. For $V/W \to \infty$ and for half band-filling ($n = N/L = 1/2$) the model can be mapped to the Ising model with antiferromagnetic exchange interaction. Thus, on bipartite lattices and for $V \gg W$ the ground state is an insulating charge-density wave for which either the A or the B lattice sites are dominantly occupied. Consequently, there must be a metal–insulator transition at half band-filling at a critical interaction V_c. As mentioned in Sect. 1.5.2, the model can be used to study the charge-ordering in the Verwey transition [5.61].

In one dimension the model is Bethe-ansatz solvable since it can be mapped onto the anisotropic Heisenberg model [5.62, 5.63]. At $V = 0$ it corresponds to the XY model (Fermi liquid), and at $V \gg t$ it becomes equivalent to an Ising model in the z direction (charge-density wave). In the case $V_\mathrm{c}^{d=1} = 2t$ the model is equivalent to the isotropic Heisenberg model. At this value the phase transition takes place from the metallic to the charge-density wave state. For dimensions $d \geq 2$ the charge-density wave is stable for sufficiently strong interactions V even for finite temperatures [5.64]. There are strong indications that the critical interaction at $T = 0$ is given by $V_\mathrm{c}^{d\geq 2} = 0^+$ on bipartite lattices for all $d \geq 2$ [5.65].

In the limit of infinite dimensions the Hartree decoupling becomes exact; see Sect. 5.3.1. This allows a detailed study of the phase diagram of the model [5.66], and a $1/d$ expansion to first order can be carried out explicitly [5.67]. Furthermore, the effects of local energetic disorder of the lattice sites can be studied [5.68], as we now discuss.

Phase Diagram in Infinite Dimensions. At half band-filling and on bipartite lattices the charge-density wave provides the solution of the Hartree equations with the lowest energy. Similar to the Hubbard model with perfect nesting the system is an insulator with a commensurate charge-density wave at all interaction strengths, $V_c^{d=\infty} = 0^+$. At finite doping $\delta = 1/2 - n$ the energetically optimal self-consistent solution of the Hartree equations is not known. The stability analysis of the translationally invariant solution against symmetry breaking shows that the model displays a surprisingly rich phase diagram as a function of doping, temperature, and interaction strength. It not only includes the charge-density wave and the homogeneous state but also phase separation between these two phases, and ordered phases with incommensurate wave vectors. Further details can be found in [5.66].

The $1/d$ corrections [5.67] quantitatively, but not qualitatively, change the phase diagram. Typically, the transition temperatures are monotonously reduced. The corrections for $d = 1$ are formally the largest, which indicates that the series will not converge there. Of course, the series expansion in $1/d$ cannot possibly provide any information on its radius of convergence. A similar problem occurs for the case of the $1/d$ expansion for variational wave functions; see Sect. 3.4.7.

Influence of Disorder. At half band-filling the system forms a charge-density wave at arbitrarily small interaction strength, $V_c^{d=\infty} = 0^+$ [5.66]. In the presence of disorder the transition is shifted to a finite value that depends on the density of states and the strength of the impurity potential [5.68]. In infinite dimensions coherent localization effects are absent, only energetic aspects of the random local impurity potential are relevant. The random distribution of the lattice site energies $\epsilon_{\boldsymbol{R}}$ shall be characterized by their second moment $\gamma^2 = \langle \epsilon_{\boldsymbol{R}}^2 \rangle$. In the charge-density wave two neighboring sites are dominantly occupied by an electron and a hole. At small $V/\widetilde{t}$ this is energetically disadvantageous since it presumes a correlation between the lattice site energies, which is actually not present. If the disorder is strong enough ($\gamma \gtrsim W$) the transition is shifted from zero in the absence of disorder to $V_c^{d=\infty}(\gamma \gtrsim W) \sim \gamma$. At small interactions the metallic state is thus found to be *stabilized* by the disorder since it suppresses the perfect nesting instability. For all finite disorder strengths $\gamma > 0$ the critical nearest-neighbor interaction is finite; for details, see [5.68].

5.5 Approximate Solutions of the Hubbard Model

In principle, every approximate treatment of the single-impurity Anderson model can be used to close formally the self-consistency equations for the infinite-dimensional Hubbard model. We start this section with a specification of our model variants, and summerize those constraints on approximate theories that we could obtain from the exactly solvable limits in the last section. Next, we discuss the merits and limitations of two numerical techniques, namely quantum Monte-Carlo calculations and exact diagonalization studies. The rest of the section is devoted to three analytical approximations to the infinite-dimensional Hubbard model, namely the iterated perturbation theory, the (iterated) non-crossing approximation, and the most recent local-moment approach.

5.5.1 Model Specifications

In the last two sections of this chapter we again focus on the Hubbard model and its metal–insulator transition at half band-filling in the absence of an external magnetic field. The infinite-dimensional Hubbard model in a magnetic field is studied in [5.69]–[5.71], and superconductivity in infinite dimensions is discussed in [5.72, 5.73]. Furthermore, the periodic Anderson model [5.20, 5.74], various other multi-band models [5.75], and the mutual influence between the electron–electron interaction and lattice vibrations [5.39, 5.42, 5.76] or lattice irregularities [5.14, 5.66, 5.77] can be treated along the same lines; for a recent review, see [5.4].

Even if we concentrate on the one-band Hubbard model and its metal-to-insulator transitions, we still have to specify the density of states for the non-interacting system, $D_{\sigma,0}(\epsilon)$. If we want to avoid (slightly) pathological effects, we should not use the Gaussian of the cubic lattice in infinite dimensions as the bare density of states. Its exponential tails may give rise to a small but finite density of states in the gap of the Mott–Hubbard insulator such that its identification becomes a more subtle problem. We cannot use a Lorentzian density of states either since the average kinetic energy is unbound in this case. Here, the non-interacting system is already rather pathological because it reduces to a model with infinite-range hopping in the thermodynamic limit. Although a closed solution for the self-consistency equations can be found, the system is always metallic [5.27] since the kinetic energy necessarily dominates the potential energy. For actual calculations the density of states of the Bethe lattice with infinite coordination number, (5.69b), is most advisable since it allows us to carry out some steps of the calculations analytically.

In addition to the non-interacting density of states we must also specify the correlation function between the kinetic energies at distance $\boldsymbol{q}$ in momentum space, $D_{\boldsymbol{q}}(\epsilon_1, \epsilon_2)$; see Sect. 5.2. The standard choices are (i) the condition for perfect nesting (5.26), and (ii) the condition for a perfectly random dispersion relation (5.28). Although case (i) allows for antiferromagnetic order, the Mott–Heisenberg insulator is excluded in the "fully frustrated" case (ii), and its phase diagram shows the metal and the Mott–Hubbard insulator only.

5.5.2 Constraints on Approximate Treatments

According to Mott's general arguments [5.60], the Hubbard model with perfectly random dispersion relation should display a transition between the metallic Fermi-liquid state and the Mott–Hubbard insulator. The value for the critical interaction strength should be of the order of the bandwidth, $U_c \approx W$; compare Sect. 1.4.6. Although the very existence of this transition is indicated by all numerical and analytical treatments of the infinite-dimensional Hubbard model, a formal proof does not exist yet.

The proof for the existence of the metallic state for $U > 0$ and thus $U_c > 0$ presented in [5.78] works only if we are able to separate low and fast time scales (high and low energy excitations), and, moreover, the single-impurity Anderson model has to scale to its "Toulouse limit", which was first studied in the connection of the Kondo model [5.79]. If these conditions are met, the iterated Toulouse limit for the Anderson model gives a self-consistent metallic solution of the infinite-dimensional Hubbard model for some finite $U > 0$. However, we may only speculate whether these conditions are ever fulfilled; see also Sect. 5.6.4.

In the same way a rigorous proof for the existence of the insulating phase for some $U > U_c \neq \infty$ has not been achieved yet. The derivation formulated in [5.29] is based on the same assumptions that lead to the alloy-analogy approximation for the Hubbard model; see Sect. 3.3.5. Hence, we are only left with the same plausibility arguments that constitute the Hubbard approximations for the description of the Mott–Hubbard insulator.

Nevertheless, we can formulate some "minimal" requirements on a good approximate theory for the Hubbard model with a perfectly random dispersion relation and a semi-elliptic density of states at half band-filling.

1. It describes a Fermi liquid for $U < U_c$. In particular, the density of states at the Fermi energy is given by its value in the absence of interactions (5.58), the imaginary part of the self-energy vanishes quadratically at the Fermi energy, and the quasi-particle renormalization factor (5.59) is finite.
2. It describes an insulator with a gap in the one-electron density of states for $U > U_c$, the entropy of the Mott–Hubbard insulator is expected to be finite in infinite dimensions; see Sect. 5.2.2.
3. For large U/W the width of both the upper and the lower Hubbard band is given by $W_{1,2} = W$ (see Sect. 5.4.2) and the atomic limit, $W = 0$, is properly reproduced.

As we will see in the next section, these conditions are necessary but far from sufficient to decide whether the transition between the metal and the Mott–Hubbard insulator is discontinuous or not.

In the presence of perfect nesting an approximate theory should also properly describe the Mott–Heisenberg insulator at half band-filling. Ideally, such a theory covers the metallic Fermi liquid, the Mott–Hubbard insulator, and the Mott–Heisenberg insulator. For a semi-elliptic density of states and in the presence of perfect nesting we may formulate the following additional criteria for a good approximate theory for half band-filling.

4. The Néel temperature behaves like $k_B T_N(U) = (W/4)^2/U - \mathcal{O}(W^3/U^2)$ for $U/W \gg 1$; see Sect. 5.1.1. For small interaction strengths, we expect $k_B T_N(U \to 0) \sim \exp(-1/(D_{\sigma,0}(0)U))$; see Sect. 3.1.2.
5. The one-electron density of states for strong interactions is given by the sum of two Hubbard ellipses of width W; see Sect. 5.4.2.

The five criteria derived here allow us to assess the quality of various numerical and analytical approaches which we discuss in the rest of this section. Explicit results for the infinite-dimensional Hubbard model are presented in the next section.

5.5.3 Quantum Monte-Carlo Calculations

Quantum Monte-Carlo calculations are the numerical standard approach to correlated electron systems. A particularly efficient algorithm for the Kondo and Anderson models was developed by Hirsch and Fye, and Gubernatis, Hirsch, and Scalapino [5.80]. This program is usually employed in the Monte-Carlo studies of the Hubbard model in infinite dimensions to solve the single-impurity problem [5.21, 5.26, 5.29], [5.81]–[5.84]. The quantum Monte-Carlo method is almost a science on its own and we will not try an attempt to discuss it here. Hence, we only sketch possibilities and limitations of this technique. The discussion follows [5.82, 5.83].

We seek the Green function of the impurity problem at imaginary times, $G_\sigma^{\text{And}}(\tau)$, iterated to convergence. From this quantity we may directly infer the free energy of the Hubbard model in infinite dimensions [5.16]. To this end we have to solve the impurity problem for a given bath Green function. For a numerical treatment the four-fermion term in the Hubbard interaction is decoupled with the help of a discrete Hubbard–Stratonovich transformation [5.85]. This results in a partition function over Ising variables on a chain with $\mathcal{L}$ sites. In this way we have split the temperature interval $[0,\beta]$ into $\mathcal{L}$ equidistant parts $\Delta\tau = \beta/\mathcal{L}$. Since the limit $\mathcal{L} \to \infty$ cannot explicitly be realized the method is principally limited to temperatures $k_B T > \mathcal{O}(W/\mathcal{L})$. The partition of the temperature interval itself is already a source for systematic errors since we may extrapolate to $\mathcal{L} \to \infty$ only. The partition function can exactly be summed for $\mathcal{L} \leq 16$. Beyond that it can only approximately be calculated within the quantum Monte-Carlo method, which provides another source of systematical and statistical errors. Fortunately, at half band-filling (symmetric Anderson model) the minus-sign problem does not play a role because of electron–hole symmetry.

After the single-impurity Anderson model has been solved numerically for one bath Green function we have to repeat the process until convergences of the self-consistency equations is obtained. Then we know the imaginary-time Green function at discrete points in the interval $[0,\beta]$. However, we are interested in $G_\sigma^{\text{And}}(\omega)$ on the real axis, or, equivalently, the self-energy $\Sigma_\sigma(\omega)$, which allows us to obtain the spectral function $A_\sigma(\boldsymbol{k},\omega)$ of the infinite-dimensional Hubbard model. The continuation of the imaginary-time Green function to the real axis is mathematically an ill-conditioned problem [5.86, 5.87] since the data set is limited and the Green function $G_\sigma^{\text{And}}(\tau)$ is without any significant structures. As the next step we thus have to use an approximation scheme, e.g., the maximum entropy algorithm, to continue the Green function to the real axis; see, e.g., [5.87]. This method is particularly problematic when we have to decide whether there is a structure in the density of states near the Fermi energy that only has a width $\Delta \ll W$. This Abrikosov–Suhl or Kondo resonance contains the essential physics and its existence or not decides whether the Hubbard model at a given interaction U describes a metal or an insulator. The approach can be stabilized, if we provide information on the behavior of the Green function at large energies from approximate theories, e.g., from NCA calculations [5.83]. Unfortunately, this can lead to "slightly" unphysical results.

These principle problems (discretization of the temperature interval, approximate evaluation of the partition function, extrapolation via maximum entropy) already occur for the calculation of the impurity Green function $G_\sigma^{\text{And}}(\tau)$ for a given bath Green function $G_{\sigma,0}^{\text{And}}(\tau)$. For the Hubbard model the whole procedure has to be iterated to convergence. Near the metal–insulator transition, however, the procedure converges only slowly. This is the typical critical slowing-down for all systems close to a phase transition. Moreover, the convergence is not monotonic, i.e., rapid oscillations between the insulating and the metallic solution occur in some interval around the critical interaction U_{c}. Some groups [5.4] interpret this behavior as coexistence of both solutions in some "critical interval" (M. Rozenberg, private communication, 1996). In contrast, other researchers find that the oscillations vanish when the numerical precision is increased (T. Pruschke, Habilitation thesis, 1995). In the vicinity of the phase transition very low temperatures thus cannot be reached because the iteration itself and not the single quantum Monte-Carlo calculation becomes too time-consuming and numerically inaccurate.

The quantum Monte-Carlo technique is a genuine finite-temperature approach. Within the restrictions discussed above, i.e., for not too low temperatures, it gives (numerically) *exact* results. Its second major advantage lies in the fact that it also allows us to calculate (local) two-electron propagators at finite temperatures, e.g., the magnetic susceptibility. This task is still rather difficult for other approximate methods. Consequently, the Néel temperature, which signals the transition into the antiferromagnetic Mott–Heisenberg insulator, can be calculated accurately and efficiently with this technique, as long as the interaction strength is not too large, $U \lesssim 1.5\,W$.

5.5.4 Exact Diagonalization Studies

Instead of the discretization of the temperature interval at finite temperatures we may discretize the single-impurity Anderson model, (5.47), in energy space [5.88, 5.89]; see also [5.90]. In this way the Anderson model is specified by $\mathcal{L}$ parameters $\Delta(\epsilon_n)$, which have to be determined self-consistently. Typically, $\mathcal{L}$ ranges from five to twelve, at the most. Obviously, the result cannot be more precise than the discretization of the bandwidth, $\epsilon_n = W/\mathcal{L}$.

The truncated impurity problem is solved exactly, and the self-consistency cycle provides the bath Green function for the next iteration, i.e., the new set of parameters $\Delta(\epsilon_n)$. However, a new problem results from the fact that the bath Green function contains more information (poles) after the iteration than before. Since the number $\mathcal{L}$ of parameters $\Delta(\epsilon_n)$ must not be changed within the same iteration procedure, we need to invent an approximation scheme to redetermine the parameter set for the next iteration from the outcome of the last iteration. To this end we may either employ Padé approximants, a Gauss quadrature as in error analysis [5.88], or an optimization of spectral moments with the help of a continued fraction expansion [5.90, 5.91]. The latter method is similar in spirit to the solution of the Hubbard model on clusters [5.92]. The full optimization of a Gauss quadrature is numerically costly for large $\mathcal{L}$, and the continued fraction expansion misses the physics at low energies. Hence, a combination of the two methods was proposed in [5.89].

In addition to the fact that a projection has to be performed after every step of the iteration procedure, this approach has the following problems: (i) the data, e.g., the eigenenergies, are discrete, and (ii) extrapolation to $\mathcal{L} \to \infty$ is not straightforward. The second point implies that larger $\mathcal{L}$ give better results but the convergence may be very slow. The first point implies that we have to "smear out" the discrete spectrum to obtain continuous curves, e.g., for the density of states. The width $\Delta\omega$ of the smearing function can be used as measure for the energy resolution and thus for the precision of the technique. As in the quantum Monte-Carlo calculations convergence problems arise in the vicinity of the Mott–Hubbard transition.

The discretization of the energy interval is numerically more efficient at zero temperature than the quantum Monte-Carlo procedure, yet it cannot be used too close to the Mott–Hubbard transition. There, the smallness of the gap and the small width of the quasi-particle peak set new small energy scales that cannot be resolved adequately by numerical methods. In the next three subsections we discuss three analytical approximations to the infinite-dimensional Hubbard model. Since they are not exact their predictions on the Mott–Hubbard and Mott–Heisenberg transition are also only approximate.

5.5.5 Iterated Perturbation Theory

The simplest way to obtain some insight into the single-impurity Anderson model at half band-filling is second-order perturbation theory in U/t relative to non-magnetic Hartree–Fock theory, i.e., we set the chemical potential $\mu = E_d = U/2$. As shown by Yamada and Yosida [5.32, 5.33] this approach works surprisingly well for the single-impurity Anderson model; for a review, see [5.31]. Therefore, we may hope to find a good first approximation to the infinite-dimensional Hubbard model as well [5.4, 5.27, 5.81]. Since the approach is not exact it cannot prove the existence of a transition at a critical interaction strength, and the results, especially in the critical region, must be analyzed with care. However, the approximation provides a first glimpse into the problems of the description of the Mott–Hubbard transition in infinite dimensions.

As already explained in Sect. 5.4.1, second-order perturbation theory relative to the paramagnetic Hartree–Fock solution reduces to the correct *atomic* limit for vanishing hopping. From this we may infer that the perturbatively motivated ansatz [5.27, 5.81]

$$\Sigma_{\sigma}^{\mathrm{And}}(\mathrm{i}\omega_n, \{G_{\sigma,0}^{\mathrm{And}}\}) = U^2 \int_0^{\beta} \mathrm{e}^{\mathrm{i}\omega_n \tau} \left[G_{\sigma,0}^{\mathrm{And}}(\tau)\right]^3 \tag{5.74}$$

for the proper (Matsubara) self-energy will *qualitatively* work both for weak and large interaction strengths. For a given bath Green function this equation defines the starting point for the iteration procedure ("iterated perturbation theory"). Other self-consistent approaches discussed in Sect. 5.4.1 use the non-interacting local Green function or the self-consistently determined lattice Green function $G_{\sigma}(\boldsymbol{k},\omega)$ (including ladder and bubble sums) instead of the bath Green function. Therefore, these approaches are different from each other. The above description is limited to the paramagnetic half-filled Hubbard model. A generalization of (5.74) to densities below half band-filling is ambiguous; see [5.93] for an example.

It is clear from the beginning that the theory will give a good approximation for small interaction strengths in the Fermi-liquid phase, and that it will not miss the existence of the upper and lower Hubbard bands. However, it is not custom-made for the Mott insulator since it does not explicitly contain the formation of local moments. Therefore, we cannot expect a good description of the transition to the antiferromagnetic Mott–Heisenberg insulator or the degenerate Mott–Hubbard insulator.

5.5.6 Non-Crossing Approximation

Another well-known approach to the Kondo model and the single-impurity Anderson model is the "non-crossing approximation" (NCA) in which a certain class of diagrams is summed; for a review, see [5.94]. The comparison with known exact results for the Kondo model from the Bethe ansatz shows that the NCA is a reliable approximation for the Kondo model, as long as the temperature/frequencies are not too small. Quantitatively, $\omega/T_{\mathrm{K}} \gtrsim 0.1$, $T/T_{\mathrm{K}} \gtrsim 0.1$ have to hold, where T_{K} is the effective low-energy scale of the system ("Kondo temperature"). This implies that the NCA misses the Fermi-liquid physics at low temperatures. Beyond that temperature/frequency range, however, the results are accurate.

The original NCA was designed for the half-filled single-impurity Anderson model at large interaction U. For our purposes we need a "generic" extension to finite interactions. Such a generalization was proposed by Pruschke and Grewe [5.95], and was subsequently used in the context of the infinite-dimensional Hubbard model [5.3, 5.21, 5.83]. Although less thoroughly tested than the NCA for the large-U limit, its extension to the case of finite interaction strength appears to be applicable in the temperature regions cited above.

In contrast to the iterated perturbation theory with its single equation (5.74) for the self-energy, the equations of the NCA are based on four self-energies for the four possible ionic configurations ($\uparrow\downarrow$, $\uparrow$, $\downarrow$, $\circ$) of the site with local energies $E_{\uparrow\downarrow} = U - 2E_d$, $E_\sigma = -E_d$, and $E_\circ = 0$. The four ionic resolvents $P_M(z)$ are written in terms of their self-energies $\Sigma_M(z)$ ($M = \uparrow\downarrow, \uparrow, \downarrow, \circ$) as [5.95]

$$P_M(z) = \frac{1}{z - E_M - \Sigma_M(z)} \tag{5.75}$$

with

$$\Sigma_\circ(z) = \sum_\sigma \int \frac{\mathrm{d}\epsilon}{\pi} \Delta(\epsilon) f_{\mathrm{FD}}(\epsilon) \Lambda_{0,\sigma}(z,\epsilon) P_\sigma(z+\epsilon) , \tag{5.76a}$$

$$\begin{aligned} \Sigma_\sigma(z) = & \int \frac{\mathrm{d}\epsilon}{\pi} \Delta(\epsilon) f_{\mathrm{FD}}(-\epsilon) \Lambda_{0,\sigma}(z-\epsilon,\epsilon) P_\circ(z-\epsilon) \\ & + \int \frac{\mathrm{d}\epsilon}{\pi} \Delta(\epsilon) f_{\mathrm{FD}}(\epsilon) \Lambda_{2,\sigma}(z+\epsilon,\epsilon) P_{\uparrow\downarrow}(z+\epsilon) , \end{aligned} \tag{5.76b}$$

$$\Sigma_{\uparrow\downarrow}(z) = \sum_\sigma \int \frac{\mathrm{d}\epsilon}{\pi} \Delta(\epsilon) f_{\mathrm{FD}}(-\epsilon) \Lambda_{2,\sigma}(z,\epsilon) P_\sigma(z-\epsilon) , \tag{5.76c}$$

where $f_{\mathrm{FD}}(\epsilon)$ denotes the Fermi–Dirac distribution and $\Delta(\epsilon)$ was defined in (5.48). In the basic version of the theory the vertex functions are given by

$$\Lambda_{0,\sigma}(z,z') = 1 + \int \frac{\mathrm{d}x}{\pi} f_{\mathrm{FD}}(x) \Delta(x) P_{-\sigma}(z+x) P_{\uparrow\downarrow}(z+z'+x) , \tag{5.77a}$$

$$\Lambda_{2,\sigma}(z,z') = 1 + \int \frac{\mathrm{d}x}{\pi} f_{\mathrm{FD}}(-x) \Delta(x) P_{-\sigma}(z-x) P_\circ(z-z'-x) , \tag{5.77b}$$

but a more complicated expression may also be used [5.95]. The approximation obtained in this way is derivable from a generating functional, and thus is conserving. For a given hybridization function $\Delta(\epsilon)$ we can calculate the ionic resolvents (5.75) from (5.76) and (5.77).

The local partition function is found as

$$\mathcal{Z}_{\mathrm{loc}} = \sum_M \oint_C \frac{\mathrm{d}z}{2\pi \mathrm{i}} P_M(z) , \tag{5.78a}$$

and the local Matsubara Green function of the impurity becomes

$$\begin{aligned} G_\sigma^{\text{And}}(\mathrm{i}\omega_n) &= \frac{1}{\mathcal{Z}_{\text{loc}}} \oint_C \frac{\mathrm{d}z}{2\pi\mathrm{i}} \mathrm{e}^{-\beta z} \Big[P_\circ(z) \Lambda_{0,\sigma}(z, \mathrm{i}\omega_n) P_\sigma(z+\mathrm{i}\omega_n) \\ &\quad + P_{-\sigma}(z) \Lambda_{2,-\sigma}(z+\mathrm{i}\omega_n, \mathrm{i}\omega_n) P_{\uparrow\downarrow}(z+\mathrm{i}\omega_n) \Big] . \end{aligned} \tag{5.78b}$$

With the help of (5.77) the Green function can be expressed explicitly in terms of the ionic resolvents and their spectral densities; see [5.95]. Obviously, even for a given hybridization function $\Delta(\epsilon)$ the evaluation of the local Green function is not simple and requires considerable numerical effort. The self-consistency is obtained via the cycle of equations (5.50), which we derived in Sect. 5.3.3. Apparently, the theory is directly applicable at all band fillings, in contrast to the iterated perturbation theory.

5.5.7 Local Moments in Hubbard's Approximations

Both the iterated perturbation theory and the non-crossing approximation use a single self-energy for the description of the Green function of the Hubbard model. In the insulator with a gap in the single-particle spectrum and, hence, two distinct Hubbard bands it is in many respects more natural to use two self-energies to describe two (separate) Hubbard bands.

A description based implicitly on local moments is already contained in Hubbard's third paper [5.96]; see Sect. 3.3. On the Bethe lattice the alloy-analogy approximation gives the following expression for the local Green function, (3.46a),

$$G_\sigma(\omega) = \frac{1}{2}\left[\frac{1}{\omega + U/2 - (W/4)^2 G_\sigma(\omega)} + \frac{1}{\omega - U/2 - (W/4)^2 G_\sigma(\omega)}\right] . \tag{5.79}$$

This can be cast into the form

$$G_\sigma(\omega) = \frac{1}{2}\left[G_{A,\sigma}(\omega) + G_{B,\sigma}(\omega)\right] , \tag{5.80a}$$

where

$$G_{A,\sigma}(\omega) = \frac{1}{\omega - (W/4)^2 G_\sigma(\omega) - \Sigma_{A,\sigma}(\omega)} , \tag{5.80b}$$

$$G_{B,\sigma}(\omega) = \frac{1}{\omega - (W/4)^2 G_\sigma(\omega) - \Sigma_{B,\sigma}(\omega)} \tag{5.80c}$$

denote the Green functions for the case in which the site was originally predominantly occupied by an $\uparrow$ electron (A) or by a $\downarrow$ electron (B) when we created/annihilated the σ electron on that site. In the paramagnetic insulator the A and B sites are equally likely to be surrounded by other A and B sites such that the Feenberg self-energy is given by $S_\sigma(\omega) = (W/4)^2 G_\sigma(\omega)$; see (5.54).

Obviously, there are two self-energies in the alloy-analogy approximation to the Hubbard model:

$$\Sigma_{A,\sigma}^{\mathrm{AAA}}(\omega) = -\sigma\frac{U}{2}, \tag{5.81a}$$

$$\Sigma_{B,\sigma}^{\mathrm{AAA}}(\omega) = +\sigma\frac{U}{2}, \tag{5.81b}$$

which correspond to the self-energies for the upper and lower Hubbard bands, shifted by the chemical potential $\mu = U/2$. As shown in Sect. 5.4.4, the alloy-analogy approximation becomes exact for the Falicov–Kimball model in infinite dimensions.

The local moments are static in the Falicov–Kimball model but *dynamic* in the Hubbard model. This has been taken partially into account in the (full) Hubbard-III approximation which can be written in the same form as (5.80) but with the frequency-dependent self-energies

$$\Sigma_{A,\sigma}^{\mathrm{H.\text{-}III}}(\omega) = -\sigma\frac{U}{2} - \frac{U(W/4)^2 G_\sigma(\omega)}{U - \sigma U/2 - 3(W/4)^2 G_\sigma(\omega)}, \tag{5.82a}$$

$$\Sigma_{B,\sigma}^{\mathrm{H.\text{-}III}}(\omega) = \sigma\frac{U}{2} + \frac{U(W/4)^2 G_\sigma(\omega)}{U + \sigma U/2 - 3(W/4)^2 G_\sigma(\omega)} \tag{5.82b}$$

(compare also [5.4, 5.97]). As shown in Sect. 3.3 for the alloy-analogy approximation and the (full) Hubbard-III approximation, the two self-energies can always be combined to give a single self-energy. In the metallic phase this combined self-energy has to fulfill the Fermi-liquid properties as discussed in Sect. 5.4. The Hubbard approximations fail to reproduce Fermi-liquid behavior in the metallic phase because the scattering is still elastic at the Fermi surface as for the impurity scattering in the Falicov–Kimball model.

It is important to note that the Fermi-liquid conditions pose restrictions onto a theory only in the vicinity of $\omega = 0$. They do not invalidate the local-moment picture at frequencies $\omega = \pm|m|U/2$ where $|m|$ is the size of the local moments.

5.5.8 Local-Moment Approach

In the preceding subsection we used Hubbard's approximations to illustrate the idea and importance of local moments for the description of the Mott–Hubbard insulator. Of course, this description is seen in the Hubbard approximations only a posteriori. Logan, Eastwood, and Tusch [5.98, 5.99] were the first to use the local-moment approach from the outset.

Mott–Heisenberg Insulator. In the case of antiferromagnetism a description with two self-energies is evidently necessary since the translational symmetry is broken down to the two sublattices. On the Bethe lattice the self-consistency equations are given by (5.55),

$$G_{A,\sigma}(\omega) = \frac{1}{\omega - (W/4)^2 G_{B,\sigma}(\omega) - \Sigma_{A,\sigma}(\omega)}, \tag{5.83a}$$

$$G_{B,\sigma}(\omega) = \frac{1}{\omega - (W/4)^2 G_{A,\sigma}(\omega) - \Sigma_{B,\sigma}(\omega)}, \tag{5.83b}$$

and we have to specify the equations for the self-energies on the two sublattices.

The essential dynamic process on a given site is the spin flip that remains when a σ electron is added to a site and the $-\sigma$ electron leaves the site afterwards. Spin-flip excitations are described by the transverse spin polarization propagator $\Pi^{\sigma,-\sigma}(\boldsymbol{R}_1, \boldsymbol{R}_2) = \langle\langle \hat{S}^{\sigma}(\boldsymbol{R}_1)\hat{S}^{-\sigma}(\boldsymbol{R}_2)\rangle\rangle$. Its RPA form for the local part reduces to

$$\Pi_{\alpha}^{\sigma,-\sigma}(\omega) = \frac{{}^{0}\Pi_{\alpha}^{\sigma,-\sigma}(\omega)}{1 - U\,{}^{0}\Pi_{\alpha}^{\sigma,-\sigma}(\omega)} \quad ; \quad \alpha = A, B \tag{5.84}$$

in infinite dimensions where

$${}^{0}\Pi_{\alpha}^{\sigma,-\sigma}\omega) = \mathrm{i}\int_{-\infty}^{\infty} \frac{\mathrm{d}\omega'}{2\pi} G_{\alpha,-\sigma}^{\text{LM-HF}}(\omega') G_{\alpha,\sigma}^{\text{LM-HF}}(\omega' - \omega) \tag{5.85}$$

is the independent-electron part of the propagator. The independent-electron Green functions are defined in terms of the "Hartree–Fock" self-energies

$$\Sigma_{A,\sigma}^{\text{LM-HF}}(\omega) = -\sigma\frac{U|m|}{2}, \tag{5.86a}$$

$$\Sigma_{B,\sigma}^{\text{LM-HF}}(\omega) = +\sigma\frac{U|m|}{2} \tag{5.86b}$$

as

$$G_{\sigma}^{\text{LM-HF}}(\omega) = \frac{1}{2}\left[G_{A,\sigma}^{\text{LM-HF}}(\omega) + G_{B,\sigma}^{\text{LM-HF}}(\omega)\right] \tag{5.87a}$$

with

$$G_{A,\sigma}^{\text{LM-HF}}(\omega) = \frac{1}{\omega - (W/4)^2 G_{B,\sigma}^{\text{LM-HF}}(\omega) - \Sigma_{A,\sigma}^{\text{LM-HF}}(\omega)}, \tag{5.87b}$$

$$G_{B,\sigma}^{\text{LM-HF}}(\omega) = \frac{1}{\omega - (W/4)^2 G_{A,\sigma}^{\text{LM-HF}}(\omega) - \Sigma_{B,\sigma}^{\text{LM-HF}}(\omega)}. \tag{5.87c}$$

The moment is self-consistently obtained from

$$|m| = \int_{-\infty}^{0} \mathrm{d}\omega \left(D_{A,\uparrow}^{\text{LM-HF}}(\omega) - D_{A,\downarrow}^{\text{LM-HF}}(\omega)\right), \tag{5.88}$$

where $D_{A,\sigma}^{\text{LM-HF}}(\omega)$ is the density of states of the corresponding Green function.

We want to scatter a σ electron off the spin-flip excitation described in terms of $\Pi_{\alpha}^{\sigma,-\sigma}(\omega)$. A spin flip occurs if a σ electron hops onto a site, and the resulting $-\sigma$ electron hops off into the surrounding medium. To correctly describe this process [5.98, 5.99] we need the Green function that avoids multiple spin-flip scatterings at that site. It is given by

$$\mathcal{G}_{\alpha,\sigma}(\omega) = \frac{1}{\left[G_{\alpha,\sigma}(\omega)\right]^{-1} + \Sigma_{\alpha,\sigma}(\omega)} \quad ; \quad \alpha = A, B \; . \tag{5.89}$$

In terms of these objects the approximation for the self-energy finally becomes

$$\Sigma_{\alpha,\sigma}(\omega) = U^2 \int_{-\infty}^{\infty} \frac{\mathrm{d}\omega'}{2\pi\mathrm{i}} \mathcal{G}_{\alpha,-\sigma}(\omega - \omega') \Pi_{\alpha}^{-\sigma,\sigma}(\omega') \quad ; \quad \alpha = A, B \; . \tag{5.90}$$

Equations (5.84)–(5.90) together with the self-consistency equations (5.83) constitute the local-moment approximation for the Mott–Heisenberg insulator in its basic form. The theory is naturally extended by using the full local Green function $G_{\alpha,\sigma}(\omega)$ or $\mathcal{G}_{\alpha,\sigma}(\omega)$ in the calculation of the spin-flip propagator in (5.85), and is robust to such renormalizations [5.98, 5.99].

Mott–Hubbard Insulator. Thus far all moments have been ordered on their corresponding sublattice sites. In agreement with Mott's general arguments presented in Sect. 1.4, local moments are always present for finite interaction strengths which, however, need not be ordered [5.60, Chap. 4.9]. Therefore, we will still have A and B sites predominantly occupied by a $\uparrow$ or $\downarrow$ electron. Hence, the above equations (5.84)–(5.90) remain valid but the self-consistency equations are not given by (5.83) as in the antiferromagnetic case but by

$$G_{A,\sigma}(\omega) = \frac{1}{\omega - (W/4)^2 G_{\sigma}(\omega) - \Sigma_{A,\sigma}(\omega)} \; , \tag{5.91a}$$

$$G_{B,\sigma}(\omega) = \frac{1}{\omega - (W/4)^2 G_{\sigma}(\omega) - \Sigma_{B,\sigma}(\omega)} \tag{5.91b}$$

with

$$G_{\sigma}(\omega) = \frac{1}{2}\left[G_{A,\sigma}(\omega) + G_{B,\sigma}(\omega)\right] \; , \tag{5.91c}$$

because every A site is surrounded by as many B sites as A sites in infinite dimensions, and the Feenberg part of the self-energy changes according to (5.53).

If a spin at some site is equally likely to be surrounded by $\uparrow$ and $\downarrow$ electrons, it does not cost any energy to flip this spin, in contrast to the case in a Néel antiferromagnet where this energy is finite. Consequently, the RPA condition

$$U^0 \Pi^{\sigma,-\sigma}(\omega_s) = 1 \tag{5.92}$$

for the spin-flip pole must give $\omega_s^{\mathrm{P}} = 0$ for the paramagnetic insulator. It can be shown that this is indeed consistent with the definition (5.88) of the local moment if we perform our calculation of the spin propagator in (5.85) using the independent-electron Green functions $G_{\alpha,\sigma}^{\mathrm{LM\text{-}HF}}(\omega)$ [5.98, 5.99].

After this brief review of present-day approximation schemes for the Hubbard model in infinite dimensions we now discuss their predictions for the Mott transition.

5.6 Metal–Insulator Transitions

In the final section of this chapter we present and compare the results of the various approximate methods discussed in the preceeding section for the Mott–Heisenberg insulator, the metallic phase, and the Mott-Hubbard insulator. The transition between the paramagnet to the Mott–Heisenberg insulator is a common thermodynamic phase transition that is well understood. On the contrary, the nature of the quantum phase transition between the Fermi-liquid ground state and the Mott–Hubbard insulator is still a matter of debate. However, we will argue that both the quasi-particle weight of the metal and the gap of the insulator vanish at a common critical interaction strength. In this way the phase diagram of the infinite-dimensional Hubbard model confirms Mott's general views on the metal–insulator transition.

5.6.1 Mott–Heisenberg Insulator

The transition from the paramagnetic to the antiferromagnetic phase is a second-order thermodynamic phase transition. It shows up as a divergence in the momentum-dependent magnetic susceptibility in the paramagnetic phase [5.26, 5.83, 5.84]. Alternatively, we can compare the free energy densities of the paramagnetic and antiferromagnetic phases in the iterated solutions to find the optimal phase [5.81, 5.82].

The limiting behavior of the Néel temperature is known for both strong and weak coupling; see Sect. 5.4. In the intermediate regime the transition temperature is comparably large, of the order of $10^{-2}\,W$, and is thus within the reach of the quantum Monte-Carlo technique. The result of such a calculation for the hypercubic lattice [5.26] is shown in Fig. 5.3. The corresponding curve for the (non-bipartite) infinite-dimensional diamond lattice can be

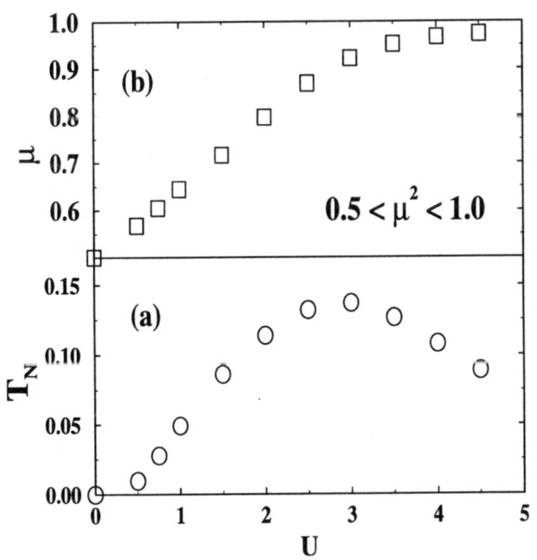

Fig. 5.3. Square of the magnetic moment in the z direction, $\mu^2 = \langle(\hat{n}_{R,\uparrow} - \hat{n}_{R,\downarrow})^2\rangle$ at the Néel temperature, and the Néel temperature T_{N} in the infinite-dimensional half-filled Hubbard model on a hypercubic lattice as a function of the interaction strength, calculated with the quantum Monte-Carlo method; the energy units are such that $W = 2\sqrt{2}$ (from [5.26]).

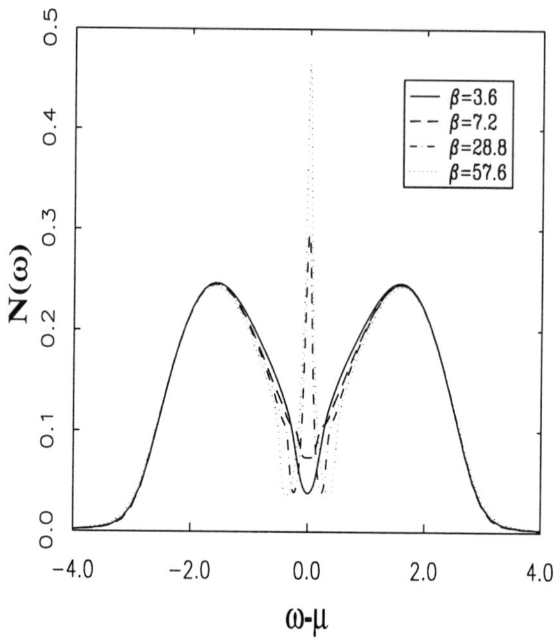

Fig. 5.4. One-electron density of states $N(\omega) \equiv D_\sigma(\omega)$ of the paramagnetic metal ($U = 3$) of the infinite-dimensional half-filled Hubbard model on a hypercubic lattice as a function of frequency for various temperatures, calculated with the quantum Monte-Carlo method; the units of energy are such that $W = 2\sqrt{2}$ (from [5.21]).

found in [5.84]. At the transition the susceptibility diverges with a power law and exponent $\nu = 1$ [5.83, 5.84]. The sublattice magnetization also displays Curie–Weiß behavior with critical exponent $\beta = 1/2$ [5.84].

The Néel temperature approaches its limiting value $k_\mathrm{B}T_\mathrm{N}(U \gg W) = (W/4)^2/U$ for large U/W; see Sect. 5.4.2. As is seen from Fig. 5.3, the agreement is already satisfactory for $U \gtrsim W$. The reason for this can be understood from the behavior of the magnetic moment in the z direction, $\mu^2 = 1 - 2\langle\hat{D}\rangle/L$, whose size as a function of the interaction strength just *above* the transition is also shown in Fig. 5.3. Above $U \approx W$ the moments have almost reached their saturation value, i.e., the density of holes and double occupancies is very small. As seen from the figure, local moments are well formed above the thermodynamic phase transition for all interaction strengths, in agreement with Mott's general arguments presented in Sect. 1.4.

The approximate analytical treatments discussed in the last section do not yet provide a detailed description of the Néel temperature as a function of the interaction strength. Exact diagonalization studies and the non-crossing approximation have not been applied to this case, and the iterated perturbation theory approach apparently fails for the antiferromagnet. As seen from Fig. 42 of [5.4], the Néel temperature as a function of the interaction strength is not well reproduced by iterated perturbation theory in its present form. The local-moment approach [5.98, 5.99] has so far been applied to the ground state. It becomes exact in the strong-coupling limit, e.g., it reproduces the known results for a single hole in an antiferromagnetic background; see Sect. 5.4.2. If the excitation energy for a single spin flip is taken as an estimate for the

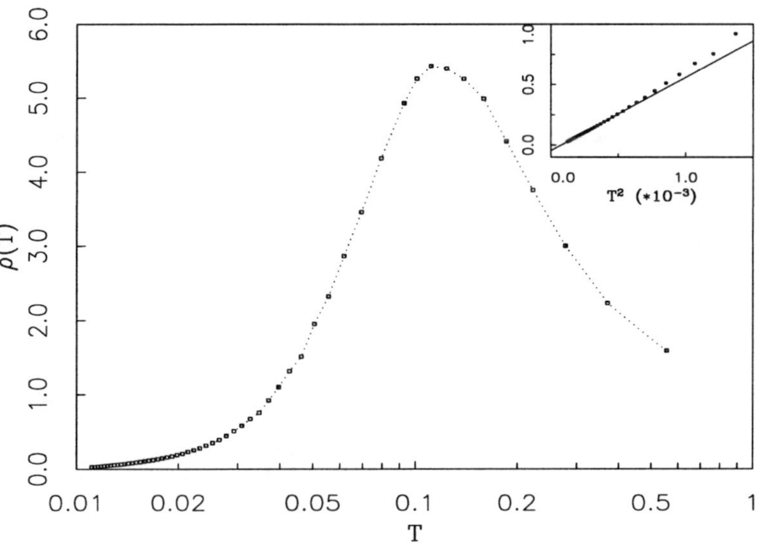

Fig. 5.5. Specific resistivity as a function of temperature in the paramagnetic metallic phase of the infinite-dimensional Hubbard model ($U = 3$) on a hypercubic lattice at half band-filling; the energy units are such that $W = 2\sqrt{2}$, the units of $\rho(T)$ are $[\pi e^2 a/(2\hbar)]^{-1} \approx 10^{-4}$ Ohm-cm (from [5.21]).

Néel temperature, a good agreement with the quantum Monte-Carlo data for $U \gtrsim W$ can be obtained [5.99]. Hence, the local-moment approach appears to capture the essential features of the Mott–Heisenberg insulator.

5.6.2 Correlated Metal

In the absence of perfect nesting we expect to find a metallic state at all temperatures for interactions that are not too strong. According to (5.46) the one-electron spectral function $A_\sigma(\epsilon, \omega)$ alone already determines the conductivity of the infinite-dimensional Hubbard model. The one-electron density of states is given by $D_\sigma(\omega) = \int d\epsilon A_\sigma(\epsilon, \omega)$. Its temperature dependence for a hypercubic lattice as obtained from quantum Monte-Carlo calculations is shown in Fig. 5.4 for $U = 3W/(2\sqrt{2}) \approx W$.

For low temperatures the density of states of the metallic phase displays the Kondo or Abrikosov–Suhl resonance, which signals the existence of Landau quasi-particles in the spectral function. The system apparently is a Fermi liquid. Nevertheless, the upper and lower Hubbard bands are already clearly visible in the metallic regime.

To reach temperatures lower than those accessible to quantum Monte-Carlo calculations we may employ the non-crossing approximation. Results for the specific resistivity are shown in Fig. 5.5. The agreement with quantum Monte-Carlo results is excellent for all temperatures for which the latter

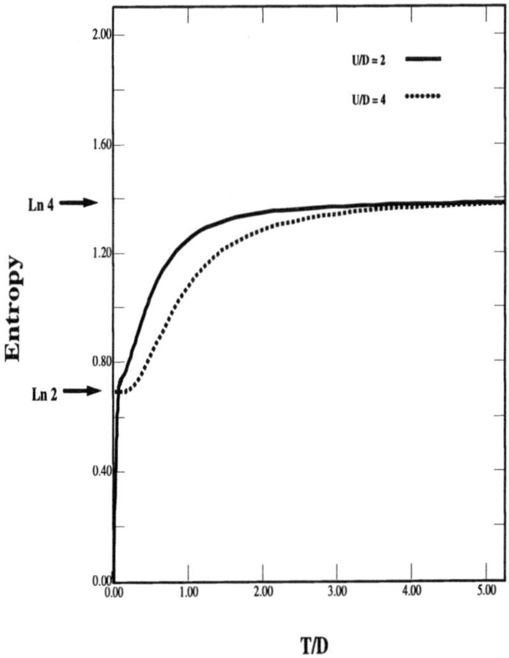

Fig. 5.6. Entropy density S/L at $U = W$ (metal) and $U = 2W$ (insulator) for the half-filled infinite-dimensional Hubbard model in iterated perturbation theory as a function of $T/(W/2)$; the Mott–Hubbard insulator displays a finite entropy at zero temperature, $S/L = \ln 2$ (from [5.81]).

technique provides reliable data. The NCA shows that the resistivity increases with a T^2 power law at small temperatures, as indeed expected for a Fermi liquid. Whereas the resistivity actually goes to zero for $T \to 0$, the non-crossing approximation for the Anderson model extrapolates to a finite (negative) intersection at $T = 0$ [5.21], as seen from the inset of Fig. 5.5. This error had to be expected; see our discussion in Sect. 5.5.6.

At zero temperature iterated perturbation theory is constructed to work well for the description of the Fermi-liquid features, and its results are reliable if we are not too close to the Mott–Hubbard transition. Preliminary results for the local-moment approach show that it, too, can be adapted to ensure Fermi-liquid properties in the metal (D. Logan, private communication, 1996).

Although all theories on the infinite-dimensional Hubbard model are only approximate, it is only the large-d approach that allows a quantitative analysis of the resistivity for correlated electrons over the whole temperature range for appreciable interaction strengths. High-temperature expansions give a decreasing electronic resistivity for $T \to \infty$, and the resistivity increases proportionally to T^2 in the regime of the validity of Fermi-liquid theory. Hence, a maximum in the resistivity is expected for some intermediate temperature. For the parameters chosen in Fig. 5.5 the Fermi-liquid regime is restricted to $T_{\mathrm{FL}} \lesssim 0.03W/(2\sqrt{2})$, which corresponds to the width of the quasi-particle peak in the density of states at zero temperature. The temperature for the maximum resistivity is $T_{\max} \approx 0.1W/(2\sqrt{2})$, and generally close to the temperature below which Fermi-liquid theory is valid. At finite hole density the system is a correlated metal for all interaction strengths; for further results for the conductivity for less than half band-filling, see [5.21, 5.93, 5.100].

Thermodynamic quantities are discussed in detail in [5.4, 5.81, 5.82]. As an example we display the entropy density as a function of temperature for two values of the interaction strengths in Fig. 5.6. The calculations were done on a Bethe lattice by using the iterated perturbation theory approximation. The figure shows that the metallic state for $U = W/2$ reaches the entropy of $S/L = \ln(2)$ on a temperature scale of $k_B T_{S=L\ln 2} \approx 0.04\,W$. Around this temperature, the charge excitations are most efficiently scattered by the spin excitations such that this temperature agrees well with the temperature T_{max} for which the specific resistivity reaches its maximum.

5.6.3 Mott–Hubbard Insulator

As already discussed in Sect. 5.4.2 the Mott–Hubbard insulator in infinite dimensions has a finite entropy density. This is also seen in Fig. 5.6 where the result of iterated perturbation theory for the Bethe lattice with infinite coordination number is shown for $U = 2W$. As discussed in Sect. 5.2.2, this result is exact for $U \gg W$ and follows naturally from the fact that every singlet is a ground state for the corresponding Heisenberg model in the absence of any nesting. Of course, a residual entropy at zero temperature is artificial and again shows that the limit of infinite dimensions can be singular as far as the interaction between localized spins is concerned.

In the insulator the Hubbard bands are separated by an energy gap. The result of quantum Monte-Carlo calculations for $U = 6W/(2\sqrt{2}) \approx 2W$ is shown in Fig. 5.7. In the Mott–Hubbard insulator the two Hubbard bands change their shape only slightly as a function of temperature. Owing to the

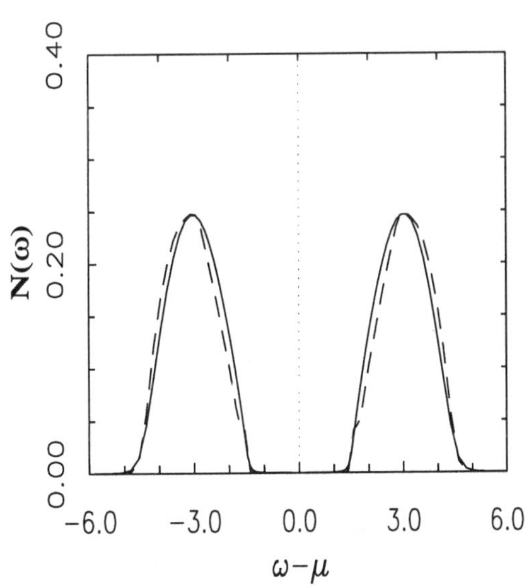

Fig. 5.7. One-electron density of states $N(\omega) \equiv D_\sigma(\omega)$ of the infinite-dimensional Hubbard model in the Mott–Hubbard insulator ($U = 6$) on a half-filled hypercubic lattice as a function of frequency at various temperatures, calculated with the quantum Monte-Carlo method; the energy units are such that $W = 2\sqrt{2}$; the temperature is $\beta = 1.44$ (*full line*) and $\beta = 28.8$ (*dashed line*) (from [5.21]).

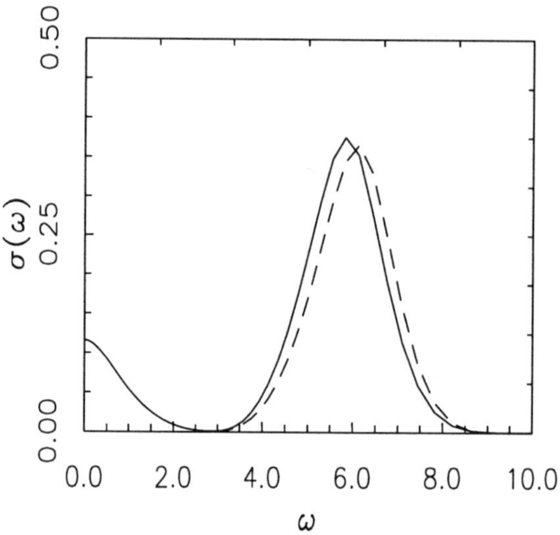

Fig. 5.8. Optical conductivity in the Mott–Hubbard insulator of the infinite-dimensional Hubbard model ($U = 6$) on a hypercubic lattice at half band-filling as a function of frequency at temperatures $\beta = 1.44$ (*full line*) and $\beta = 28.8$ (*dashed line*); the energy units are such that $W = 2\sqrt{2}$, the units for the conductivity are $\pi e^2 a/(2\hbar) \approx 10^4\,(\text{Ohm-cm})^{-1}$ (from [5.21]).

exponential tails in the Gaussian bare density of states, $D_{\sigma,0}(\epsilon)$, the density of states, $D_\sigma(\epsilon)$, exactly vanishes only at $\omega = 0$. It is also seen that the Hubbard bands display their full width, $W_{1,2} \approx W$, which has to hold in the limit of large interaction strengths; see Sect. 5.4.2. Various approximate treatments fulfill this criterion for a good approximation to the Mott–Hubbard insulating phase of the infinite-dimensional Hubbard model. We showed this explicitly for the Hubbard-III approximation in Sect. 3.3.6. Furthermore, both the iterated perturbation theory and the local-moment approach predict the full bandwidth $W_{1,2} = W$ for the upper and lower Hubbard bands for $U \gg W$. The proofs in these cases are lengthy and bare little relation to the fact that these approaches are exact in the atomic limit [5.99]. Hence, the large-U limit alone, although non-trivial to obtain, does not allow us to discriminate between the various analytical approximation schemes.

Finally, Fig. 5.8 displays the optical conductivity $\sigma(\omega)$ in the insulating phase at $U = 6W/(2\sqrt{2})$ as a function of temperature. At high temperatures the conductivity is finite at $\omega = 0$ but rapidly goes to zero as the temperature is lowered: a Drude peak as expected for a metal, (1.3), does not form, the system is insulating. For a Gaussian density of states the energy gap in the one-electron density of states does not fully develop. Therefore, thermal excitations always contribute to an exponentially small conductivity at $\omega = 0$ for $T > 0$, but they are not visible on the scale of the figure. As already seen in the metal, the position of the broad maximum around $\omega \approx U$ (optical excitations between the lower and upper Hubbard bands) only slightly shifts as a function of temperature.

5.6.4 Mott–Hubbard Transition

As pointed out in Sect. 5.6.1 we observe a standard thermodynamic phase transition into the antiferromagnetically ordered Mott–Heisenberg insulator at the Néel temperature, and the transition can easily be monitored by quantum Monte-Carlo calculations. The situation is different for the Mott–Hubbard transition, which is a quantum phase transition at zero temperature. Since an exact solution does not exist we will discuss two different conceivable scenarios in the following.

Scenario I: Discontinuous Transition. In this scenario one assumes that there is a region of interaction strengths for which the self-consistency equations give *both* a metallic *and* an insulating solution. This scenario is supported by the iterated perturbation theory approximation; see [5.4] for details. If we begin the iteration with a "metallic" input guess, we find a converged solution with a finite quasi-particle weight $Z_{\mathrm{F}} = m/m^* \geq 0$ up to a critical value $U_{\mathrm{c}}^{\mathrm{M}\to\mathrm{I}}$ at which Z_{F} goes to zero. On the other hand, if we start with an "insulating" input guess, we find a converged solution with a finite gap Δ down to and including a critical value $U_{\mathrm{c}}^{\mathrm{I}\to\mathrm{M}}$ with $U_{\mathrm{c}}^{\mathrm{M}\to\mathrm{I}} > U_{\mathrm{c}}^{\mathrm{I}\to\mathrm{M}}$, and thus a "coexistence region" $U_{\mathrm{c}}^{\mathrm{I}\to\mathrm{M}} \leq U \leq U_{\mathrm{c}}^{\mathrm{M}\to\mathrm{I}}$. Within iterated perturbation theory and on the Bethe lattice with infinite coordination number the critical interaction strengths are found in [5.4] as $U_{\mathrm{c,IPT}}^{\mathrm{I}\to\mathrm{M}} \approx 1.3\,W$ and $U_{\mathrm{c,IPT}}^{\mathrm{M}\to\mathrm{I}} \approx 1.7\,W$. Actually, the critical value for the transition from the metallic into the insulating state could be much higher. The numerical analysis of iterated perturbation theory at zero temperature by Macarie and d'Ambrumenil [5.101] suggests that $U_{\mathrm{c,IPT}}^{\mathrm{M}\to\mathrm{I}} \gtrsim 3\,W$. Thus, it is conceivable that the coexistence region is rather large, and possibly infinite.

In the case of a coexistence region the absolute energy determines the true ground-state phase. As an illustration let us assume that the transition happens at some U within the critical interval. Coming from the metallic side we would then observe that the quasi-particle weight jumps to zero, and likewise the gap opens discontinuously. Furthermore, the energy would be continuous but its first derivative with respect to U, the mean number of double occupancies, would show a jump discontinuity.

If the transition occurs at one of the boundaries, it can be of second order at best. We will find that either the quasi-particle weight drops to zero discontinuously, if $U_{\mathrm{c}} = U_{\mathrm{c}}^{\mathrm{I}\to\mathrm{M}}$, or the gap jumps to a finite value, if $U_{\mathrm{c}} = U_{\mathrm{c}}^{\mathrm{M}\to\mathrm{I}}$. In either case the transition is still a discontinuous one. Within the iterated perturbation theory some numerical and analytical arguments are presented in [5.4] which locate the transition at $U_{\mathrm{c,IPT}} = U_{\mathrm{c}}^{\mathrm{M}\to\mathrm{I}}$; see, however, [5.101]. It would imply that the gap is already pre-formed and the energy scales for the metal (width of the Kondo resonance in the one-electron density of states) and the insulator (size of the gap) are indeed well separated: close to the transition the density of states in the metallic phase is strictly zero between the Kondo resonance and the upper (lower) Hubbard band.

A finite coexistence region at zero temperature would have severe consequences for *finite* temperatures. In this case the phase diagram would display a line of first-order transitions $U_{\mathrm{c}}(T \leq T_{\mathrm{CP}})$ in the thermodynamic sense, which terminates in a second-order critical point T_{CP} [5.4].

Scenario II: Continuous Transition. In this scenario one assumes that the quasi-particle weight Z_{F} vanishes at the same critical interaction strength where the gap opens. Thus the transition is continuous and (at least) of second order. This scenario agrees with our concept of the Mott–Hubbard transition as a quantum phase transition at zero temperature. At finite temperatures a phase transition in the thermodynamic sense does not exist. Instead, there is a crossover region from the paramagnetic metal to the paramagnetic Mott–Hubbard insulator; see Sect. 1.2.1.

The solution of the $1/r$ Hubbard model supports the idea of the Mott–Hubbard transition as a continuous quantum phase transition. Unfortunately, an exact solution of the infinite-dimensional Hubbard model does not seem to be within reach. Although we cannot rule out a discontinuous transition, we will argue in the following that there is no convincing evidence in favor of this scenario in the infinite-dimensional Hubbard model; the scenario of a continuous phase transition appears to be the more natural one.

A discontinuous transition at zero temperature and a "first-order" phase transition at finite temperatures do not seem to be compatible with Mott's general ideas on the effect of a short-range potential for the Mott transition. Furthermore, it is difficult to reconcile such behavior with our concept of the Mott–Hubbard transition as a quantum phase transition at zero temperature; see Sect. 1.2.1. In principle, the limit of infinite dimensions could produce new, unphysical effects but this appears to be unlikely since the interaction becomes even more "local" in this limit. After these preliminary remarks we discuss the situation for zero and finite temperature in more detail.

Zero Temperature. The exactly solvable $1/r$ Hubbard model supplies an example where an approximate solution would predict a first-order transition. As discussed in detail in Sect. 4.2.4, the exact solution shows that the Mott–Hubbard transition in this model at $U_{\mathrm{c}} = W$ is of third order, i.e., only the third derivative of the ground-state energy with respect to the interaction strength displays a jump discontinuity. In addition, the transition is continuous because the gap opens continuously at the transition. However, the approximate variational theory in Sect. 3.4.10 would predict a "coexistence region" and a first-order transition. The variational metallic and insulating wave functions give $U_{\mathrm{c,var}}^{\mathrm{M}\to\mathrm{I}} = \infty$ and $U_{\mathrm{c,var}}^{\mathrm{I}\to\mathrm{M}} = 0$, respectively, and a first-order transition at $U_{\mathrm{c,var}} = W$ with a jump discontinuity in the number of double occupancies follows. Both variational states are excellent for large and small interaction strengths but their deviations from the exact wave function are crucial near the transition.

The example of the variational approach to the Mott–Hubbard transition indicates that approximate theories readily predict coexistence regions which, however, might be absent in the exact solution. In this case, the magnitude of the coexistence interval might be taken as the region for which the approximation is unreliable. These remarks also apply to the results of iterated perturbation theory. Fourth-order iterated perturbation theory at half band-filling shows that the overall features of the self-energy change only slightly [5.102], but it is also seen that (iterated) perturbation theory as a whole becomes unreliable for $U \gtrsim W$.

In the absence of long-range order all numerical and analytical calculations indicate that the Mott-Hubbard transition should occur in the region $W \lesssim U_\mathrm{c} \lesssim 1.5\,W$. As seen from Fig. 5.3, the local moments are well developed at the transition, and the local-moment approach of Sect. 5.5.8 should be applicable. In this case $U_{\mathrm{c,LM}}^{\mathrm{I}\to\mathrm{M}} = 1.21\,W$ is found, at which the gap closes linearly, i.e., the gap exponent is $\nu = 1$ [5.99]. This is consistent with the predictions from exact diagonalization studies [5.4] which, however, are not too reliable close to the transition, i.e, they cannot really access the critical regime. The local-moment approach is presently under investigation in the metallic regime. It remains to be seen whether or not a (large) coexistence region will be found, although this appears to be unlikely (D. Logan, private communication, 1996).

As we have seen, approximate theories at zero temperature cannot be used to decide whether or not a coexistence region exists. We may ask whether there are arguments in favor of a discontinuous transition which are independent of the details of the microscopic description. If one *assumes* that the high and low energy scales are well separated in the metallic phase close to the transition, one can indeed show that $U_\mathrm{c}^{\mathrm{M}\to\mathrm{I}} > U_\mathrm{c}^{\mathrm{I}\to\mathrm{M}}$, i.e., the transition is discontinuous. Unfortunately, this "projective self-consistent technique" [5.4, 5.103] only proves that the basic *assumption* of separate energy scales is self-consistently fulfilled. The method cannot exclude, however, that there is always a finite spectral weight between the Kondo resonance and the upper (lower) Hubbard band, which leads to a continuous transition.

As pointed out by Nozières (P. Nozières, private communication, 1996), the electrons in the Kondo resonance have to screen the local moments to guarantee Fermi-liquid behavior; otherwise the metallic state cannot persist. If we assume a preformed gap we have to face Nozières exhaustion problem when we approach $U_\mathrm{c}^{\mathrm{M}\to\mathrm{I}}$: there will not be enough mobile electrons to completely screen the local moments. Hence, there are severe problems to stabilize the Fermi-liquid phase for $U < U_\mathrm{c}^{\mathrm{M}\to\mathrm{I}}$. The exhaustion problem does not occur if we abandon the idea that we can strictly separate itinerant and local features of the electrons. This implies, however, that the density of states in the metallic phase is finite everywhere between the upper (lower) Hubbard band, and a preformed gap in the metallic phase does not exist.

Finite Temperatures. At finite temperatures quantum Monte-Carlo calculations are exact within the limitations discussed in Sect. 5.5.3. In principle, it should be possible to decide with this method whether or not a finite coexistence region exists at finite temperatures. Indeed, as used in [5.4], persistent oscillations between the insulating and the metallic solution in the self-consistency procedure are taken as indication for a (large) coexistence region (M. Rozenberg, private communication, 1996).

These oscillations could be reproduced by others but were seen to disappear when the numerical accuracy was increased (T. Pruschke, Habilitation thesis, 1995). The non-crossing approximation and the fully converged numerically stable quantum Monte-Carlo calculations suggest that the Mott–Hubbard transition occurs at $U_c \approx 1.3\,W$, with only a small coexistence region of much less than a tenth of the bandwidth, if there is any (T. Pruschke, Habilitation thesis, 1995, and private communication, 1996); see also Fig. 5.9. Hence, neither the quantum Monte-Carlo data nor the non-crossing approximation provide evidence for a coexistence region at temperatures accessible to these approaches.

It is seen that the arguments from quantum Monte-Carlo calculations in favor of a discontinuous Mott–Hubbard transition in the infinite-dimensional Hubbard model are, at best, inconclusive.

5.6.5 Comparison with Basic Theoretical Concepts

Despite the fact that the details of the Mott–Hubbard transition have not fully been resolved for the infinite-dimensional Hubbard model we can compare its main features with the basic theoretical concepts presented in Chap. 1.

Zero Temperature. In the absence of perfect nesting, Mott's general concepts on the Mott–Hubbard transition can be tested because the antiferromagnetic phase does not interfere. However, as pointed out in Sect. 5.4.2 the Mott–Hubbard insulator in infinite dimensions has a finite entropy, which shows that the neglect of ordering is not a realistic approximation at low temperatures. Nevertheless, the charge gap in the Mott–Hubbard insulator can be large compared to the Néel temperature such that the insulating signatures of the Mott–Hubbard insulator prevail when the ordering of the moments ceases to exist. Therefore, the quantum phase transition between the metal and the Mott–Hubbard insulator is an essential concept that allows us to understand the overall phase diagram even in the presence of antiferromagnetism at low temperatures.

The exact results for the infinite-dimensional Hubbard model in the limits of small and large interactions completely agree with Mott's predictions: the half-filled system is a correlated metal (Fermi liquid) at small interaction strengths, but it is a Mott–Hubbard insulator with a gap of size $\Delta(U \gg W) = U - W$ for large interaction strengths. In addition, the upper and lower Hubbard bands display the full bandwidth.

Unfortunately, different scenarios have been proposed for the Mott–Hubbard transition (see Sect. 5.6.4) and it is not yet clear which of them will prevail. A discontinuous transition could be an artefact of the approximation scheme, and a continuous quantum phase transition seems to be more likely: the metallic and insulating states do not coexist, the gap in the Mott–Hubbard insulator opens continuously at U_c where the quasi-particle

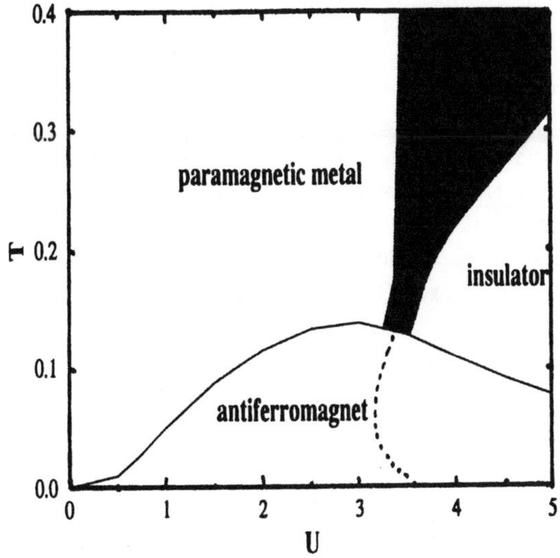

Fig. 5.9. Phase diagram of the infinite-dimensional Hubbard model on a hypercubic lattice at half band-filling as a function of interaction strength calculated with the quantum Monte-Carlo method; the energy units are such that $W = 2\sqrt{2}$ (from [5.83]).

weight of the metallic phase goes to zero. This scenario is certainly in better agreement with Mott's views and is also in harmony with exact results in one dimension. Owing to the lack of an exact solution of the infinite-dimensional Hubbard model, however, this issue cannot be settled now.

Phase Diagram at Half Band-Filling. The phase diagram for an infinite-dimensional hypercubic lattice is shown in Fig. 5.9. Two thermodynamic phases are found: the paramagnetic phase above the Néel temperature and the antiferromagnetically ordered state below T_{N}. As has been shown by Fig. 5.3 of Sect. 5.6.1, preformed moments order at the thermodynamic phase transition. Thus, the antiferromagnetic state is characterized as a Mott–Heisenberg insulator.

Above the Néel temperature the paramagnetic state is conducting but the conductivity is large ("metallic") for "small" U/W and (very) small ("insulating") for "large" U/W. In between there is a large crossover regime (shaded black in the figure). The boundaries of the crossover regime in the figure have been determined qualitatively from numerical instabilities in the quantum Monte-Carlo calculations (left boundary) and a vanishingly small compressibility in the NCA calculations (right boundary) [5.83]. Thus we may roughly identify the shaded part of the phase diagram as the crossover region between semimetal and semiconductor; see Fig. 1.1 of Sect. 1.2.1. Of course, its limits may not be understood as true phase boundaries at finite temperatures.

In the numerical simulation the phase boundaries can naturally be extended into the antiferromagnetic phase because we may choose a reference system with the same density of states but an energy correlation function without the perfect nesting property. In this case the crossover region extends down to zero temperature and, most likely, terminates in a single critical point $U_c \approx 1.3\,W$, the critical value for the quantum phase transition between the Fermi-liquid metal and the Mott–Hubbard insulator. In the presence of antiferromagnetism the Mott–Hubbard quantum phase transition is hidden behind the Mott–Heisenberg phase. However, the gap for current-carrying excitations in the Mott Hubbard insulator *persists* for $T > T_N$, beyond the ordering of the moments. Without the concept of a hidden quantum phase transition this insulating behavior could not be understood.

All three essential phases discussed in Sect. 1.4 can actually be found in the phase diagram of the infinite-dimensional Hubbard model: the antiferromagnetic Mott–Heisenberg insulator, the paramagnetic correlated metal (good conductor at $T > 0$), and the paramagnetic Mott–Hubbard insulator (bad conductor at $T > 0$). In addition, the crossover regime between semimetallic and semiconducting behavior could be estimated.

As noted in [5.4, 5.81] the dominant features of the phase diagram of vanadium sesquioxide (see Fig. 1.12 of Sect. 1.5.3) are qualitatively reproduced by the infinite-dimensional Hubbard model, Fig. 5.9. As discussed in Sect. 1.5.3 this observation confirms that Mott's views on the metal–insulator transition are correct. Furthermore, the comparison of the phase diagrams shows that the simple Hubbard model is indeed able to describe basically the Mott transition. However, as we pointed out in Sect. 1.1, all other interactions of the electrons may and will modify the nature of the Mott transition in real materials. For example, the lattice degrees of freedom in effect transmute the Mott–Hubbard quantum phase transition in V_2O_3 to a thermodynamic phase transition for $T_N < T \leq T_c^{\text{lattice}}$ where the lattice distortion plays the role of the order parameter [5.104]. The description of this transition is beyond the scope of the purely electronic Hubbard model.

To gain further insight into metal–insulator transitions due to electron–electron interactions we study various other models for interacting electrons in the next chapter.

6. Further Models with Hubbard Interaction

In this chapter we collect further models that contain the Hubbard interaction and seem to be suitable to describe a metal–insulator transition. In view of the vast literature on the subject, the given list of models necessarily cannot be exhaustive.

In the first section we complete our study of the Hubbard model with a short discussion on generic model extensions. First, we investigate some problems of the Bethe-ansatz approach for degenerate bands. Next, we list some exact results for the Hubbard model with an additional nearest-neighbor interaction. Finally, we briefly discuss the Hubbard model with general nearest-neighbor interactions, which has simple ground states in certain parameter regions.

In the second section we exclusively discuss models in which the number of double occupancies is dynamically conserved. Thus, these models are possible candidates for the description of high-temperature superconductivity. The Harris–Lange model, which can be derived from a large-U expansion of the Hubbard Hamiltonian, and the Montorsi–Rasetti model as its "simplified" version are all prototypes. Both the Harris Lange model and the $U(2|2)$-supersymmetric Hubbard model can be solved exactly in one dimension.

In the third section we are concerned with three exactly solvable models for which the operators for the kinetic energy and the potential energy commute. First, the Hubbard model with infinite-range hopping is found to be trivial in the thermodynamic limit. Next, a reduced version of the $1/r$ Hubbard model is shown to exactly reproduce Stoner's theory of ferromagnetism ("$1/r$ Stoner model"). Lastly, the interaction in the "Landau–Hubbard model" is reduced to momentum-independent forward scattering only, which makes it exactly solvable in all dimensions. It displays a metal–insulator transition at $U_{\rm c} = W$, and illustrates Mott's concept of the splitting of the Hubbard bands.

6.1 Degenerate and Extended Hubbard Models

In this section we first discuss the $SU(N_f)$-symmetric Bethe-ansatz equations. For half band-filling and $N_f > 2$, they describe a first-order metal–insulator transition. Unfortunately, the Hubbard-type model that corresponds to these equations has not yet been identified. Next, we present exact statements on the extended Hubbard model that contains a nearest-neighbor interaction besides the Hubbard on-site term. Finally, we study the most general nearest-neighbor Hubbard model, and determine the region in parameter space for which a given (simple) variational wave function is an exact ground state. These statements are (almost) independent of lattice structure and dimension, and mostly apply to models with a conserved number of double occupancies, which we discuss in the next section.

6.1.1 Band-Degenerate Hubbard Model in One Dimension

A natural extension of the Bethe-ansatz equations for the Hubbard model can be achieved by their $SU(N_f)$-invariant formulation in the case of a N_f-fold band degeneracy. The discussion follows [6.1].

The degenerate Bethe-ansatz equations were first formulated in [6.2], and are written in the diction of [6.1] as

$$\begin{aligned} 2\pi I_j + \sum_{\alpha=1}^{M_1} \Theta\big(2\sin(k_j) - 2\Lambda_\alpha^{(1)}\big) &= k_j L\,, \\ & \qquad j = 1,\dots,N\,, \qquad (6.1\text{a}) \\ 2\pi J_\alpha^{(m)} + \sum_{j=1}^{M_m} \Theta\big(2\Lambda_\alpha^{(m)} - 2\Lambda_\beta^{(m)}\big) &= \sum_{\beta=1}^{M_m-1} \Theta\big(\Lambda_\alpha^{(m)} - \Lambda_\beta^{(m-1)}\big) \\ &\quad + \sum_{\beta=1}^{M_m+1} \Theta\big(\Lambda_\alpha^{(m)} - \Lambda_\beta^{(m+1)}\big)\,, \\ & \qquad \alpha = 1,\dots,M_m\,, \qquad (6.1\text{b}) \end{aligned}$$

where m takes the values $m = 1,\dots,N_f$, and we assign $M_0 = N$, $M_{N_f} = 0$, $\Lambda_j^0 = \sin(k_j)$, and $\Theta(x) = -2\arctan\big[x/(U/(2t))\big]$. Depending on the integer parity of the total electron number N and the number of electrons per band M_m, the quantum numbers I_j and $J_\alpha^{(m)}$ are integers or half-odd integers. The analysis of (6.1) proceeds as described in Sect. 4.1. As well as the ground-state properties [6.2, 6.3] one may also obtain all critical exponents [6.1].

The main problem of the $SU(N_f)$-generalized Lieb–Wu equations (6.1) is the fact that one cannot specify which Hubbard-type model they actually describe [6.4, 6.5]. The Hamiltonian must dynamically exclude interactions between more than two electrons per lattice site. Otherwise the validity of the Yang–Baxter equations cannot be guaranteed [6.1, 6.3]. In the continuum limit the equations belong to a model in which the electrons interact with a $1/(\sinh r)^2$ potential. Furthermore, the model transforms into the $SU(N_f)$ Heisenberg model in the limit of strong interactions and half band-filling. For $U = 0$ the equations describe a Fermi gas with N_f degenerate bands. In the limit $N_f \to \infty$ one recovers a model that corresponds to a Bose gas with local interactions. Also the analysis of small systems does not allow the unambiguous identification of the Hamiltonian, which apparently contains Hubbard X terms and long-range interactions [6.1].

The study of (6.1) offers two advantages. First, the equations contain a metal–insulator transition at half band-filling where the critical coupling is finite for $N_f > 2$: $U_c(N_f > 2) > 0$ [6.1]–[6.6]. Since the transition occurs at a finite interaction strength it can be studied in detail [6.1]. In particular, it is found to be discontinuous in so far as the gap jumps from zero to a finite value at the critical value [6.1]. Second, merits and failures of $1/N_f$ expansions can be studied for these equations; see [6.4].

Nevertheless it would be highly desirable to know the quantum-mechanical model whose ground-state properties are in fact described by (6.1). Otherwise we cannot even qualitatively assess whether the model and its features are generic for real systems with degenerate bands.

6.1.2 Extended Hubbard Model

The Hubbard model with an additional, spin-independent nearest-neighbor interaction is usually called the "extended Hubbard model". The Hamilton operator thus reads

$$\hat{H}_{\text{ext}} = \hat{T} + U\hat{D} + \frac{V}{Z}\sum_{\boldsymbol{R},\boldsymbol{\tau}}(\hat{n}_{\boldsymbol{R}} - 1)(\hat{n}_{\boldsymbol{R}+\boldsymbol{\tau}} - 1) , \tag{6.2}$$

where $\boldsymbol{\tau}$ denotes the vectors to the Z nearest-neighbor sites and $\hat{n}_{\boldsymbol{R}} = \hat{n}_{\boldsymbol{R},\uparrow} + \hat{n}_{\boldsymbol{R},\downarrow}$ is the operator for the electron density at site $\boldsymbol{R}$. The scaling is performed in such a way that the model is properly defined in the limit $Z \to \infty$. $\hat{T}$ and $\hat{D}$ are, as before, the operators for the kinetic energy and the number of double occupancies; see Sect. 2.1, (2.20b).

As we discussed in Sect. 2.4, the model (6.2) could possibly describe essential electronic properties of one-dimensional metallic or semiconducting materials such as polymers, metal halides, or charge transfer salts where U, V are positive (repulsive interactions). Unfortunately, there are far fewer exact statements possible on the physical properties of the model $\hat{H}_{\text{ext}}$ in (6.2) than for the Hubbard model itself. In principle, the extended Hubbard model can be treated with all the approximation schemes that were presented in Chap. 3. In addition, one can employ the simplifications that arise in the limit of infinite dimensions; see Chap. 5. We do not elaborate these interesting approaches here, but concentrate on exact statements on the extended Hubbard model in one dimension [6.7], and the U-V phase diagram at half band-filling.

Exact Results in One Dimension. In one dimension and for $U = \infty$ the extended Hubbard model reduces to the problem of spinless fermions with nearest-neighbor interaction, which has been solved exactly [6.8]; see also

Sect. 5.4. Now, the quarter-filled band, $n = 1/2$, is the interesting case for the metal–insulator transition since it corresponds to a half-filled band for spinless fermions. We then obtain a metal–insulator transition at $V = 2t$. For $V < U = \infty$ and $n \neq 1/2$, $n \neq 1$ the system is metallic.

For quarter-filled bands another point in the U-V phase diagram allows an exact solution, namely $V = \infty$ [6.9]. In this case double occupancies cannot decay, and the spin of the unpaired electrons is irrelevant since they cannot be nearest neighbors. Hence, the problem again reduces to free spinless fermions, which move on a shortened chain. Thus, the $V = \infty$ model can be solved for all densities. One obtains a phase transition between a metal and an insulator for $n = 1/2$ at $(V = \infty, U = 4t)$ [6.9]. The general low-energy properties of the extended Hubbard model can be studied within the g-ology model [6.10]; see Sect. 4.4.1. We shall not elaborate further on this approach here.

Phase Diagram at Half Band-Filling. In the absence of the kinetic energy (atomic limit) we are left with a model of classical statistical mechanics [6.11]. For $V < U/2$, all sites are singly occupied in the ground state. If a finite kinetic energy was present, the unique ground state in $d > 2$ dimensions would be a spin-density wave state. Therefore, the phase for $V < U/2$ is called the "spin-density wave". For $V > U/2$, and on a half-filled bipartite lattice, the ground state is a charge-density wave with a period of twice the lattice distance since the additional V term implies an attractive interaction between double occupancies and holes. The line $v = V/U = 1/2$ in the U-V parameter space thus separates the spin- and charge-density wave states.

Within perturbation theory we may now investigate how far the inclusion of the kinetic energy will influence the phase separation line between the charge- and spin-density wave states [6.12]. For weak interactions, $U, V \to 0$, there is no shift in the phase separation line, $v_{\mathrm{c}}(U \to 0, V \to 0, d) = 1/2$. For strong interactions, $U, V \to \infty$, the spin-density wave is slightly stabilized over the charge-density wave in all finite dimensions d due to the exchange interaction. In infinite dimensions the phase separation line stays at its value for the atomic problem, i.e., $v_{\mathrm{c}}(U \to \infty, V \to \infty, d = \infty) = 1/2$. In general, $v_{\mathrm{c}}(U \to \infty, V \to \infty, d < \infty) > 1/2$, but the deviations are rather small. The results imply that the influence of the kinetic energy on the stability of both phases is rather weak.

The extended Falicov–Kimball model for which one spin species is immobile ($\hat{T}_{\downarrow} \equiv 0$) is exactly solvable in infinite dimensions [6.13]. The phase separation line in this model is given by $v_{\mathrm{c}}^{\mathrm{FK}}(U, V, d = \infty) = 1/2$ for all V/U. The transition between the two phases is continuous, in contrast to the case of the Hubbard model.

Our analytical results for the extended Hubbard models are seen to be rather limited. As we discuss in the rest of this chapter, we may alternatively investigate other Hubbard-type models which, by construction, allow for a more detailed analysis. However, the relevance of these models for real systems is less well established.

6.1.3 Generalized Hubbard Models

We close this section with a brief look at exact results for the most general Hubbard model with nearest-neighbor terms. The model may be written as $\hat{H}^{\text{gen}} = \sum_{\boldsymbol{R},\boldsymbol{T}} \hat{H}^{\text{gen}}_{\boldsymbol{R},\boldsymbol{R}+\boldsymbol{T}}/2$ [6.14, 6.15] with the two-site term ($m \equiv \boldsymbol{R}_m$)

$$\begin{aligned}
\hat{H}^{\text{gen}}_{l,m} &= \sum_{\sigma} \left(\hat{c}^{+}_{l,\sigma}\hat{c}_{m,\sigma} + \text{h.c.}\right)\Big[-t + X\left(\hat{n}_{l,-\sigma} + \hat{n}_{m,-\sigma}\right)\Big] + \frac{\mu}{Z}\left(\hat{n}_l + \hat{n}_m\right) \\
&\quad + \frac{U}{Z}\Big((\hat{n}_{l,\uparrow} - 1/2)(\hat{n}_{l,\downarrow} - 1/2) + (l \leftrightarrow m)\Big) + V(\hat{n}_l - 1)(\hat{n}_m - 1) \\
&\quad + Y\left(\hat{c}^{+}_{l,\uparrow}\hat{c}^{+}_{l,\downarrow}\hat{c}_{m,\downarrow}\hat{c}_{m,\uparrow} + \text{h.c.}\right) + \frac{J_{\text{xy}}}{2}\left(\hat{S}^{+}_l\hat{S}^{-}_m + \text{h.c.}\right) + J_{\text{z}}\hat{S}^{z}_l\hat{S}^{z}_m \,,
\end{aligned} \tag{6.3}$$

where Z is the number of nearest neighbors. Usually, the analysis is restricted to half band-filling where electron–hole symmetry simplifies the discussion.

Instead of looking for the exact ground state for some fixed set of couplings, we may select a (simple) trial state, and determine the region in parameter space for which it is a (possibly degenerate) ground state. For example, we may ask under which conditions the Brinkman–Rice insulator

$$|\text{BR}\rangle = \prod_{\boldsymbol{R}_l \in \mathcal{A}} \hat{c}^{+}_{\boldsymbol{R}_l,\uparrow} \prod_{\boldsymbol{R}_m \in \mathcal{A}'} \hat{c}^{+}_{\boldsymbol{R}_m,\downarrow}|\text{vacuum}\rangle \tag{6.4}$$

is a ground state of $\hat{H}^{\text{gen}}$ for arbitrary disjoint sets of lattice points $\mathcal{A}$ and $\mathcal{A}'$. Since the hopping term would create double occupancies, we have to demand $X = t$ from the outset.

We want to determine the region in parameter space for which a state $|\psi_0\rangle$ is a ground state. The idea is to show that our chosen state not only gives an upper bound for the ground-state energy, but also provides a lower bound. The latter condition generally imposes constraints on the model parameters. This allows us to determine analytically parameter regions in (6.3) in which the physical properties of the system at zero temperature are known. Unfortunately, the corresponding ground states, like the Brinkman–Rice insulator, are very simple, and we may very well conclude that the electron dynamics is not particularly interesting in these parameter regions.

There are several ways to establish the lower bound property for $\hat{H}^{\text{gen}}$ for a given state $|\psi_0\rangle$. First, one may try to write the Hamiltonian as a sum over a part $\hat{H}^{\text{gen}}_0$, for which the chosen $|\psi_0\rangle$ is an eigenstate, and a positive semidefinite part that annihilates $|\psi_0\rangle$ [6.16]. Obviously, the approach is ingenious but not very systematic.

The second method makes use of Gerschgorin's theorem for $n \times n$ matrices: if one sums the absolute values of all matrix entries in row i, the maximum sum of all n rows is an upper bound for the absolute value of the largest eigenvalue. Bounds obtained with this method are, in general, superior to those of the first method [6.17].

The third method is the "optimal ground state" approach [6.15]. If we split the total Hamiltonian $\hat{H}^{\text{gen}}$ into a sum of sub-Hamiltonians $\hat{H}^{\text{gen}}_{l,m}/2$, the sum over the exact ground-state energies of the sub-Hamiltonians is a lower bound to the ground-state energy of the total Hamiltonian [6.18]. For the two-site Hamiltonians $\hat{H}^{\text{gen}}_{l,m}/2$ the eigenstates and eigenenergies are readily determined [6.15]. From the eigenstates of the two-site problem one may construct (variational) states $|\psi_0\rangle$ for the full Hamiltonian. If the upper bound obtained from $|\psi_0\rangle$ agrees with the lower bound, $|\psi_0\rangle$ is called an "optimal" ground state. In general, the bounds obtained in this approach are superior to the previous two, and the method is very systematic.

For example, the paramagnetic states $|\sigma, \sigma'\rangle$ are two-site ground states for $X = t$, $J_{\text{xy}} = J_{\text{z}} = 0$, and have the two-site energy $E_{\text{para}} = -U/(2Z) + 2\mu/Z$. They are the lowest of all eight different two-site eigenenergies, if the inequality

$$U/Z \geq \text{Max}\left(2|t| + \frac{2\mu}{Z}, V + |Y|, -V + \frac{2|\mu|}{Z}\right) \tag{6.5}$$

holds. Accordingly, the state $|\text{BR}\rangle$ in (6.4) is found to be an optimal ground state, if (6.5) is fulfilled [6.15]. This case includes the insulating phase of the Harris–Lange model for $V = Y = \mu = 0$ and $U \geq 2|t|Z$ for all lattice types. We will discuss these models in the next section.

6.2 Models with Conserved Double Occupancies

In this section we discuss various models that dynamically conserve local electron pairs. For this reason they could possibly be relevant for the explanation of high-temperature superconductivity.

The number of double occupancies remains unchanged during a hopping process if we include terms for the correlated hopping of electrons. They were called "Hubbard X" or "bond-charge" terms in Sect. 2.4. These terms naturally appear within a large-U expansion of the repulsive Hubbard model on the operator level. This expansion was systematically performed for the first time by Harris and Lange [6.19]. For this reason we will denote the resulting model as the "Harris–Lange model".

The Harris–Lange model is exactly solvable in one dimension and most closely resembles Mott's Gedanken example of Sect. 1.4.6. It describes the lower and upper Hubbard charge bands with the dispersion relation of non-interacting electrons. The Harris–Lange model displays a transition from the Brinkman–Rice insulator for $U > U_{\text{c}} = W$ to a "superconductor" in which "η"-paired electrons appear. This transition also occurs in higher dimensions.

The Montorsi–Rasetti model is the "simplified" version of the Harris–Lange model. Although now one spin species is immobile, an analytical solution of this model has not yet been obtained. One can show, however, that there is again an insulating state for half band-filling above $U_c = W$, for an arbitrary lattice structure.

As last model with a conserved number of double occupancies we discuss the supersymmetric Hubbard model. In one dimension it can be solved exactly via the Bethe ansatz. Its ground state shows long-range order of paired electrons ("η pairing") and thus has superconducting properties.

6.2.1 Bond–Charge Interactions and Harris–Lange Model

In the atomic limit the occupation of the lattice sites is conserved. The number of double occupancies is a good quantum number, and the one-electron Green function displays two poles at the energies zero and U. If we allow for the kinetic energy as a perturbation, the two energy levels broaden into the upper and lower Hubbard bands; see Sect. 1.4. In Sect. 3.3 we used Green functions to discuss the upper and lower Hubbard bands. Of course, these two bands can also be described in terms of operators. Harris and Lange [6.19] were the first to carry out such an operator perturbation theory in the kinetic energy $\hat{T}$, which we now discuss.

We split the Fermi annihilation operator into a part that destroys an electron on a singly occupied site and another part that destroys an electron on a doubly occupied site:

$$\hat{c}_{\boldsymbol{R},\sigma} = (1 - \hat{n}_{\boldsymbol{R},-\sigma})\hat{c}_{\boldsymbol{R},\sigma} + \hat{n}_{\boldsymbol{R},-\sigma}\hat{c}_{\boldsymbol{R},\sigma} \; . \tag{6.6}$$

The corresponding creation operator can be treated accordingly. We may split the kinetic energy operator into

$$\hat{T} = \hat{T}^0_{\text{LHB}} + \hat{T}^0_{\text{UHB}} + \hat{T}^+ + \hat{T}^- \; , \tag{6.7a}$$

$$\hat{T}^0_{\text{LHB}} = (-t) \sum_{\boldsymbol{R},\boldsymbol{T},\sigma} (1 - \hat{n}_{\boldsymbol{R},-\sigma})\, \hat{c}^+_{\boldsymbol{R},\sigma} \hat{c}_{\boldsymbol{R}+\boldsymbol{T},\sigma} \,(1 - \hat{n}_{\boldsymbol{R}+\boldsymbol{T},-\sigma}) \; , \tag{6.7b}$$

$$\hat{T}^0_{\text{UHB}} = (-t) \sum_{\boldsymbol{R},\boldsymbol{T},\sigma} \hat{n}_{\boldsymbol{R},-\sigma} \hat{c}^+_{\boldsymbol{R},\sigma} \hat{c}_{\boldsymbol{R}+\boldsymbol{T},\sigma} \hat{n}_{\boldsymbol{R}+\boldsymbol{T},-\sigma} \; , \tag{6.7c}$$

$$\hat{T}^+ = (-t) \sum_{\boldsymbol{R},\boldsymbol{T},\sigma} \hat{n}_{\boldsymbol{R},-\sigma} \hat{c}^+_{\boldsymbol{R},\sigma} \hat{c}_{\boldsymbol{R}+\boldsymbol{T},\sigma} \,(1 - \hat{n}_{\boldsymbol{R}+\boldsymbol{T},-\sigma}) = \left(\hat{T}^-\right)^+ . \tag{6.7d}$$

The zeroth-order operator $\hat{T}^0_{\text{LHB}}$ for the lower Hubbard band describes the hopping of holes. Doubly occupied sites can move in the upper Hubbard band via $\hat{T}^0_{\text{UHB}}$. Their number is conserved by both hopping processes. The operator $\hat{T}^+$ ($\hat{T}^-$) increases (decreases) the number of double occupancies by one.

As with the Foldy–Wouthuysen transformation for the Dirac equation [6.20, Chap. 20] we apply a canonical transformation that eliminates the operators $\hat{T}^{\pm}$ to a given order in t/U,

$$\hat{c}_{\boldsymbol{R},\sigma} = e^{\mathrm{i}\hat{S}(\bar{c})} \bar{c}_{\boldsymbol{R},\sigma} e^{-\mathrm{i}\hat{S}(\bar{c})} \tag{6.8}$$

with $\left(\hat{S}(\bar{c})\right)^{+} = \hat{S}(\bar{c})$. As shown by Harris and Lange [6.19] the transformation operator to lowest order in t/U reads

$$\hat{S}(\bar{c}) = \frac{\mathrm{i}t}{U}\left(\hat{T}^{+}(\bar{c}) - \hat{T}^{-}(\bar{c})\right) , \tag{6.9}$$

which can easily be verified since $[\hat{D},\hat{T}^{\pm}]_{-} = \pm\hat{T}^{\pm}$.

For all physical operators that do not contain a factor of U/t we may make the replacement

$$\hat{c}_{\boldsymbol{R},\sigma} = \bar{c}_{\boldsymbol{R},\sigma} \tag{6.10}$$

because the error is only of order (t/U) and we do not have to distinguish between the operators $\hat{c}_{\boldsymbol{R},\sigma}$ and $\bar{c}_{\boldsymbol{R},\sigma}$ to lowest order in t/U. The energies obtained from the Harris–Lange model thus agree with those of the Hubbard model to order $t(t/U)^{-1}$ and $t(t/U)^{0}$. If we neglect all correction terms to order t/U and higher, the transformed Hubbard Hamiltonian becomes the Harris–Lange model,

$$\hat{H}_{\mathrm{HL}}(\gamma) = \hat{T}^{0}_{\mathrm{LHB}} + (\gamma - 1)\hat{T}^{0}_{\mathrm{UHB}} + \hat{D} . \tag{6.11}$$

Since $\hat{T}_{\mathrm{LHB}}$ and $\hat{T}_{\mathrm{UHB}}$ separately conserve the number of double occupancies we included the real parameter γ as a model modification. The expansion of the Hubbard model around the atomic limit gives $\gamma = 2$. The operator $\hat{T}^{0}_{\mathrm{LHB}} = \hat{P}_{D=0}\hat{T}\hat{P}_{D=0}$ is the Gutzwiller-projected kinetic energy operator and just describes the $\hat{T}$ model; see Sect. 2.5, (2.40).

The Harris–Lange model with $\gamma = 2$ is the natural starting point for a t/U expansion of the Hubbard Hamiltonian. The Hamiltonian for $\gamma = 0$ may be viewed as a special version of the Hubbard model with an additional "bond-charge" interaction $\hat{H}_X$

$$\begin{aligned} \hat{H}_{\mathrm{HL}}(\gamma = 0) &\equiv \hat{H}_{\mathrm{BC}}(t, X = t, U) \\ &= \hat{T} + t\hat{H}_X + U\hat{D} , \end{aligned} \tag{6.12a}$$

$$\hat{H}_X = \sum_{\boldsymbol{R},\boldsymbol{T},\sigma} \hat{c}^{+}_{\boldsymbol{R},\sigma}\hat{c}_{\boldsymbol{R}+\boldsymbol{T},\sigma}\left(\hat{n}_{\boldsymbol{R},-\sigma} + \hat{n}_{\boldsymbol{R}+\boldsymbol{T},-\sigma}\right) , \tag{6.12b}$$

where $\hat{H}_X$ is just the Hubbard X term; see Sect. 2.4, (2.21). The number of double occupancies is conserved for $X = t$, $[\hat{T} + t\hat{H}_X, \hat{D}]_{-} = 0$ but this is no longer the case for general values of X/t. In the context of high-temperature superconductivity $\hat{H}_{\mathrm{BC}}(t, X, U)$ has been introduced and studied by Hirsch [6.21]; hence it is sometimes called the "Hirsch model".

6.2.2 Harris–Lange Model in One Dimension

In one dimension one can exactly solve the Harris–Lange model for $\gamma = 0$ and $\gamma = 2$ [6.17, 6.22, 6.23]. The value $\gamma = 1$ corresponds to the exactly solvable $\hat{T}$ model on chain segments. Bariev solved another model with a similar structure via the Bethe-ansatz technique [6.24]. Other one-dimensional models with correlated hopping can be found in [6.25, 6.26]. In the following we concentrate on the Harris–Lange model.

As in the $\hat{T}$ model, which covers the sector without any double occupancies, we can recover the spectrum of the Harris–Lange model for $\gamma = 0$ and $\gamma = 2$ from a model of spinless fermions. In the thermodynamic limit the spectrum for $N_{\rm C}$ charges (holes and double occupancies) is given by

$$E = \sum_{j=1}^{N_{\rm C}} \epsilon(k_j) + UD \tag{6.13}$$

in both cases. Here, $\epsilon(k)$ is the bare dispersion relation, D is the number of double occupancies, and k_j are true momenta which are quantized in units of $(2\pi/L)$. Since the orientation of the spins on the singly occupied sites plays no role, each energy level is 2^{N-2D}-fold degenerate.

The result for the ground-state energy density in the thermodynamic limit is given by [6.22, 6.23]

$$E_0(U;\overline{d})/L = -(2t/\pi)\sin[(n-2\overline{d})\pi] + U\overline{d}\,. \tag{6.14}$$

This energy has to be minimized with respect to the density of double occupancies $\overline{d} = D/L$. Concepts based on variational theories such as the Gutzwiller-correlated wave function (see Sect. 3.4) could become exact for this model. In the thermodynamic limit the average number of double occupancies is given by ($W = 4t$ is the bandwidth)

$$\overline{d} = \begin{cases} n/2 & \text{for} \quad U \le -W \\ n/2 - [\arccos(-U/W)]/(2\pi) & \text{for} \quad -W \le U \le U_{\rm c}(n) \\ 0 & \text{for} \quad U_{\rm c}(n) \le U \end{cases} , \tag{6.15}$$

with $U_{\rm c}(n) = -W\cos(n\pi)$. The model displays local pairs (double occupancies) and thus superconducting correlations for all $U < U_{\rm c}(n)$. For $1/2 < n < 3/2$ this phase extends to positive values of U, $U_{\rm c}(n) > U > 0$, where the density of local pairs continuously decreases to zero at $U = U_{\rm c}(n)$. For $U > U_{\rm c}(n)$ the ground state does not contain any double occupancies, and one recovers the physics of the one-dimensional $\hat{T}$ model.

For $\gamma = 0$ one of the highly degenerate states for attractive interaction and an even electron number is the fully "η-paired state",

$$|\Psi_0(U \le -W)\rangle = \left(\hat{C}^+\right)^{N/2} |\text{vacuum}\rangle\,, \tag{6.16}$$

where $\hat{C}^+ = \sum_l \hat{c}_{l,\uparrow}^+ \hat{c}_{l,\downarrow}^+ = \sum_k \hat{c}_{k,\uparrow}^+ \hat{c}_{-k,\downarrow}^+$ creates local double occupancies without center-of-mass momentum. This state displays long-range order in the BCS pair amplitude and is thus characterized as "superconducting". In the region $-W < U < U_{\rm c}(n)$ the ground state still possesses $2D$ paired electrons:

$$|\Psi_0(-W < U < U_{\rm c}(n)\rangle = \left(\hat{C}^+\right)^{D} |\hat{T}\,\text{model}\rangle_{N-2D}\,. \tag{6.17}$$

$|\hat{T}\,\text{model}\rangle_{N-2D}$ denotes the ground state of the $\hat{T}$ model with the $N-2D<N$ remaining electrons in the subspace of zero double occupancies. Hence we obtain paired electrons even for a formally repulsive interaction. It should be kept in mind, however, that the Harris–Lange model stems from a large-U expansion, and the validity of the expansion in t/U terminates at larger values of U than the critical value $U_{\rm c}(n)$.

At half band filling, $n=1$, we obtain a transition from a superconductor to a Brinkman–Rice insulator; see Sect. 3.4 and 6.1.3. The number of double occupancies vanishes together with the Drude weight of the electrical conductivity at the transition at $U_{\rm c}(n=1)=W$ in a continuous but not differentiable manner [6.22],

$$\overline{d}(n=1, U\to W) = \frac{1}{2\pi}\sqrt{\frac{2(W-U)}{W}}\;. \tag{6.18}$$

The ground-state energy per lattice site near the transition behaves like $E_0(n=1, U\to W)\sim (1-U/W)^{3/2}$.

The overall phase diagram remains qualitatively the same in higher dimensions [6.22]. This can be proven with the help of the techniques discussed in Sect. 6.1.3. For example, (6.5) shows that the Brinkman–Rice insulator (6.4) is the exact ground state for $U\geq W$ for all lattice types.

The one-dimensional Harris–Lange model is the model that comes closest to Mott's Gedanken example of Sect. 1.4.6. In fact, we can *interpret* the spectrum in (6.13) in terms of an upper and a lower Hubbard band [6.27]. The spin degrees of freedom are spectators only and give rise to a large degeneracy of the eigenenergies. The charges are described by two Hubbard bands with dispersion relations $\epsilon_{\rm LHB}=\epsilon(k)$ and $\epsilon_{\rm UHB}(k)=U+\epsilon(k)$, where k is the momentum of a charge excitation. The width of both bands is $W_1=W_2=W$. At half band-filling the system is an insulator since the charge gap is finite for $U\geq W$, $\Delta\mu=U-W$ and closes at $U_{\rm c}=W$, in accordance with Mott's ideas. As already discussed above, there is an insulator-to-superconductor transition at $U_{\rm c}$, but in this regime the model no longer resembles Mott's example since there the number of double occupancies is (approximately) conserved only for $U\gg W$.

6.2.3 Montorsi–Rasetti Model

A special case of the Harris–Lange model is the Montorsi–Rasetti model [6.28]. In this model the spin is not conserved during a hopping process. The Hamilton operator reads

$$\begin{aligned}\hat{H}_{\rm MR}(\gamma) &= -t\sum_{\boldsymbol{R},\boldsymbol{T}}\sum_{\sigma,\sigma'}\hat{H}^0_{\boldsymbol{R},\sigma;\boldsymbol{R}+\boldsymbol{T},\sigma'}(\gamma)+U\hat{D}-\mu\hat{N}\;, \qquad (6.19)\\ \hat{H}^0_{\boldsymbol{R}_1,\sigma;\boldsymbol{R}_2,\sigma'}(\gamma) &= \hat{c}^+_{\boldsymbol{R}_1,\sigma}\hat{c}_{\boldsymbol{R}_2,\sigma'}\Big(1-\hat{n}_{\boldsymbol{R}_1,-\sigma}\\ &\qquad -\hat{n}_{\boldsymbol{R}_2,-\sigma'}+\gamma\hat{n}_{\boldsymbol{R}_1,-\sigma}\hat{n}_{\boldsymbol{R}_2,-\sigma'}\Big)\;.\end{aligned}$$

As in the case of the Harris–Lange model the hopping of the electrons preserves the number of double occupancies for all γ.

The hopping amplitudes for processes with and without a spin flip are equal. Hence we may introduce two new fermion operators,

$$\hat{d}_{\boldsymbol{R}} = \left(\hat{c}_{\boldsymbol{R},\uparrow} + \hat{c}_{\boldsymbol{R},\downarrow}\right)/\sqrt{2}\,, \tag{6.20a}$$

$$\hat{f}_{\boldsymbol{R}} = \left(\hat{c}_{\boldsymbol{R},\uparrow} - \hat{c}_{\boldsymbol{R},\downarrow}\right)/\sqrt{2}\,, \tag{6.20b}$$

which allows us to transform the model onto a simplified Harris–Lange model for which one electron species is immobile:

$$\begin{aligned}\hat{H}_{\mathrm{MR}}(\gamma) &= (-t)\sum_{\boldsymbol{R},\boldsymbol{T}} \hat{M}^{f}_{\boldsymbol{R},\boldsymbol{R}+\boldsymbol{T}}(\gamma)\hat{d}^{+}_{\boldsymbol{R}}\hat{d}_{\boldsymbol{R}+\boldsymbol{T}} \\ &\quad + U\sum_{\boldsymbol{R}} \hat{n}^{d}_{\boldsymbol{R}}\hat{n}^{f}_{\boldsymbol{R}} - \mu\sum_{\boldsymbol{R}}\left(\hat{n}^{d}_{\boldsymbol{R}} + \hat{n}^{f}_{\boldsymbol{R}}\right)\,, \end{aligned} \tag{6.21}$$

with $\hat{n}^{f}_{\boldsymbol{R}} = \hat{f}^{+}_{\boldsymbol{R}}\hat{f}_{\boldsymbol{R}}$ and $\hat{n}^{d}_{\boldsymbol{R}} = \hat{d}^{+}_{\boldsymbol{R}}\hat{d}_{\boldsymbol{R}}$. The f electrons are not dynamically mobile and interact with the d electrons via a local repulsion. In addition, the hopping of the d electrons via the operators

$$\hat{M}^{f}_{\boldsymbol{R}_1,\boldsymbol{R}_2}(\gamma) = 1 - \hat{n}^{f}_{\boldsymbol{R}_1} - \hat{n}^{f}_{\boldsymbol{R}_1} + \gamma\hat{n}^{f}_{\boldsymbol{R}_1}\hat{n}^{f}_{\boldsymbol{R}_2} \tag{6.22}$$

is modified in such a way that the number of double occupancies between f and d electrons remains conserved. The Montorsi–Rasetti model and the Harris–Lange model share the same relationship as the Falicov–Kimball model and the Hubbard model.

Since the Falicov–Kimball model can be solved exactly only in infinite dimensions, the Montorsi–Rasetti model cannot analytically be solved in general [6.29], in contrast to the claims in [6.28]. Nevertheless, we can deduce some analytical statements using the methods described in Sect. 6.1.3. For example, the model contains a transition to the Brinkman–Rice insulator at half band-filling for a finite interaction strength, $U_{\mathrm{c}} = W$, for arbitrary lattices [6.16, 6.17].

In both the Falicov–Kimball model and the Montorsi–Rasetti model only one electron species is dynamically mobile. Consequently, there is no minus-sign problem in quantum Monte-Carlo calculations, and numerical studies can most efficiently be carried out. Detailed numerical studies to the Falicov–Kimball and Montorsi–Rasetti model can be found in [6.30].

6.2.4 Supersymmetric Hubbard Model

In Chap. 4 we viewed (modified) Hubbard models as systematic extensions of Heisenberg models. We identified the exchange coupling with squares of hopping matrix elements, which allowed us to construct exactly solvable one-dimensional Hubbard models. Naturally, there are many other ways to generalize Heisenberg models. In the approach discussed below the global symmetry group of the model is enlarged.

Heisenberg models have two fermionic degrees of freedom per lattice site and are $SU(2)$-invariant. Obviously, we may add one bosonic degree of freedom for holes in a suitable, i.e., supersymmetric way. In this manner we obtain the tJ model (see Sect. 2.5) at its $SU(1|2)$-supersymmetric point $J = 2t$. It is exactly solvable in one dimension [6.31].

Correspondingly, we may add the double occupancies as another bosonic degree of freedom, and obtain the supersymmetric Hubbard model with $SU(2|2)$ symmetry, which can be extended into $U(2|2)$ with sixteen generators if we add the operator for the Hubbard interaction

$$\hat{D}^{\times} = \sum_{\boldsymbol{R}} \hat{D}^{\times}_{\boldsymbol{R}} = \sum_{\boldsymbol{R}} (\hat{n}_{\boldsymbol{R},\uparrow} - 1/2)(\hat{n}_{\boldsymbol{R},\downarrow} - 1/2) \, . \tag{6.23}$$

With a term for the chemical potential μ and for an external magnetic field $h = g\mu_B \mathcal{H}_0/2$ the Hamilton operator explicitly reads [6.32]

$$\begin{aligned} \hat{H}_{\text{susy}} &= t \sum_{\boldsymbol{R},\boldsymbol{T}} \hat{H}^0_{\boldsymbol{R},\boldsymbol{R}+\boldsymbol{T}} + U\hat{D}^{\times} - (\mu - h)\hat{N}_{\uparrow} - (\mu + h)\hat{N}_{\downarrow} \, , \\ \hat{H}^0_{\boldsymbol{R}_1,\boldsymbol{R}_2} &= \hat{\boldsymbol{S}}_{\boldsymbol{R}_1} \cdot \hat{\boldsymbol{S}}_{\boldsymbol{R}_2} - \hat{\boldsymbol{C}}_{\boldsymbol{R}_1} \cdot \hat{\boldsymbol{C}}_{\boldsymbol{R}_2} - \hat{D}^{\times}_{\boldsymbol{R}_1} \\ & \quad - \sum_{\sigma} \left[\hat{L}^{+}_{\boldsymbol{R}_1,\sigma} \hat{L}_{\boldsymbol{R}_2,\sigma} - \hat{U}^{+}_{\boldsymbol{R}_1,\sigma} \hat{U}_{\boldsymbol{R}_2,\sigma} \right] \, , \end{aligned} \tag{6.24}$$

where $\hat{\boldsymbol{S}}_{\boldsymbol{R}}$ and $\hat{\boldsymbol{C}}_{\boldsymbol{R}}$ denote the Heisenberg vector operators for the $SU(2)$ spin and charge transformations; see Sect. 2.5. In addition, the operators

$$\hat{L}_{\boldsymbol{R},\sigma} = \hat{c}_{\boldsymbol{R},\sigma}(1 - \hat{n}_{\boldsymbol{R},-\sigma}) \, , \tag{6.25a}$$

$$\hat{U}_{\boldsymbol{R},\sigma} = \hat{c}_{\boldsymbol{R},\sigma}\hat{n}_{\boldsymbol{R},-\sigma} \tag{6.25b}$$

are used to describe the hopping of holes in the lower and the hopping of double occupancies in the upper Hubbard band, respectively. Their number is conserved during a hopping process between two lattice sites. The model is a special case of an even more general model that can be derived from symmetry considerations on a one-band model [6.14].

It does not come as a surprise that the model in (6.24) is Bethe-ansatz solvable in one dimension since the operator $\hat{H}^0_{\boldsymbol{R},\boldsymbol{R}+\boldsymbol{T}}$ acts as the permutation operator between the 4×4 states of two neighboring lattice sites. The corresponding Bethe-ansatz equations for $\hat{H}_{\text{susy}}$ are a systematic generalization of the Lieb–Wu equations and can be found in [6.32].

The phase diagram in the ground state as a function of the density and the interaction strength U/t is rather similar to the one-dimensional Harris–Lange model at $\gamma = 0$; see Sect. 6.2.2. For example, the unique ground state for attractive interaction and an even electron number is solely formed by paired charges ("η pairing"),

$$|\Psi_0(U < 0)\rangle = \left(\hat{C}^{+}\right)^{N/2} |\text{vacuum}\rangle \, , \tag{6.26}$$

where $\hat{C}^+ = \sum_l \hat{c}^+_{l,\uparrow}\hat{c}^+_{l,\downarrow} = \sum_k \hat{c}^+_{k,\uparrow}\hat{c}^+_{-k,\downarrow}$ creates local double occupancies with zero center-of-mass momentum. This state displays long-range order in the BCS pairing amplitude and is thus characterized as "superconducting".

In analogy to the Harris–Lange model there is a region $0 \leq U \leq U_c(n,d)$ in the parameter space of the supersymmetric Hubbard model for which the ground state still contains $2D$ paired electrons,

$$|\Psi_0(0 \leq U \leq U_c(n,d))\rangle = \left(\hat{C}^+\right)^D |\mathrm{tJ\ model}\rangle_{N-2D} \,. \tag{6.27}$$

$|\mathrm{tJ\ model}\rangle_{N-2D}$ denotes the ground state of the supersymmetric tJ model for the $N-2D<N$ remaining electrons in the subspace of no double occupancies [6.31]. In one dimension one finds $U_c^{\max}(n,d=1) = U_c(n=1,d=1) = 4t\ln 2$. Thus we obtain paired electrons for a formally repulsive interaction strength, in analogy to the Harris–Lange model.

The phase diagram should qualitatively stay the same in higher dimensions. In particular, η pairing at repulsive interaction occurs, but bounds on the precise value for $U_c(n,d\geq 2)$ have not been given yet. Nevertheless, the critical interaction should be positive around half band-filling.

The evidence for superconductivity in a Hubbard model with formally repulsive interactions is weakened by the observation that there are also parts hidden in $\hat{H}^0$ that describe an attraction between electrons. The effective Hubbard interaction therefore is $U^{\mathrm{eff}} = U - 2t$ such that the superconducting regime is effectively much smaller. In addition, the local pairs can move as a whole. The energy gain due to correlated hopping facilitates the formation of these pairs when the on-site repulsion is not too strong.

6.3 Models with Commuting Operators

In this section we discuss three models for which the operators for the kinetic and the potential energy commute. Our first example is the Hubbard model with infinite-range hopping, which reduces to the atomic Hubbard model in the thermodynamic limit, i.e., it describes an insulator for all U/t. Next, we study a reduced version of the one-dimensional $1/r$ Hubbard model. For this model the Stoner theory of ferromagnetism becomes exact and it thus provides an example of itinerant ferromagnetism. As our last model with commuting operators we discuss the Landau–Hubbard Hamiltonian for which the interaction is diagonal in momentum space. This implies the assumption of an (unphysical) infinite-range interaction and correlated hopping.

The metal–insulator transition in models with commuting operators is necessarily the result of the crossing of energy levels ("level crossing transition"), since an exact eigenstate of both operators remains an exact eigenstate independent of the relative mixing of both operators, i.e., the interaction strength. In general, the ground-state energy as a function of the interaction strength is not differentiable at the transition ("first-order" phase transition).

6.3.1 Hubbard Model with Infinite-Range Hopping

In Sect. 5.3 we introduced a ferromagnetic Heisenberg model with infinite-range interactions. If the couplings between two arbitrary lattice sites are identical, $J(\boldsymbol{R}_1 - \boldsymbol{R}_2) = J/L$, the mean-field theory gives the exact thermodynamics of the model. We may now ask in how far the Hubbard model simplifies, if we allow for an infinite-range hopping,

$$t(\boldsymbol{R}_1 - \boldsymbol{R}_2) = -t \ . \tag{6.28a}$$

The dispersion relations becomes

$$\epsilon(\boldsymbol{k}) = -tL\delta_{\boldsymbol{k},\boldsymbol{0}} \ , \tag{6.28b}$$

which fully specifies the Hubbard model with infinite-range hopping. A Kondo lattice model with infinite-range hopping is discussed in [6.33].

The model was first completely solved in [6.34]. It is obvious that, in the thermodynamic limit, the orbital $\boldsymbol{k} = \boldsymbol{0}$ will be occupied with unit probability as soon as at least two electrons are in the system. All other $\boldsymbol{k}$ orbitals have no kinetic energy and $\boldsymbol{k} = \boldsymbol{0}$ is only a single point in $\boldsymbol{k}$ space. Hence, the kinetic and potential energy operators commute in the thermodynamic limit,

$$[\hat{T}, \hat{D}]_- = 0 + \mathcal{O}(1/L) \ . \tag{6.29}$$

The kinetic and potential energy operators can thus be simultaneously diagonalized. In this sense the model is trivial [6.34] since the system always behaves as if it was in the atomic limit.

We introduce the fugacity $z = \exp(\beta\mu)$ to write the free energy density in the thermodynamic limit as

$$f = f_{\text{kin}} + f_{\text{pot}} = -2t - \frac{1}{\beta} \ln\left(1 + 2z + z^2 \mathrm{e}^{-\beta U}\right) \tag{6.30}$$

since we may separately diagonalize the kinetic and potential energy; see Sect. 2.5.4. The average number of double occupancies $\overline{d}(n, U, T)$ is given by (2.38)

$$\mathrm{e}^{-\beta U} = \frac{(1 - n + \overline{d})\overline{d}}{(n/2 - \overline{d})^2} \tag{6.31}$$

and thus has the form of a law of mass action between the singly occupied sites (occupation probability $(n/2 - \overline{d})$ for $\uparrow$ and $\downarrow$ spins), and holes (occupation probability $(1 - n + \overline{d})$) and double occupancies (occupation probability $\overline{d}$). A comparison with (3.58c) and its generalization to finite temperatures [6.35] shows that the Gutzwiller approximation (see Sect. 3.4) becomes exact for the Hubbard model with infinite-range hopping.

The ground state for $n \leq 1$ has no double occupancy as follows from (6.31), and each lattice site is at most singly occupied. The energy for the annihilation of an electron is given by $\mu^-(n \leq 1) = 0$, while the addition of an electron for at least half band-filling requires the energy $\mu^+(n \geq 1) = U$.

Thus, at half band-filling and for finite interactions, the system is an insulator with an energy gap $\Delta\mu(n=1)=U$. Obviously, the model is unphysical since it describes a "metallic" state for $n\neq 1$ only in the sense that the energy gap vanishes, $\Delta\mu(n<1)=\Delta\mu(n>1)=0$. In contrast, the system is *always* insulating since the electrons at $\boldsymbol{k}=\boldsymbol{0}$ cannot be excited, and all other electrons do not move at all. The system displays only the properties of the atomic limit.

The model is also pathologic in several other aspects. For example, the limit of strong interactions (states at $\boldsymbol{k}=\boldsymbol{0}$ are emptied) and the thermodynamic limit (states at $\boldsymbol{k}=\boldsymbol{0}$ are always occupied) do not commute. Furthermore, the thermodynamic pressure does not vanish even for vanishing electron density ($n\to 0$) if we first perform the thermodynamic limit, since the two electrons at $\boldsymbol{k}=\boldsymbol{0}$ produce a finite energy density. The only significance of the model lies in the fact that the (thermal) Gutzwiller approximation provides the exact results for the average number of double occupancies for this very special case.

6.3.2 Stoner Model with Long-Range Hopping

As discussed in Sect. 2.5.5 it is difficult to construct Hubbard-type models that display genuine itinerant ferromagnetism. The Lieb–Mattis theorem forbids one-dimensional itinerant ferromagnetism for nearest-neighbor hopping, and models with a more complicated dispersion relation are often unsatisfactory in other respects. For example, flat-band models become ferromagnetic at arbitrarily small interaction strength, or Tasaki's split-band model describes an insulator beyond the ferromagnetic transition. Below we discuss a one-dimensional model with long-range hopping and Coulomb interactions for which Stoner's theory of ferromagnetism becomes exact. It thus provides an exactly solvable example of one-dimensional itinerant ferromagnetism. However, as an indirect consequence of the Lieb–Mattis theorem, this ferromagnetism is not free of pathologies either.

The model construction starts from the $1/r$ Hubbard model of Sect. 4.2, whose hopping amplitude decays like the inverse of the distance r between two sites. Its dispersion relation is linear, $\epsilon(k)=t(k \bmod 2\pi)$, within the Brillouin zone. The Hamilton operator reads

$$\begin{aligned}\hat{H}_{\text{Stoner}} &= \hat{T}+U\hat{D}^{0}\,.,\\ \hat{T} &= \sum_{|k|<\pi,\sigma}\epsilon(k)\hat{c}^{+}_{k,\sigma}\hat{c}_{k,\sigma}\,,\\ \hat{D}^{0} &= \hat{N}_{\uparrow}\hat{N}_{\downarrow}/L+\sum_{0<q\leq 2\pi-\Delta}\hat{\rho}^{\text{N}}_{\uparrow}(q)\hat{\rho}^{\text{N}}_{\downarrow}(-q)+\hat{\rho}^{\text{U}}_{\uparrow}(q)\hat{\rho}^{\text{U}}_{\downarrow}(-q)\,.\end{aligned}\tag{6.32}$$

We set $\Delta=2\pi/L$ and the bandwidth is $W=2\pi t$. The operators $\hat{\rho}^{\text{N,U}}_{\sigma}(q)$ were defined in (4.54). The operator $\hat{\rho}^{\text{N}}_{\sigma}(q)$ scatters an electron inside the

Brillouin zone ("normal process"), whereas $\hat{\rho}_\sigma^{\mathrm{U}}(q)$ scatters an electron out of the Brillouin zone ("umklapp process"). The kinetic energy of a state will not be changed by the action of $\hat{D}^0$,

$$[\hat{T}, \hat{D}^0]_- = 0 \,. \tag{6.33}$$

Since the model describes long-range hopping of the electrons and also contains long-range hopping contributions in the interaction term, the Lieb–Mattis theorem [6.36] does not apply. Hence the ground state of this one-dimensional model could be ferromagnetic.

Although a simultaneous diagonalization of both operators in $\hat{H}_{\mathrm{Stoner}}$ is possible in principle we have not succeeded in the explicit construction of all eigenstates of this model. The spectrum, however, can be conjectured from the effective Hamilton operator we have already derived for the $1/r$ Hubbard model in Sect. 4.2. To this end we have only to neglect those parts in $\hat{H}_1^{\mathrm{eff}}$, (4.59), by which the two configurations of a doublet mix. In this way we obtain the effective Hamilton operator for the $1/r$ Stoner model

$$\hat{H}^{\mathrm{eff}}_{\mathrm{Stoner}} = \hat{H}_0^{\mathrm{eff}} + \hat{H}_2^{\mathrm{eff}} \,, \tag{6.34}$$

$$\begin{aligned}\hat{H}_2^{\mathrm{eff}} = \sum_{-\pi<\mathcal{K}<\pi-\Delta} \Bigg\{ & \left[\frac{U}{2} - \left(\pi t - \frac{U}{4\pi}(2\mathcal{K}+\Delta)\right)\right] \hat{d}^+_{\mathcal{K}}\hat{d}_{\mathcal{K}}\hat{e}^+_{\mathcal{K}+\Delta}\hat{e}_{\mathcal{K}+\Delta} \\ & + \left[\frac{U}{2} + \left(\pi t - \frac{U}{4\pi}(2\mathcal{K}+\Delta)\right)\right] \hat{s}^+_{\mathcal{K},\uparrow}\hat{s}_{\mathcal{K},\uparrow}\hat{s}^+_{\mathcal{K}+\Delta,\downarrow}\hat{s}_{\mathcal{K}+\Delta,\downarrow}\Bigg\}\end{aligned}$$

with $\hat{H}_0^{\mathrm{eff}}$ from (4.58).

If we add an external magnetic field ($\mathcal{H}_0$) and the chemical potential $\mu_\sigma = \mu - \sigma(g\mu_B\mathcal{H}_0)/2$ we may compactly write

$$\begin{aligned}\hat{H}^{\mathrm{eff}}_{\mathrm{Stoner}} = \sum_{-\pi<\mathcal{K}<\pi} \Bigg\{ & \sum_\sigma h^s_{\mathcal{K},\sigma}\hat{n}^s_{\mathcal{K},\sigma} + h^d_{\mathcal{K}}\hat{n}^d_{\mathcal{K}} + h^e_{\mathcal{K}}\hat{n}^e_{\mathcal{K}} \\ & + \widetilde{J}_{\mathcal{K}}\left[\hat{n}^d_{\mathcal{K}-\Delta}\hat{n}^e_{\mathcal{K}} - \hat{n}^s_{\mathcal{K}-\Delta,\uparrow}\hat{n}^s_{\mathcal{K},\downarrow}\right]\Bigg\}\end{aligned} \tag{6.35a}$$

with $h^s_{\mathcal{K},\sigma} = (t\mathcal{K})/2 - \mu_\sigma$, $h^d_{\mathcal{K}} = -(t\mathcal{K})/2 - 2\mu + U$, $h^e_{\mathcal{K}} = -(t\mathcal{K})/2$, and the new coupling

$$\widetilde{J}_{\mathcal{K}} = \frac{W+U}{2}\left(\frac{2\mathcal{K}-\Delta}{2\pi} - 1\right) \le 0 \,. \tag{6.35b}$$

Here we also introduced the notations $\hat{n}^s_{\mathcal{K},\sigma} = \hat{s}^+_{\mathcal{K},\sigma}\hat{s}_{\mathcal{K},\sigma}$ etc. The coupling $\widetilde{J}_{\mathcal{K}}$ is now ferromagnetic whereas it was antiferromagnetic in the $1/r$ Hubbard model; see (4.60).

As for the $1/r$ Hubbard model we obtain an effective Ashkin–Teller model whose thermodynamics can be determined with the help of the transfer-matrix method. The analysis of the ground state of the model in (6.35) is simple. As for the $1/r$ Hubbard model we may focus on two electrons each in neighboring orbitals. For two electrons we occupy the neighboring orbitals $\mathcal{K}, \mathcal{K}+\Delta$ with the configuration $\boxed{\bullet \circ}_{\mathcal{K},\mathcal{K}+\Delta}$ if $U \leq W$ holds; otherwise, if $U > W$, we choose one of the three energetically equivalent spin-triplet configurations $\uparrow\uparrow_{\mathcal{K},\mathcal{K}+\Delta}$, $\downarrow\uparrow_{\mathcal{K},\mathcal{K}+\Delta}$, or $\downarrow\downarrow_{\mathcal{K},\mathcal{K}+\Delta}$. For $U > W$ the singlet configuration is higher in energy than the triplet configuration for all orbitals.

Since these statements are independent of the orbital $\mathcal{K}$ the symbolic ground state only contains $\boxed{\bullet \circ}_{\mathcal{K},\mathcal{K}+\Delta}$ pairs in the regime $-\pi < \mathcal{K} < \mathcal{K}_{\mathrm{F}} = \pi(2n-1)$ and holes $\circ_{\mathcal{K}}$ in the regime $\mathcal{K}_{\mathrm{F}} < \mathcal{K} < \pi$, if $U \leq W$. For $U > W$ we obtain a $3^{N/2}$-fold degenerate ground state of the outer product of $N/2$ pairs of spin-triplets for $-\pi < \mathcal{K} < \mathcal{K}_{\mathrm{F}} = \pi(2n-1)$ and holes $\circ_{\mathcal{K}}$ for $\mathcal{K}_{\mathrm{F}} < \mathcal{K} < \pi$. In this class of states we also find the fully polarized ferromagnetic state. For the $1/r$ Hubbard model the ground state could be interpreted as a product over singlet pairs (resonating valence bond or RVB picture). The ground state of the $1/r$ Stoner model for $U > W$ is seen to be a product over RVB triplet pairs, and thus is vastly degenerate.

The ground-state energy in the thermodynamic limit is ready calculated as

$$(E_0/L)(n, U/W) = \begin{cases} W(n^2-2n)/4 + Un^2/4 & \text{for} \quad U \leq W \\ W(n^2-n)/2 & \text{for} \quad U \geq W \end{cases} . \tag{6.36}$$

These are precisely the results of Stoner's theory; see Sect. 3.1, (3.12). The ground-state energy for $U \leq W$ is composed of the energy of the Fermi gas, $(E_0/L)(n, U=0) = W(n^2-n)/4$, and the Hartree term, $(E_0/L)^{\mathrm{Hartree}} = Un^2/4$. After the transition at $U_{\mathrm{c}}^{\mathrm{Stoner}} = W$ we find the ground-state energy of the fully ferromagnetically polarized state.

The density of states of the non-interacting system is constant, $D_\sigma(\epsilon) = 1/W$ for $|\epsilon| \leq W/2$, such that the critical coupling strength $U_{\mathrm{c}}^{\mathrm{Stoner}}$ indeed obeys the Stoner criterion (3.11),

$$U_{\mathrm{c}}^{\mathrm{Stoner}} D_{\sigma,0}(0) = 1 \; . \tag{6.37}$$

This justifies the term "Stoner model" for the toy model of (6.32). Due to the particular dispersion relation, however, the magnetization does not gradually reach its saturation value for U values above the transition. Instead, the fully ferromagnetically polarized state is among the huge class of states that can be formed from $N/2$ triplet pairs. In this sense the Lieb–Mattis theorem is not fully spoiled because a ground-state singlet with spin $S = 0$ is still degenerate with the ferromagnetic state with spin $S = N/2$ for electron numbers $N \bmod 4 = 0$.

The transition at $n = 1$ also is a metal–insulator transition because the fully ferromagnetically polarized state at half band-filling is an insulator. The transition is of "first-order" since the ground-state energy is not differentiable with respect to the interaction parameter U. The reason for that lies in the nature of this level crossing transition. The paramagnetic Fermi sea remains eigenstate also for $U > W$ but it has a higher energy than the ferromagnetic state. The transition is pathologic for yet another reason. At arbitrarily weak magnetic field we obtain a transition from a paramagnetic state with zero magnetization into the fully polarized ferromagnetic state.

The significance of this model lies in the fact that we can give an explicit example for Stoner ferromagnetism for itinerant electrons without infinite-range interactions and/or hopping ranges. Unfortunately, the model is much too special to formulate general criteria for the existence of ferromagnetism in a one-band system. However, it confirms the tendencies of the Lieb–Mattis theorem that the hopping has to be long-ranged and that a lattice (finite bandwidth) are necessary conditions for the occurrence of itinerant ferromagnetism.

6.3.3 Landau–Hubbard Model

Landau's Fermi-liquid theory provides the standard description of electron–electron interactions in metals at low temperatures; see Sect. 1.4.1. Moreover, it can be linked to an exactly solvable model since the Landau quasi-particles keep their momentum during a scattering process. This can be seen from Landau's energy functional of a normal Fermi liquid, which gives the change in internal energy due to the presence of quasi-particle excitations of density $\delta n_{\boldsymbol{p},\sigma}$ [6.37]

$$E - E_0 = \sum_{\boldsymbol{p},\sigma} (\epsilon(\boldsymbol{p}) - \mu)\delta n_{\boldsymbol{p},\sigma} + \frac{1}{2} \sum_{\boldsymbol{p},\sigma;\boldsymbol{p}',\sigma'} f(\boldsymbol{p},\sigma;\boldsymbol{p}',\sigma')\delta n_{\boldsymbol{p},\sigma}\delta n_{\boldsymbol{p}',\sigma'} \, . \quad (6.38)$$

E_0 is the ground-state energy of the interacting Fermi system, and terms of the order $(\delta n_{\boldsymbol{p},\sigma})^3$ in the Taylor expansion of $E - E_0$ have been neglected. Both terms in (6.38) are of second order in the density of quasi-particle excitations since $(\epsilon(\boldsymbol{p}) - \mu)$ itself is of order $(\delta n_{\boldsymbol{p},\sigma})$. Indeed, the energy functional (6.38) describes a dilute gas of quasi-particles whose momentum is conserved, i.e., only forward scattering with non-singular amplitudes $f(\boldsymbol{p},\sigma;\boldsymbol{p}',\sigma')$ is allowed. In this case, the physics of Fermi-liquid theory can be worked out in all detail [6.37].

Of course, Landau's model only describes low-lying quasi-particle excitations of the filled Fermi sea. Nevertheless, we may readily generalize (6.38) to a model for interacting electrons with pure forward scattering,

$$\hat{H}_{\text{forward}} = \sum_{\boldsymbol{k},\sigma} \epsilon(\boldsymbol{k})\hat{n}_{\boldsymbol{k},\sigma} + \sum_{\boldsymbol{k},\sigma;\boldsymbol{k}',\sigma'} V(\boldsymbol{k},\sigma;\boldsymbol{k}',\sigma')\hat{n}_{\boldsymbol{k},\sigma}\hat{n}_{\boldsymbol{k}',\sigma'} \, . \quad (6.39)$$

Obviously, the operators for the kinetic and potential energy commute with each other, i.e., the model (6.39) is a classical, Ising-type model. Unfortunately, the spin-1/2 Ising model cannot be solved explicitly for general interaction $V(\boldsymbol{k},\sigma;\boldsymbol{k}',\sigma')$ and orbital-dependent "magnetic fields" $\epsilon(\boldsymbol{k})$. Hence, we have to introduce a further approximation to obtain a completely solvable model.

In analogy to the derivation of the Hubbard model in Sect. 2.3.2 we neglect all terms except those that are independent of the (momentum) indices in $V(\boldsymbol{k},\sigma;\boldsymbol{k}',\sigma')$. This defines the Landau–Hubbard model as

$$\hat{H}_{\mathrm{LH}} = \sum_{\boldsymbol{k},\sigma} \epsilon(\boldsymbol{k})\hat{n}_{\boldsymbol{k},\sigma} + U \sum_{\boldsymbol{k}} \hat{n}_{\boldsymbol{k},\uparrow}\hat{n}_{\boldsymbol{k},\downarrow} , \tag{6.40}$$

where we absorbed some terms in the chemical potential. The model was apparently independently introduced in [6.38]–[6.40]. The simplification of a momentum-independent interaction term allows an explicit solution since now the $\boldsymbol{k}$ orbitals are independent of each other. Their ordering and occupation, however, are determined by the dispersion relation and the interaction strength.

The model is formally equivalent to the atomic limit of the Hubbard model with orbital-dependent energies. Consequently, the free energy density reads

$$f(U,T,\mu) = -(T/L)\sum_{\boldsymbol{k}} \ln\left(1 + 2\mathrm{e}^{-\beta(\epsilon(\boldsymbol{k})-\mu)} + \mathrm{e}^{-\beta(2\epsilon(\boldsymbol{k})+U-2\mu)}\right) \tag{6.41}$$

[see (2.38), Sect. 2.5.4]. All thermodynamic quantities are readily derived from (6.41) [6.40].

In the ground state those $\boldsymbol{k}$ levels with energies $\epsilon(\boldsymbol{k}) \leq \mu - U/2$ are doubly occupied and those with $\mu - U/2 \leq \epsilon(\boldsymbol{k}) \leq \mu$ are singly occupied with either an $\uparrow$ or $\downarrow$ electron. This already shows that the model does not describe a Fermi liquid in the metallic phase for $U \neq 0$. In particular, the ground state is highly spin-degenerate for all finite interaction strengths. Below half band-filling we always find an empty $\boldsymbol{k}$ orbital left for the addition of an extra electron, and this energy level is thermodynamically close in energy to the last occupied one. Hence, the system is metallic for $n < 1$, even at $U = \infty$. In one dimension, and for $U = \infty$, the spectrum and its degeneracies agree with the result for the Hubbard model, i.e., the $\hat{T}$ model. Furthermore, the non-interacting limit and the atomic limit of the Hubbard model are properly reproduced.

At half band-filling, and for $U < W$, there are some singly occupied and empty $\boldsymbol{k}$ orbitals, and the chemical potential is continuous. For $U > W$, however, all orbitals are singly occupied in the ground state, and the chemical potentials are given by $\mu^- = W/2$, $\mu^+ = U - W/2$, respectively. The gap is given by $\Delta\mu = \mu^+ - \mu^- = U - W$, and the model describes a metal–insulator transition at $U_{\mathrm{c}} = W$.

Certainly, the Landau–Hubbard model is the simplest model conceivable that illustrates the band-splitting effect of the Coulomb interaction: in the presence of a $-\sigma$ electron in the orbital $\boldsymbol{k}$ the dispersion of the σ electron is $\epsilon(\boldsymbol{k}) + U$, otherwise it is $\epsilon(\boldsymbol{k})$. These are the two Hubbard bands that, within the model, have the full bandwidth W, independent of U, and split for *all* $\boldsymbol{k}$ orbitals for arbitrarily small $U > 0$. The gap $\Delta\mu = U - W$ opens at $U_{\mathrm{c}} = W$ when the two bands cease to overlap.

The model is rather unrealistic since we expect Fermi-liquid theory to hold at small excitation energies. Clearly, a momentum-dependent coupling $U(\boldsymbol{k})$ with $U(\boldsymbol{k}) \to 0$ for $\epsilon(\boldsymbol{k}) \to E_{\mathrm{F}}$ would allow a more realistic description of the metallic phase. The Hubbard bands as signatures for the local-moment formation should only show up at higher energies, at least as long as the interaction is small. The present model displays different behavior because the interaction is of infinite range,

$$U \sum_{\boldsymbol{k}} \hat{n}_{\boldsymbol{k},\uparrow} \hat{n}_{\boldsymbol{k},\downarrow} = (U/L) \sum_{\boldsymbol{R}_1,\boldsymbol{R}_2,\boldsymbol{R}_3,\boldsymbol{R}_4} \delta_{\boldsymbol{R}_1+\boldsymbol{R}_2,\boldsymbol{R}_3+\boldsymbol{R}_4} \hat{c}^{+}_{\boldsymbol{R}_1,\uparrow} \hat{c}_{\boldsymbol{R}_2,\uparrow} \hat{c}^{+}_{\boldsymbol{R}_3,\downarrow} \hat{c}_{\boldsymbol{R}_4,\downarrow} , \tag{6.42}$$

and also contains terms that describe an infinite-range (correlated) electron hopping. Nevertheless, the model may serve as an instructive toy model to illustrate the concept of the split upper and lower Hubbard bands. In addition, it is possibly the simplest model that can be cast into the concept of free particles with a "statistical interaction" [6.41]. We will not further elaborate on this interesting new development [6.42].

With these considerations on various modifications on the Hubbard model and their metal–insulator transitions we close our presentation of analytical approaches to the metal–Mott insulator transition.

7. Conclusions

In this book we have reviewed our present-day understanding of the Mott transition as a genuine correlation effect. We used exact and approximate treatments of model Hamiltonians to support our basic concepts on the canonical phase diagram. We summarize our results in this last chapter.

Basic Concepts: The Canonical Phase Diagram

As discussed in Chap. 1, the interaction between electrons leads to magnetic moments. In general, as pointed out by Mott [7.1, 7.2], these moments have formed at temperatures far above their ordering (Néel) temperature $T_{\rm N}(U)$ below which the undoped system is a (antiferromagnetic) Mott–Heisenberg insulator. The sublattice magnetization plays the role of the order parameter in this thermodynamic phase transition.

The more subtle and interesting question is the conducting behavior at temperatures above $T_{\rm N}(U)$. A material can effectively remain insulating (Mott–Hubbard insulator) even above the Néel temperature where the order of the moments ceases to exist. Such behavior can essentially be understood as the consequence of a quantum phase transition at zero temperature: even if the long-range order of the moments was absent, the electron correlations will drive the metal into the Mott–Hubbard insulator at some finite interaction strength $U_{\rm c}$.

The canonical (U,T) phase diagram for an undoped system then displays three generic phases and their respective transitions; an (antiferromagnetic) Mott–Heisenberg insulator below the Néel temperature $T_{\rm N}(U)$, a correlated metal above $T_{\rm N}(U)$ and for sufficiently low interaction strengths, and finally a paramagnetic Mott–Hubbard insulator above $T_{\rm N}(U)$ and larger interaction strengths. Note that there is no sharp distinction between the paramagnetic metal and the paramagnetic Mott–Hubbard insulator for $T > T_{\rm N}(U) \geq 0$ since these are not two different phases in the sense of statistical physics. Instead, there is a crossover as a function of the interaction strength at finite temperatures.

The phase diagram of vanadium sesquioxide (Fig. 1.12, p. 45) displays these three canonical phases, which suggests that our basic concepts are correct. Furthermore, it confirms that a correlated electron approach is the proper starting point for theoretical investigations of these materials. A complicated band structure, band degeneracies, the influence of impurities and the electron–phonon interaction often blur the basic mechanism behind the Mott metal–insulator transition. For example, the electron–phonon interaction is a relevant perturbation in the crossover region from the metal to the Mott–Hubbard insulator, and gives rise to a line of first-order transitions in the phase diagram of V_2O_3. Therefore, and to bring out the essential underlying physics, we study simple model Hamiltonians which capture the essential elements of the Mott transition.

Models and Methods: Results for the Hubbard Hamiltonian

In Chap. 2 we derived the Hubbard model, conceivably the simplest model to describe interacting itinerant lattice electrons. It thus is the natural starting point to study the Mott transition.

The common approximation schemes discussed in Chap. 3 provide a complementary view on the various metallic and insulating phases. For example, the Hartree–Fock approach focuses on possible types of long-range order in the Mott–Heisenberg insulator, Hubbard's Green-function decoupling scheme approximates the Mott–Hubbard insulator reasonably well, and Gutzwiller-correlated variational wave functions and Slave Boson approaches reasonably describe the metallic Fermi-liquid regime. Hence, the available approximate methods for correlated electrons provide strong evidence for the correlated metal, the Mott–Hubbard insulator, and the Mott–Heisenberg insulator as essential parts of the canonical phase diagram of the Hubbard model at half band-filling. However, a satisfactory theory for the Mott transition requires an equally good description of all three canonical phases.

Exact solutions for all interaction strengths can be obtained for the Hubbard model in one dimension (Chap. 4). Unfortunately, special properties are often the price we have to pay for integrability. For example, perfect nesting is the generic case in one dimension, which implies that the half-filled Hubbard model describes a Mott–Hubbard insulator at all finite interaction strengths. Nesting is absent for the $1/r$ Hubbard model: in complete agreement with our scenario for the Mott–Hubbard transition its exact solution displays a quantum phase transition from the metal into the paramagnetic Mott–Hubbard insulator at a finite interaction strength.

The combined analytical and numerical results in infinite dimensions (Chap. 5) clearly demonstrate that the Hubbard model is able to describe the generic (U, T) phase diagram at half band-filling and the corresponding Mott transition(s). However, more research efforts are necessary to clarify important details of the Mott–Hubbard transition. For example, it remains an open question whether this zero-temperature quantum phase transition is discontinuous or continuous, and whether or not there is a line of first-order transitions at finite temperatures.

In the course of this book it has become clear that the description of the Mott–Hubbard quantum phase transition requires systematic and controlled approaches to interacting many-electron systems. Substantial progress in the understanding of the Mott transition has been made but a lot remains to be learned. Heavy fermion materials, high-temperature superconductors, and relativistic many-particle systems [7.3] will certainly stimulate further intensive research on models and methods for correlated electron systems.

Appendix

A.1 Calculations for Hartree–Fock Theory

In this first section we present some details on the antiferromagnetic Hartree–Fock theory for the Hubbard model. We diagonalize the Hartree–Fock Hamiltonian for a simple cubic lattice, and discuss the resulting gap at half band-filling.

A.1.1 Diagonalization of the Hartree–Fock Hamiltonian

For simple cubic lattices in d dimensions we transform the Hartree–Fock Hamiltonian $\hat{H}^{\mathrm{HF}}$, (3.2), into momentum space,

$$\begin{aligned}\hat{H}^{\mathrm{HF}} &= \frac{UL(n^2+m^2)}{4} + \sum_{\boldsymbol{k}\in\mathrm{MBZ},\sigma}\Big[\epsilon(\boldsymbol{k})\hat{n}_{\boldsymbol{k},\sigma} + \epsilon(\boldsymbol{k}+\boldsymbol{Q})\hat{n}_{\boldsymbol{k}+\boldsymbol{Q},\sigma} \quad \text{(A.1.1)}\\ &\qquad - \frac{Um\lambda_{\sigma,\uparrow}}{2}\left(\hat{c}^{+}_{\boldsymbol{k}+\boldsymbol{Q},\sigma}\hat{c}_{\boldsymbol{k},\sigma} + \mathrm{h.c.}\right)\Big]\,,\end{aligned}$$

where $\boldsymbol{Q} = \boldsymbol{G}/2 = (\pi,\ldots,\pi)/a$ denotes half a reciprocal lattice vector. MBZ is the magnetic Brillouin zone in which the two vectors $\boldsymbol{k}$ and $\boldsymbol{k}+\boldsymbol{Q}$ from the first Brillouin zone are identical, and $\lambda_{\sigma,\sigma} = +1$, $\lambda_{\sigma,-\sigma} = -1$ hold. We introduce the (Bogoliubov) quasi-particles on the A and B sublattice via the new fermion operators

$$\hat{a}_{\boldsymbol{k},\sigma} = \cos\phi_{\boldsymbol{k},\sigma}\hat{c}_{\boldsymbol{k},\sigma} + \sin\phi_{\boldsymbol{k},\sigma}\hat{c}_{\boldsymbol{k}+\boldsymbol{Q},\sigma}\,, \tag{A.1.2a}$$

$$\hat{b}_{\boldsymbol{k},\sigma} = -\sin\phi_{\boldsymbol{k},\sigma}\hat{c}_{\boldsymbol{k},\sigma} + \cos\phi_{\boldsymbol{k},\sigma}\hat{c}_{\boldsymbol{k}+\boldsymbol{Q},\sigma}\,, \tag{A.1.2b}$$

which has the inverse transformation

$$\hat{c}_{\boldsymbol{k},\sigma} = \cos\phi_{\boldsymbol{k},\sigma}\hat{a}_{\boldsymbol{k},\sigma} - \sin\phi_{\boldsymbol{k},\sigma}\hat{b}_{\boldsymbol{k},\sigma}\,, \tag{A.1.2c}$$

$$\hat{c}_{\boldsymbol{k}+\boldsymbol{Q},\sigma} = \sin\phi_{\boldsymbol{k},\sigma}\hat{a}_{\boldsymbol{k},\sigma} + \cos\phi_{\boldsymbol{k},\sigma}\hat{b}_{\boldsymbol{k},\sigma}\,. \tag{A.1.2d}$$

The new quasi-particles are defined such that $\hat{a}^{+}_{\boldsymbol{k},\sigma}$ ($\hat{b}^{+}_{\boldsymbol{k},\sigma}$) creates σ particles predominantly on the A (B) lattice sites.

The Hartree–Fock Hamiltonian becomes diagonal in the new operators if the following relations hold:

$$\tan(2\phi_{\boldsymbol{k},\sigma}) = \frac{Um\lambda_{\sigma,\uparrow}}{\epsilon(\boldsymbol{k})-\epsilon(\boldsymbol{k}+\boldsymbol{Q})}\,, \tag{A.1.3a}$$

$$\sin^2\phi_{\boldsymbol{k},\sigma} = \frac{1}{2}\left(1-\lambda_{\sigma,\uparrow}\frac{\epsilon(\boldsymbol{k})-\epsilon(\boldsymbol{k}+\boldsymbol{Q})}{\sqrt{[\epsilon(\boldsymbol{k})-\epsilon(\boldsymbol{k}+\boldsymbol{Q})]^2+(Um)^2}}\right)\,. \tag{A.1.3b}$$

It finally becomes

$$\begin{aligned}\hat{H}^{\mathrm{HF}} &= \frac{UL(n^2+m^2)}{4}+\sum_{\boldsymbol{k}\in\mathrm{MBZ}}\Big[E^-(\boldsymbol{k})\left(\hat{a}^+_{\boldsymbol{k},\uparrow}\hat{a}_{\boldsymbol{k},\uparrow}+\hat{b}^+_{\boldsymbol{k},\downarrow}\hat{b}_{\boldsymbol{k},\downarrow}\right) \qquad \text{(A.1.4a)}\\ &\qquad + E^+(\boldsymbol{k})\left(\hat{a}^+_{\boldsymbol{k},\downarrow}\hat{a}_{\boldsymbol{k},\downarrow}+\hat{b}^+_{\boldsymbol{k},\uparrow}\hat{b}_{\boldsymbol{k},\uparrow}\right)\Big]\,,\end{aligned}$$

$$E^\pm(\boldsymbol{k}) = \frac{1}{2}\left[\epsilon(\boldsymbol{k})+\epsilon(\boldsymbol{k}+\boldsymbol{Q})\pm\sqrt{[\epsilon(\boldsymbol{k})-\epsilon(\boldsymbol{k}+\boldsymbol{Q})]^2+(Um)^2}\right]\,. \tag{A.1.4b}$$

At finite temperatures the expectation values for the occupation number operators of the quasi-particles read

$$\langle\hat{n}^a_{\boldsymbol{k},\uparrow}\rangle = \langle\hat{n}^b_{\boldsymbol{k},\downarrow}\rangle = f_{\mathrm{FD}}(E^-(\boldsymbol{k})-\mu^{\mathrm{HF}})\,, \tag{A.1.5a}$$

$$\langle\hat{n}^b_{\boldsymbol{k},\uparrow}\rangle = \langle\hat{n}^a_{\boldsymbol{k},\downarrow}\rangle = f_{\mathrm{FD}}(E^+(\boldsymbol{k})-\mu^{\mathrm{HF}})\,, \tag{A.1.5b}$$

where $f_{\mathrm{FD}}(x)=(\exp(\beta x)+1)^{-1}$ denotes the Fermi–Dirac distribution. We note in passing that the Hartree–Fock chemical potential μ^{HF} will, in general, differ from the exact chemical potential. We have to calculate μ^{HF} according to its definition

$$n = \sum_{\boldsymbol{k}\in\mathrm{MBZ},\sigma}\left[f_{\mathrm{FD}}(E^+(\boldsymbol{k})-\mu^{\mathrm{HF}})+f_{\mathrm{FD}}(E^-(\boldsymbol{k})-\mu^{\mathrm{HF}})\right]\,. \tag{A.1.6}$$

The self-consistency equation for the sublattice magnetization m is given by

$$\begin{aligned} m &= |\langle\hat{n}_{\boldsymbol{R},\uparrow}\rangle-\langle\hat{n}_{\boldsymbol{R},\downarrow}\rangle|\\ &= \frac{2}{L}\sum_{\boldsymbol{k}\in\mathrm{MBZ}}\frac{Um}{\sqrt{[\epsilon(\boldsymbol{k})-\epsilon(\boldsymbol{k}+\boldsymbol{Q})]^2+(Um)^2}} \qquad \text{(A.1.7)}\\ &\qquad\times\left(f_{\mathrm{FD}}(E^-(\boldsymbol{k})-\mu^{\mathrm{HF}})-f_{\mathrm{FD}}(E^+(\boldsymbol{k})-\mu^{\mathrm{HF}})\right)\,.\end{aligned}$$

A discussion of the thermodynamics can be found, e.g., in [A.1, A.2].

A.1.2 Ground-State Properties for Half Band-Filling

In the following we concentrate on the half-filled Hubbard model on a simple cubic lattice at zero temperature. In this case the dispersion relation obeys the "perfect nesting" condition, $\epsilon(\boldsymbol{k})=-\epsilon(\boldsymbol{k}+\boldsymbol{Q})$ for all $\boldsymbol{k}$ from the magnetic

Brillouin zone. Equation (A.1.6) gives $\mu^{\mathrm{HF}} = 0$ (for all temperatures) since $(\mathrm{e}^{x}+1)^{-1} + (\mathrm{e}^{-x}+1)^{-1} = 1$.

To study the insulating properties we have to determine the Hartree–Fock band gap for $T = 0$. At half-filling the lower band is completely filled, $f_{\mathrm{FD}}(E^{-}(\boldsymbol{k})) = 1$, and the upper band is empty, $f_{\mathrm{FD}}(E^{+}(\boldsymbol{k})) = 0$. Then, we can rewrite (A.1.7) for the antiferromagnetic solutions ($m \neq 0$) as

$$\frac{1}{U} = \int_{-W/2}^{W/2} \frac{\mathrm{d}\epsilon D_{\sigma,0}(\epsilon)}{\sqrt{(Um)^2 + 4\epsilon^2}} \tag{A.1.8}$$

where $W = 4dt$ and $D_{\sigma,0}(\epsilon)$ denote the bandwidth and the density of states for a single spin species in d dimensions, respectively. The Hartree–Fock gap for an excitation is thus calculated as

$$\Delta\mu^{\mathrm{HF}} = \min_{\{\boldsymbol{k}\}} \left[E^{+}(\boldsymbol{k}) - E^{-}(\boldsymbol{k})\right] = \min_{\{\boldsymbol{k}\}} \left[\sqrt{4\left[\epsilon(\boldsymbol{k})\right]^2 + (Um)^2}\right] = Um\,. \tag{A.1.9}$$

According to (A.1.9) a band gap opens at the Fermi surface, $\epsilon(\boldsymbol{k}) = 0$, for $m > 0$.

The size of the Hartree–Fock gap can easily be estimated for large and small interaction strengths. For small U/t one obtains

$$\frac{1}{U} \approx \int_{Um/2}^{E_{\mathrm{c}}'} \frac{\mathrm{d}\epsilon D_{\sigma,0}(\epsilon)}{\epsilon} + \text{regular contributions} \tag{A.1.10a}$$

$$\approx D_{\sigma,0}(Um/2)\ln\left[2E_{\mathrm{c}}'/(Um)\right] + \text{regular contributions,} \tag{A.1.10b}$$

where E_{c}' is a cut-off energy. If the density of states is finite at $\epsilon = 0$, as in one or three dimensions, we may replace $D_{\sigma,0}(Um/2)$ by $D_{\sigma,0}(0)$ and solve the equation for $\Delta\mu^{\mathrm{HF}} = Um$. In two dimensions the density of states is itself logarithmically divergent near zero energy (van-Hove singularity), which results in another logarithmic factor in (A.1.10b). Finally we obtain the Hartree–Fock energy gap $\Delta\mu^{\mathrm{HF}} = Um$ as [A.3]

$$\Delta\mu^{\mathrm{HF}}(U \to 0, d=1) \sim \exp\left(-\frac{2\pi t}{U}\right), \tag{A.1.11a}$$

$$\Delta\mu^{\mathrm{HF}}(U \to 0, d=2) \sim \exp\left(-\sqrt{\frac{E_{\mathrm{c}}}{U}}\right), \tag{A.1.11b}$$

$$\Delta\mu^{\mathrm{HF}}(U \to 0, d=3) \sim \exp\left(-\frac{1}{UD_{\sigma,0}(0)}\right). \tag{A.1.11c}$$

We see that the gap opens for arbitrarily small interaction strengths. For situations with "perfect nesting" we thus find $U_{\mathrm{c}}^{\mathrm{HF}}(n=1) = 0^{+}$ as the critical interaction strength within Hartree–Fock theory.

For large interactions the Hartree–Fock ground state of the Hubbard model smoothly transforms into that of the Heisenberg model. For $U = \infty$

we thereby obtain the Néel state. It is apparent that this classical spin state lacks all quantum-mechanical spin fluctuations. The neglect of these contributions implies in particular that low-energy spin excitations (spin waves) are not contained in the theory. For $U \gg W$ the gap for the charge excitations, (A.1.8) and (A.1.9), becomes

$$\Delta\mu^{\rm HF}(U\to\infty,d) \approx U - \frac{2}{U}\int_{W/2}^{W/2} {\rm d}\epsilon\epsilon^2 D_{\sigma,0}(\epsilon)\;. \tag{A.1.12}$$

In Hartree–Fock theory there is *no* bandwidth contribution to the gap since the particles are fixed onto their respective sublattice sites.

The ground-state energy is readily calculated as

$$\begin{aligned}(E_0^{\rm HF}/L) &= \frac{U(1+m^2)}{4} + 2\sum_{E^-(\boldsymbol{k})\le\mu^{\rm HF}} E^-(\boldsymbol{k}) \\ &= \frac{U(1+m^2)}{4} - \int_{-W/2}^{0} {\rm d}\epsilon D_{\sigma,0}(\epsilon)\sqrt{(Um)^2+4\epsilon^2}\;. \end{aligned}\tag{A.1.13}$$

In the limits of small and large interaction strengths one obtains

$$(E_0^{\rm HF}/L)(U\to 0) = E_{\text{Fermi sea}} + \frac{U}{4} + \text{exp. small terms}\;, \tag{A.1.14a}$$

$$(E_0^{\rm HF}/L)(U\to \infty) = -\frac{1}{U}\int_{W/2}^{W/2} {\rm d}\epsilon\epsilon^2 D_{\sigma,0}(\epsilon)\;. \tag{A.1.14b}$$

For lattices without "perfect nesting" one finds that m becomes finite above a critical interaction strength, $U_{\rm c}^{\rm HF} > 0$. Typically, the energy gap is not direct but indirect, i.e., the maximum of $E^-(\boldsymbol{k})$ and the minimum of $E^+(\boldsymbol{k})$ are not at the same but at different momenta $\boldsymbol{k}$. The calculations for a one-dimensional example are carried out in [A.4].

A.2 Green Functions and Hubbard-I Approximation

In the first subsection we recall some basic properties of Green functions that are suitable for the description of the physics of many-particle systems. More details of this standard many-particle language can be found in [A.5]–[A.7]. In the second subsection we present the Green-function decoupling method in its simplest form to derive the Hubbard-I approximation.

A.2.1 Green Functions

We start with a list of definitions and elementary relationships [A.8]; see also [A.6, Chap. 3].

The thermodynamic average over an operator $\hat{X}$ is given by

$$\langle \hat{X} \rangle = \frac{1}{\mathcal{Z}} \mathrm{Tr}\{X \mathrm{e}^{-\beta(\hat{H}-\mu\hat{N})}\} \quad , \quad \mathcal{Z} = \mathrm{Tr}\{\mathrm{e}^{-\beta(\hat{H}-\mu\hat{N})}\} \, . \tag{A.2.1}$$

Heisenberg operators $\hat{X}(t)$ are defined by $\hat{X}(t) = \exp(\mathrm{i}\hat{H}t)\hat{X}\exp(-\mathrm{i}\hat{H}t)$. In Hubbard's notation [A.8, A.9] the retarded ("+") and advanced ("−") Green function of two fermion operators $\hat{A}$ and $\hat{B}$ are defined by

$$\langle\langle \hat{A}(t); \hat{B}(t')\rangle\rangle^{\pm} = \mp \mathrm{i}\theta\{\pm(t-t')\}\langle[\hat{A}(t), \hat{B}(t')]_{+}\rangle \, , \tag{A.2.2}$$

where $[\hat{A}, \hat{B}]_{\pm} = \hat{A}\hat{B} \pm \hat{B}\hat{A}$ and $\theta(x)$ denote the (anti-)commutator and the step function, respectively.

Since the system is homogeneous in time it is advantageous to introduce the energy-dependent Green function via the Fourier transformation

$$\langle\langle \hat{A}; \hat{B}\rangle\rangle_{E}^{\pm} = \int_{-\infty}^{\infty} \mathrm{d}t \, \mathrm{e}^{\mathrm{i}Et} \, \langle\langle \hat{A}(t); \hat{B}(0)\rangle\rangle^{\pm} \, . \tag{A.2.3}$$

For complex E the retarded (advanced) Green function is regular in the upper (lower) half-plane. Besides the real axis the Green function in energy representation,

$$\langle\langle \hat{A}; \hat{B}\rangle\rangle_{E} = \begin{cases} \langle\langle \hat{A}; \hat{B}\rangle\rangle_{E}^{+} \, , \text{ if } \mathrm{Im}(E) > 0 \\ \langle\langle \hat{A}; \hat{B}\rangle\rangle_{E}^{-} \, , \text{ if } \mathrm{Im}(E) < 0 \end{cases} \quad \text{(Im: imaginary part)} \, , \tag{A.2.4}$$

is a regular function in the complex plane. The Fourier transform of the time-ordered Green function $\langle\langle \hat{A}; \hat{B}\rangle\rangle_{E}$ obeys the following equation of motion

$$E\langle\langle \hat{A}; \hat{B}\rangle\rangle_{E} = \langle[\hat{A}, \hat{B}]_{+}\rangle + \langle\langle[\hat{A}, \hat{H}]_{-}; \hat{B}\rangle\rangle_{E} \, , \tag{A.2.5}$$

which can be derived by taking the time-derivative of the Green function and a Fourier transformation afterwards.

For translationally invariant systems it is natural to investigate the one-electron Green function in its energy-momentum representation,

$$\begin{aligned} G_{\sigma}(\boldsymbol{k}; E) &= \frac{1}{L} \sum_{\boldsymbol{R}_1, \boldsymbol{R}_2} \exp\left(-\mathrm{i}\boldsymbol{k}(\boldsymbol{R}_1 - \boldsymbol{R}_2)\right) G_{\sigma}(\boldsymbol{R}_1 - \boldsymbol{R}_2; E) \\ &= \frac{1}{L} \sum_{\boldsymbol{R}_1, \boldsymbol{R}_2} \exp\left(-\mathrm{i}\boldsymbol{k}(\boldsymbol{R}_1 - \boldsymbol{R}_2)\right) \langle\langle \hat{c}_{\boldsymbol{R}_1,\sigma}; \hat{c}^{+}_{\boldsymbol{R}_2,\sigma}\rangle\rangle_{E} \, . \end{aligned} \tag{A.2.6}$$

This Green function provides the one-electron spectral function, $A_{\sigma}(\boldsymbol{k}, E)$, the one-electron density of states for σ electrons, $D_{\sigma}(E)$, and the momentum distribution, $n_{\sigma}(\boldsymbol{k})$ (at zero temperature, $T = 0$),

$$A_{\sigma}(\boldsymbol{k}, E) = \frac{1}{2\pi} \lim_{\eta \to 0^{+}} \mathrm{i}\left[G_{\sigma}(\boldsymbol{k}; E + \mathrm{i}\eta) - G_{\sigma}(\boldsymbol{k}; E - \mathrm{i}\eta)\right] \, , \tag{A.2.7a}$$

$$D_{\sigma}(E) = \frac{1}{L} \sum_{\boldsymbol{k}} A_{\sigma}(\boldsymbol{k}, E) \, , \tag{A.2.7b}$$

$$n_{\sigma}(\boldsymbol{k}) = \int_{-\infty}^{\infty} \mathrm{d}E A_{\sigma}(\boldsymbol{k}, E) \, . \tag{A.2.7c}$$

With the help of the spectral function one may calculate further physical quantities such as the electron density and the average kinetic and potential energy; see [A.6, Chaps. 4,7]. The one-electron Green and spectral function contain the whole information about the one-electron excitations of the system.

The quantities introduced in this appendix are very helpful for the discussion of physical properties of many-particle systems and will be used for the derivation of the Hubbard-I approximation.

A.2.2 Hubbard-I Approximation

In his first article on the Hubbard model [A.8] Hubbard formulated an (uncontrolled) approximate solution for his model. This "Hubbard-I" approximation is based on the Green-function decoupling technique, which had successfully been applied to spin systems [A.10].

If we start with the equation of motion (A.2.5) for the one-electron Green function, we immediately realize that we have to calculate a two-electron Green function,

$$\Gamma_\sigma(\boldsymbol{R}_1, \boldsymbol{R}_2; E) = \langle\langle \hat{n}_{\boldsymbol{R}_1,-\sigma} \hat{c}_{\boldsymbol{R}_1,\sigma}; \hat{c}^+_{\boldsymbol{R}_2,\sigma} \rangle\rangle_E \,. \tag{A.2.8}$$

If we set up an equation of motion for this Green function we obtain yet another higher-order Green function, and the hierarchy does not terminate. The Green-function decoupling technique is based on a physically motivated reduction of the N-electron Green function to N'-electron Green functions with $N' \leq N-1$. This approach can be justified only if there is an explicit expansion parameter λ, e.g., if the N-electron Green function contains terms of the order $\mathcal{O}(\lambda^N)$ at the most. In all other cases the approach is basically uncontrolled.

Similar to the Hartree–Fock decoupling of the Hubbard interaction the simplest approximation is the following reduction of the higher-order Green functions [A.8],

$$\langle\langle \hat{n}_{\boldsymbol{R}_1,-\sigma} \hat{c}_{\boldsymbol{R}_1,\sigma}; \hat{c}^+_{\boldsymbol{R}_2,\sigma} \rangle\rangle_E \quad \mapsto \quad \langle \hat{n}_{\boldsymbol{R}_1,-\sigma} \rangle \langle\langle \hat{c}_{\boldsymbol{R}_1,\sigma}; \hat{c}^+_{\boldsymbol{R}_2,\sigma} \rangle\rangle_E \,. \tag{A.2.9}$$

The equations of motions for the one-electron Green function

$$EG_\sigma(\boldsymbol{R}_1, \boldsymbol{R}_2; E) = \delta_{\boldsymbol{R}_1,\boldsymbol{R}_2} + \sum_{\boldsymbol{R}} t(\boldsymbol{R}_1, \boldsymbol{R}) G_\sigma(\boldsymbol{R}, \boldsymbol{R}_2; E) + U\Gamma_\sigma(\boldsymbol{R}_1, \boldsymbol{R}_2; E) \tag{A.2.10}$$

now close. After Fourier transformation it follows that

$$G_\sigma(\boldsymbol{k}; E) = \frac{1}{E - \epsilon(\boldsymbol{k}) - Un_{-\sigma}} \tag{A.2.11}$$

if we assume translational invariance. We thus find that the decoupling according to (A.2.9) reproduces the Hartree theory.

It seems to be unsystematic to terminate the hierarchy at this point since the equation of motion for $\Gamma_\sigma(\boldsymbol{R}_1, \boldsymbol{R}_2; E)$ contains only two-electron Green functions,

$$\begin{aligned} E\Gamma_\sigma(\boldsymbol{R}_1, \boldsymbol{R}_2; E) &= \delta_{\boldsymbol{R}_1,\boldsymbol{R}_2}\langle\hat{n}_{\boldsymbol{R}_1,-\sigma}\rangle + U\Gamma_\sigma(\boldsymbol{R}_1, \boldsymbol{R}_2; E) \\ &\quad + \sum_{\boldsymbol{R}\neq\boldsymbol{R}_1} t(\boldsymbol{R}_1, \boldsymbol{R})\langle\langle\hat{n}_{\boldsymbol{R}_1,-\sigma}\hat{c}_{\boldsymbol{R},\sigma}; \hat{c}^+_{\boldsymbol{R}_2,\sigma}\rangle\rangle_E \qquad \text{(A.2.12)} \\ &\quad + \sum_{\boldsymbol{R}\neq\boldsymbol{R}_1} t(\boldsymbol{R}_1, \boldsymbol{R})\Big\{\langle\langle\hat{c}^+_{\boldsymbol{R}_1,-\sigma}\hat{c}_{\boldsymbol{R},-\sigma}\hat{c}_{\boldsymbol{R}_1,\sigma}; \hat{c}^+_{\boldsymbol{R}_2,\sigma}\rangle\rangle_E \\ &\qquad - \langle\langle\hat{c}^+_{\boldsymbol{R},-\sigma}\hat{c}_{\boldsymbol{R}_1,-\sigma}\hat{c}_{\boldsymbol{R}_1,\sigma}; \hat{c}^+_{\boldsymbol{R}_2,\sigma}\rangle\rangle_E\Big\} . \end{aligned}$$

Hence we have to supply the decoupling rule (A.2.9) with the terms

$$\langle\langle\hat{c}^+_{\boldsymbol{R}_1,-\sigma}\hat{c}_{\boldsymbol{R},-\sigma}\hat{c}_{\boldsymbol{R}_1,\sigma}; \hat{c}^+_{\boldsymbol{R}_2,\sigma}\rangle\rangle_E \mapsto \langle\hat{c}^+_{\boldsymbol{R}_1,-\sigma}\hat{c}_{\boldsymbol{R},-\sigma}\rangle\langle\langle\hat{c}_{\boldsymbol{R}_1,\sigma}; \hat{c}^+_{\boldsymbol{R}_2,\sigma}\rangle\rangle_E , \qquad \text{(A.2.13a)}$$

$$\langle\langle\hat{c}^+_{\boldsymbol{R},-\sigma}\hat{c}_{\boldsymbol{R}_1,-\sigma}\hat{c}_{\boldsymbol{R}_1,\sigma}; \hat{c}^+_{\boldsymbol{R}_2,\sigma}\rangle\rangle_E \mapsto \langle\hat{c}^+_{\boldsymbol{R},-\sigma}\hat{c}_{\boldsymbol{R}_1,-\sigma}\rangle\langle\langle\hat{c}_{\boldsymbol{R}_1,\sigma}; \hat{c}^+_{\boldsymbol{R}_2,\sigma}\rangle\rangle_E . \qquad \text{(A.2.13b)}$$

If we assume translational invariance, the last term of the right-hand side of (A.2.12) vanishes and we obtain

$$\Gamma_\sigma(\boldsymbol{k}, E) = \frac{n_{-\sigma}}{E-U}\left[1 + \epsilon(\boldsymbol{k})G_\sigma(\boldsymbol{k}; E)\right] , \qquad \text{(A.2.14a)}$$

$$G_\sigma(\boldsymbol{k}; E) = \frac{Z^+_\sigma(\boldsymbol{k})}{E - E^+_\sigma(\boldsymbol{k})} + \frac{Z^-_\sigma(\boldsymbol{k})}{E - E^-_\sigma(\boldsymbol{k})} \qquad \text{(A.2.14b)}$$

with the quasi-particle energies and weights

$$E^\pm_\sigma(\boldsymbol{k}) = \frac{1}{2}\left[\epsilon(\boldsymbol{k}) + U \pm \sqrt{(\epsilon(\boldsymbol{k}) - U)^2 + 4Un_{-\sigma}\epsilon(\boldsymbol{k})}\right] , \qquad \text{(A.2.14c)}$$

$$Z^\pm_\sigma(\boldsymbol{k}) = \pm\frac{E^\pm_\sigma(\boldsymbol{k}) - U(1 - n_{-\sigma})}{E^+_\sigma(\boldsymbol{k}) - E^-_\sigma(\boldsymbol{k})} . \qquad \text{(A.2.14d)}$$

The spectral function apparently becomes

$$A_\sigma(\boldsymbol{k}, E) = Z^-_\sigma(\boldsymbol{k})\delta(E - E^-_\sigma(\boldsymbol{k})) + Z^+_\sigma(\boldsymbol{k})\delta(E - E^+_\sigma(\boldsymbol{k})) . \qquad \text{(A.2.14e)}$$

Equations (A.2.14) constitute the Hubbard-I approximation for the one-electron Green function. It describes undamped quasi-particles with energies $E^\pm_\sigma(\boldsymbol{k})$ (upper and lower Hubbard bands) and spectral weights $Z^\pm_\sigma(\boldsymbol{k})$.

The one-electron density of states (A.2.7b)

$$D_\sigma(E) = \sum_\pm \int \mathrm{d}x\delta\left(E - E^\pm(x)\right) D_{\sigma,0}(x)Z^\pm_\sigma(x) \qquad \text{(A.2.15)}$$

can be rewritten as [A.8]

$$D_\sigma(E) = D_{\sigma,0}\left(g(E,n_{-\sigma})\right) , \tag{A.2.16a}$$

$$g(E,n_{-\sigma}) = E - Un_{-\sigma} - \frac{U^2 n_{-\sigma}(1-n_{-\sigma})}{E-U(1-n_{-\sigma})} . \tag{A.2.16b}$$

The quasi-particle bands can thus be obtained from a projection of the unperturbed band structure.

A.3 Generalized Lieb–Wu Integral Equations

In this section we collect some calculations for the $1/\sinh r$ Hubbard model. First, we derive the generalized Lieb–Wu integral equations from the asymptotic Bethe-ansatz equations, and solve them for half band-filling. Next, we evaluate the integral equations for the quasi-particle dispersions. Finally, we address the limiting case $\kappa \to 0$ at half band-filling.

A.3.1 Derivation of the Integral Equations

The asymptotic Bethe-ansatz equations for the $1/\sinh r$ Hubbard model read (see Sect. 4.3)

$$k_j L = 2\pi I_j + \sum_{\alpha=1}^{M} \Theta\bigl(2H(k_j) - 2\Lambda_\alpha\bigr) \; ; \; j = 1,\ldots,N , \tag{A.3.1a}$$

$$-\sum_{\beta=1}^{N_\downarrow} \Theta\bigl(\Lambda_\alpha - \Lambda_\beta\bigr) = -2\pi J_\alpha + \sum_{j=1}^{N} \Theta\bigl(2H(k_j) - 2\Lambda_\alpha\bigr) \; ; \; \alpha = 1,\ldots,N_\downarrow , \tag{A.3.1b}$$

where the function $\Theta(x) = -2\arctan\bigl[x/(U/(2t))\bigr]$ was introduced in (4.12c). Furthermore, we used

$$H(k) = \frac{1}{2t}\left[-\eta(k) + \frac{\epsilon(k)(\epsilon(k)-U)}{2t\sinh(\kappa)}\right] , \tag{A.3.2a}$$

$$\epsilon(k) = -2t\sinh(\kappa)\sum_{n=1}^{\infty}(-1)^n \sin(kn)/\sinh(\kappa n) , \tag{A.3.2b}$$

$$\eta(k) = 2t\sinh(\kappa)\sum_{n=1}^{\infty}(-1)^n \cos(kn)\cosh(\kappa n)/\sinh^2(\kappa n) . \tag{A.3.2c}$$

For $\kappa \to \infty$ we have $H(k) = -\eta(k)/(2t) = \cos(k)$ which becomes $H(k) = \sin(k_{\mathrm{HM}})$ for the "standard" Hubbard model after the shift $k = k_{\mathrm{HM}} - \pi/2$.

In the ground state the distance between the quantum numbers I_j (J_α) is unity. Hence we obtain from (A.3.1a)

$$1=\frac{2\pi}{L(k_{j+1}-k_j)}+\sum_{\alpha=1}^{N_\downarrow}\frac{\Theta\big(2H(k_{j+1})-2\Lambda_\alpha\big)-\Theta\big(2H(k_j)-2\Lambda_\alpha\big)}{L(k_{j+1}-k_j)}\,. \tag{A.3.3}$$

In the thermodynamic limit, $L\to\infty$, we may set

$$\frac{1}{L(k_{j+1}-k_j)} = \rho_0(k)\,, \tag{A.3.4a}$$

$$\frac{1}{L(\Lambda_{\alpha+1}-\Lambda_\alpha)} = \sigma_0(\Lambda)\,, \tag{A.3.4b}$$

$$\frac{1}{L}\sum_{\alpha=1}^{N_\downarrow} = \int_{B_1}^{B_2}\frac{\mathrm{d}\Lambda}{L(\Lambda_{\alpha+1}-\Lambda_\alpha)} = \int_{B_1}^{B_2}\mathrm{d}\Lambda\sigma_0(\Lambda)\,, \tag{A.3.4c}$$

$$\frac{\Theta\big(2H(k_{j+1})-2\Lambda_\alpha\big)-\Theta\big(2H(k_j)-2\Lambda_\alpha\big)}{k_{j+1}-k_j} = \Theta'\big(2H(k)-2\Lambda\big)2H'(k)\,, \tag{A.3.4d}$$

where the prime indicates the derivative of the function. From (A.3.3) the first generalized Lieb–Wu equation follows as

$$2\pi\rho_0(k)=1+H'(k)\int_{B_1}^{B_2}\mathrm{d}\Lambda\sigma_0(\Lambda)K\left[H(k)-\Lambda;U/(4t)\right] \tag{A.3.5}$$

with the condition $\int_{B_1}^{B_2}\mathrm{d}\Lambda\sigma(\Lambda)=N_\downarrow/L$. The integral kernel obeys $K[x;y]=2y/(x^2+y^2)=\int_{-\infty}^{\infty}\mathrm{d}p\exp(-y|p|)\exp(\mathrm{i}px)$.

Correspondingly, from (A.3.1b) we obtain the second generalized Lieb–Wu equation by an expansion in $(\Lambda_{\alpha+1}-\Lambda_\alpha)/L$ as

$$\begin{aligned}2\pi\sigma_0(\Lambda) &= \int_{Q_1}^{Q_2}\mathrm{d}k\rho_0(k)K\left[H(k)-\Lambda;U/(4t)\right]\\ &\quad -\int_{B_1}^{B_2}\mathrm{d}\Lambda'\sigma_0(\Lambda')K\left[\Lambda-\Lambda';U/(2t)\right]\end{aligned} \tag{A.3.6}$$

with the condition $\int_{Q_1}^{Q_2}\mathrm{d}k\rho_0(k)=N/L$.

In the absence of an external magnetic field we have $N_\downarrow=N/2$ and it immediately follows that $B_2=\infty=-B_1$ if we integrate (A.3.6) over Λ. We introduce the Fourier transform via

$$\overline{\sigma}_0(\omega) = \int_{-\infty}^{\infty}\mathrm{d}\Lambda\mathrm{e}^{-\mathrm{i}\omega\Lambda}\sigma_0(\Lambda)\,, \tag{A.3.7a}$$

$$\sigma_0(\Lambda) = \frac{1}{2\pi}\int_{-\infty}^{\infty}\mathrm{d}\omega\mathrm{e}^{\mathrm{i}\omega\Lambda}\overline{\sigma}_0(\omega) \tag{A.3.7b}$$

and find from (A.3.6) that

$$\overline{\sigma}_0(\omega)=\int_{Q_1}^{Q_2}\mathrm{d}k\rho_0(k)\frac{\mathrm{e}^{-\mathrm{i}\omega H(k)}}{2\cosh(\omega U/(4t))}\,. \tag{A.3.8}$$

The integral equation (A.3.5) becomes

$$\rho_0(k) = \frac{1}{2\pi} + H'(k)\frac{4t}{U}\int_{Q_1}^{Q_2} \mathrm{d}p\rho_0(p)R\left(\frac{4t}{U}\Big[H(k) - H(p)\Big]\right) \qquad \text{(A.3.9a)}$$

with the kernel [A.11]

$$R(x) = \int_{-\infty}^{\infty} \frac{\mathrm{d}y}{2\pi}\frac{\mathrm{e}^{\mathrm{i}xy}}{1+\mathrm{e}^{2|y|}} = \frac{1}{\pi}\sum_{n=1}^{\infty}(-1)^{n+1}\frac{2n}{x^2+(2n)^2} \ . \qquad \text{(A.3.9b)}$$

Unfortunately, away from half band-filling this integral equation can be solved only numerically; for the Hubbard model, see [A.11].

A.3.2 Solution for Half-Filled Bands

For half band-filling we find $Q_2 = \pi = -Q_1$ which can easily be proven. We integrate (A.3.5) over k and find that the integral over Λ vanishes after the integral over k has been performed since $H(\pi) = H(-\pi)$. Next we multiply (A.3.9a) with $\exp(-\mathrm{i}\omega H(k))$ and integrate over all k. The double integral on the right-hand side vanishes again because of $H(\pi) = H(-\pi)$. We obtain the simple result

$$\int_{-\pi}^{\pi} \mathrm{d}k\rho_0(k)\mathrm{e}^{-\mathrm{i}\omega H(k)} = \int_{-\pi}^{\pi}\frac{\mathrm{d}k}{2\pi}\mathrm{e}^{-\mathrm{i}\omega H(k)} \ . \qquad \text{(A.3.10)}$$

Now, (A.3.8) for $\overline{\sigma}_0(\omega)$ can be solved explicitly,

$$\overline{\sigma}_0(\omega) = \int_{-\pi}^{\pi}\frac{\mathrm{d}k}{2\pi}\frac{\mathrm{e}^{-\mathrm{i}\omega H(k)}}{2\cosh(\omega U/(4t))} \ . \qquad \text{(A.3.11a)}$$

The final result for $\rho(k)$ from (A.3.5) reads

$$\rho_0(k) = \frac{1}{2\pi} + H'(k)\int_{-\pi}^{\pi}\frac{\mathrm{d}p}{2\pi}\int_{-\infty}^{\infty}\mathrm{d}\omega\frac{\mathrm{e}^{\mathrm{i}\omega[H(k)-H(p)]}}{1+\mathrm{e}^{|\omega U|/(2t)}} \ . \qquad \text{(A.3.11b)}$$

We introduce the functions

$$\widetilde{J}_0^{\binom{c}{s}}(\omega) \;=\; \begin{pmatrix}\mathrm{Re}\\ \mathrm{Im}\end{pmatrix}\int_{-\pi}^{\pi}\frac{\mathrm{d}k}{2\pi}\mathrm{e}^{\mathrm{i}\omega H(k)} \ , \qquad \text{(A.3.12a)}$$

$$\widetilde{J}_1^{\binom{c}{s}}(\omega) \;=\; \begin{pmatrix}\mathrm{Re}\\ \mathrm{Im}\end{pmatrix}\int_{-\pi}^{\pi}\frac{\mathrm{d}k}{2\pi}\mathrm{e}^{\mathrm{i}\omega H(k)}\epsilon(k)H'(k) \qquad \text{(A.3.12b)}$$

to arrive at a more compact representation of the quasi-particle densities in the ground state

$$\sigma_0(\Lambda) \;=\; \int_0^{\infty}\frac{\mathrm{d}\omega}{2\pi\cosh(\omega U/(4t))}\Big[\cos(\omega\Lambda)\widetilde{J}_0^c(\omega) + \sin(\omega\Lambda)\widetilde{J}_0^s(\omega)\Big] \ , \qquad \text{(A.3.13a)}$$

$$\begin{aligned}\rho_0(k) &= \frac{1}{2\pi} + H'(k)\int_0^\infty \frac{\mathrm{d}\omega}{\pi(1+\mathrm{e}^{\omega U/(2t)})} \\ &\quad \times \left[\cos(\omega H(k))\widetilde{J}_0^c(\omega) + \sin(\omega H(k))\widetilde{J}_0^s(\omega)\right] .\end{aligned} \tag{A.3.13b}$$

For $\kappa \to \infty$ we have $\widetilde{J}_0^{\mathrm{c}}(\omega) \to J_0(\omega)$, $\widetilde{J}_0^{\mathrm{s}}(\omega) \to 0$, and $\widetilde{J}_1^{\mathrm{c}}(\omega) \to -2J_1(\omega)/\omega$, $\widetilde{J}_1^{\mathrm{s}}(\omega) \to 0$, where $J_0(\omega)$ and $J_1(\omega)$ are Bessel functions. Since $H(k) = \sin(k_{\mathrm{HM}})$ for $\kappa = \infty$, we see that we recover the result obtained by Lieb and Wu [A.12] for the quasi-particle densities for the "standard" Hubbard model in that limit.

The ground-state energy density is readily calculated as

$$\begin{aligned}(E_0/L)(N=L,U) &= \int_{-\pi}^{\pi} \mathrm{d}k\rho_0(k)\epsilon(k) \\ &= \int_0^\infty \frac{2\,\mathrm{d}\omega}{1+\mathrm{e}^{\omega U/(2t)}}\left[\widetilde{J}_0^{\mathrm{c}}(\omega)\widetilde{J}_1^{\mathrm{c}}(\omega) + \widetilde{J}_0^{\mathrm{s}}(\omega)\widetilde{J}_1^{\mathrm{s}}(\omega)\right] .\end{aligned} \tag{A.3.14}$$

For $\kappa \to \infty$ we again recover the Lieb–Wu result [A.12].

A.3.3 Quasi-Particle Dispersions

Frahm and Korepin [A.13] have shown that the physical properties of a Hubbard-type model that can be solved by (asymptotic) Bethe ansatz are most efficiently analyzed with the help of the quasi-particle dispersions. For the $1/\sinh r$ Hubbard model the integral equations for these dispersions read

$$\tilde{\epsilon}_{\mathrm{c}}^{\mathrm{QP}}(k) = \epsilon_{\mathrm{c}}^0(k) + \int_{B_1}^{B_2} \frac{\mathrm{d}\Lambda}{2\pi} K[H(k)-\Lambda;U/(4t)]\tilde{\epsilon}_{\mathrm{s}}^{\mathrm{QP}}(\Lambda) , \tag{A.3.15a}$$

$$\begin{aligned}\tilde{\epsilon}_{\mathrm{s}}^{\mathrm{QP}}(\Lambda) &= \int_{Q_1}^{Q_2} \frac{\mathrm{d}k}{2\pi} H'(k)K[H(k)-\Lambda;U/(4t)]\tilde{\epsilon}_{\mathrm{c}}^{\mathrm{QP}}(k) \\ &\quad - \int_{B_1}^{B_2} \frac{\mathrm{d}\Lambda'}{2\pi} K[\Lambda-\Lambda';U/(2t)]\tilde{\epsilon}_{\mathrm{s}}^{\mathrm{QP}}(\Lambda') ,\end{aligned} \tag{A.3.15b}$$

where $\epsilon_{\mathrm{c}}^0(k) = -\mu + \epsilon(k)$, and k denotes a quasi-momentum that is not identical to the physical momentum of the excitation.

The conditions

$$\tilde{\epsilon}_{\mathrm{c}}^{\mathrm{QP}}(Q_1) = \tilde{\epsilon}_{\mathrm{c}}^{\mathrm{QP}}(Q_2) , \tag{A.3.16a}$$

$$\tilde{\epsilon}_{\mathrm{s}}^{\mathrm{QP}}(B_1) = \tilde{\epsilon}_{\mathrm{s}}^{\mathrm{QP}}(B_2) \tag{A.3.16b}$$

are necessary for the $1/\sinh r$ Hubbard model to fix the respective integration boundaries Q_1, Q_2, and B_1, B_2 for all densities since these boundaries are not necessarily symmetric around zero.

The chemical potential for the interacting model μ^- follows from the condition

$$\max_{\{k\}}[\tilde{\epsilon}_{\mathrm{c}}^{\mathrm{QP}}(k)] = 0 . \tag{A.3.17}$$

The ground-state energy can be obtained from the quasi-particle dispersions as

$$(E_0 - \mu N)/L = \int_{Q_1}^{Q_2} \frac{\mathrm{d}k}{2\pi} \tilde{\epsilon}_{\mathrm{c}}^{\mathrm{QP}}(k) . \tag{A.3.18}$$

The knowledge of the quasi-particle dispersions will be sufficient for the calculation of the ground-state energy and the chemical potential.

For half band-filling we can again solve the integral equations analytically using the Fourier transformation since $B_2 = \infty = -B_1$ and $Q_2 = \pi = -Q_1$. With the help of the definitions

$$\tilde{\epsilon}_{\mathrm{s}}^{\mathrm{QP}}(\omega) = \int_{-\infty}^{\infty} \mathrm{d}\Lambda \mathrm{e}^{-\mathrm{i}\omega\Lambda} \tilde{\epsilon}_{\mathrm{s}}^{\mathrm{QP}}(\Lambda) , \tag{A.3.19a}$$

$$\tilde{\epsilon}_{\mathrm{s}}^{\mathrm{QP}}(\Lambda) = \int_{-\infty}^{\infty} \frac{\mathrm{d}\omega}{2\pi} \mathrm{e}^{\mathrm{i}\omega\Lambda} \tilde{\epsilon}_{\mathrm{s}}^{\mathrm{QP}}(\omega) \tag{A.3.19b}$$

it follows from (A.3.15b) that

$$\tilde{\epsilon}_{\mathrm{s}}^{\mathrm{QP}}(\omega) = \frac{E_{\mathrm{c}}(\omega)}{2\cosh(\omega U/(4t))} , \tag{A.3.20a}$$

$$E_{\mathrm{c}}(\omega) = \int_{-\pi}^{\pi} \mathrm{d}k \tilde{\epsilon}_{\mathrm{c}}^{\mathrm{QP}}(k) H'(k) \mathrm{e}^{-\mathrm{i}\omega H(k)} . \tag{A.3.20b}$$

We insert this result into (A.3.15a) and express $\tilde{\epsilon}_{\mathrm{c}}^{\mathrm{QP}}(k)$ as a function of $E_{\mathrm{c}}(\omega)$ as

$$\tilde{\epsilon}_{\mathrm{c}}^{\mathrm{QP}}(k) = \epsilon_{\mathrm{c}}^{0}(k) + \int_{-\infty}^{\infty} \frac{\mathrm{d}\omega}{2\pi} \frac{\mathrm{e}^{\mathrm{i}\omega H(k)} E_{\mathrm{c}}(\omega)}{1 + \mathrm{e}^{|\omega U/(2t)|}} . \tag{A.3.21}$$

This relation is now inserted into the definition (A.3.20b) of $E_{\mathrm{c}}(\omega)$. The resulting double integral over k and ω vanishes since the k integrand is a complete differential. Its integration gives zero because $H(\pi) = H(-\pi)$. Only the first part coming from the non-interacting dispersion relation remains such that the explicit result for $E_{\mathrm{c}}(\omega)$ reads

$$E_{\mathrm{c}}(\omega) = \int_{-\pi}^{\pi} \mathrm{d}k \epsilon_{\mathrm{c}}^{0}(k) H'(k) \mathrm{e}^{-\mathrm{i}\omega H(k)} . \tag{A.3.22}$$

Therefore, we have analytically solved the integral equations.

With the help of the functions defined in (A.3.12) we can represent our results in a compact way. We finally find

$$\tilde{\epsilon}_{\rm s}^{\rm QP}(\Lambda) = \int_0^\infty \frac{{\rm d}\omega}{\cosh(\omega U/(4t))} \left[\cos(\omega\Lambda)\widetilde{J}_1^{\rm c}(\omega) + \sin(\omega\Lambda)\widetilde{J}_1^{\rm s}(\omega)\right], \tag{A.3.23a}$$

$$\begin{aligned}\tilde{\epsilon}_{\rm c}^{\rm QP}(k) = \epsilon_{\rm c}^{0}(k) + 2\int_0^\infty \frac{{\rm d}\omega}{1+{\rm e}^{\omega U/(2t)}} \\ \times \left[\cos(\omega H(k))\widetilde{J}_1^{\rm c}(\omega) + \sin(\omega H(k))\widetilde{J}_1^{\rm s}(\omega)\right].\end{aligned} \tag{A.3.23b}$$

It can be further analyzed for weak and strong interactions. For $U=0$ we may carry out the integral over ω after we have inserted the definitions (A.3.12). For $\tilde{\epsilon}_{\rm c}^{\rm QP}(k)$ the ω integral gives the contribution $\delta(H(k)-H(k'))$. Now that $H(k)$ is a symmetric function with a maximum at $k=0$ we can easily perform the remaining integral over k', which gives the contribution $\pm\epsilon_{\rm c}^{0}(k)$ for $-\pi<k<0$ and $0<k<\pi$, respectively. We obtain $\tilde{\epsilon}_{\rm c}^{\rm QP}(k)=2\epsilon_{\rm c}^{0}(k)$ for $-\pi<k<0$, and $\tilde{\epsilon}_{\rm c}^{\rm QP}(k)=0$ for $0<k<\pi$, as it should. For large U/t we find instead that $\tilde{\epsilon}_{\rm c}^{\rm QP}(k)=\epsilon_{\rm c}^{0}(k)$, because the integral over ω is only of the order $\mathcal{O}(t/U)$. This follows after the substitution $z=U\omega$.

A.3.4 Limit of Long-Range Hopping

As an example we analyze the limit $\kappa\to 0$ for the quasi-particle dispersion of the charges for half band-filling. To obtain the correct results of Sect. 4.2, we have to take this limit *after* the thermodynamic limit.

According to (A.3.21) and (A.3.22) the quasi-particle dispersion for the charges obeys ($\Delta\tilde{\epsilon}_{\rm c}^{\rm QP}(k)=\tilde{\epsilon}_{\rm c}^{\rm QP}(k)-\epsilon_{\rm c}^{0}(k)$)

$$\Delta\tilde{\epsilon}_{\rm c}^{\rm QP}(k) = \int_{-\infty}^{\infty} \frac{{\rm d}\omega}{2\pi}\frac{1}{1+\exp(|\omega U/(2t)|)} \int_{-\pi}^{\pi} {\rm d}p\,\epsilon_{\rm c}^{0}(p)H'(p){\rm e}^{{\rm i}\omega[H(p)-H(k)]} \tag{A.3.24}$$

for half band-filling. The interesting regime is $|k|<\pi-\kappa$ since it covers the whole Brillouin zone for $\kappa=0$. To perform the integral over p we have to distinguish between two cases: (i) $|p|<\pi-\kappa$ and (ii) $|p|>\pi-\kappa$. The integration region (ii) is only proportional to κ; the integrand, however, diverges like $1/\kappa$ such that there is a finite contribution from region (ii).

Region (i). For small κ we may approximate $H(p)\approx\pi^2/(12\kappa)-Up/(4\kappa t)$ and $\epsilon_{\rm c}^{0}(p)=-\mu+tp$. After the substitution $z=\omega/\kappa$ and the limit $\kappa\to 0$ the integral over z becomes elementary. The result is proportional to $\delta(p-k)$, and the remaining integral over p is trivial. The contribution of the region (i) is simply given by (i) $=-tk/2$.

Region (ii). In this region we set ($u=U/t$)

$$H(|p|>\pi-\kappa) \approx \frac{(\pi/\kappa)(\pi-p)\left((\pi/\kappa)(\pi-p)-u\right)-\pi^2}{4\kappa} \tag{A.3.25}$$

and $\epsilon_c^0(p) = -\mu + (\pi t/\kappa)(\pi - p)$. After the substitution $z = \omega/\kappa$ and $x = (\pi/\kappa)(\pi - p)$ we obtain finite expressions in which we may set $\kappa = 0$. The integral over z yields

$$\frac{1}{2}\int_{-\pi}^{\pi} \mathrm{d}x \frac{x(u-2x)}{4} \delta\left[(uk - \pi^2 + x(x-u))/4\right] . \tag{A.3.26}$$

The remaining integral is elementary and the second contribution is (ii) $= \left[U/2 - \sqrt{(U/2)^2 + (\pi t)^2 - Utk}\right]/2$.

In sum we find ($W = 2\pi t$ is the bandwidth)

$$\tilde{\epsilon}_c^{\mathrm{QP}}(k) = -\mu^- + \left[Wk/\pi + U - \sqrt{U^2 + W^2 - 2UWk/\pi}\right]/4 . \tag{A.3.27}$$

We insert this result in (A.3.18) and calculate the ground-state energy density as

$$(E_0/L)(N = L, U) = \begin{cases} (U - W)/4 - U^2/(12W) & \text{for } U \le W \\ -W^2/(12U) & \text{for } U \ge W \end{cases} , \tag{A.3.28}$$

in agreement with (4.61) of Sect. 4.2. The results for the energy gap, (4.63), agree as well.

The quasi-particle dispersion $\tilde{\epsilon}_c^{\mathrm{QP}}(k)$, which we have calculated here, describes the *lower, filled* Hubbard band, $\epsilon_{\mathrm{LHB}}(k) = \mu^- + \tilde{\epsilon}_c^{\mathrm{QP}}(k)$. The dispersion $\epsilon_{\mathrm{UHB}}(k)$, which we derived in (4.70) of Sect. 4.2, describes charge excitations of the ground state into the *upper* Hubbard band. We thus find $\epsilon_{\mathrm{UHB}}(k) = \mu^+ - \tilde{\epsilon}_c^{\mathrm{QP}}(k)$. If we take into account that $\mu^+ + \mu^- = U$ we find that this expression agrees with (4.70).

The results of Sect. 4.2 show that the $1/r$ Hubbard model allows an equal treatment of charge and spin excitations via the dispersion relations $\epsilon_c(k)$ and $\epsilon_s(k)$ for charge and spin excitations. Such a representation seems to be possible only for the $1/r$ Hubbard model because it is only for this limiting model that the scattering matrix between quasi-particles is unity, i.e., the quasi-particles do not interact with each other.

References

Chapter 1

1.1 N. F. Mott, *Metal–Insulator Transitions*, 2nd edition (Taylor and Francis, London, 1990).

1.2 T. V. Ramakrishnan in *The Metallic and Non-Metallic States of Matter*, ed. by P. P. Edwards and C. R. N. Rao (Taylor and Francis, London, 1985), p. 23, and in *Metal–Insulator Transitions Revisited*, ed. by P. P. Edwards and C. R. N. Rao (Taylor and Francis, London, 1995), p. 293.

1.3 M. Imada, J. Phys. Soc. Jpn **62**, 1105 (1993); *ibid.* **63**, 851, 3059, 4294 (1994); *ibid.* **64**, 2954 (1995); in *Spectroscopy of Mott Insulators and Correlated Metals*, ed. by A. Fujimori and Y. Tokura, Springer Series in Solid-State Sciences **119** (Springer, Berlin, 1995), p. 3; J. Low Temp. Phys. **99**, 437 (1995).

1.4 G. D. Mahan, *Many-Particle Physics*, 2nd edition (Plenum Press, New York, 1990).

1.5 N. W. Ashcroft and N. D. Mermin, *Solid-State Physics* (Holt–Saunders, Philadelphia, 1976).

1.6 R. M. Wentzcovitch, W. W. Schulz, and P. B. Allen, Phys. Rev. Lett. **72**, 3389 (1994); T. M. Rice, H. Launois, and J. P. Pouget, *ibid.* **73**, 3042 (C) (1994); R. M. Wentzcovitch, W. W. Schulz, and P. B. Allen, *ibid.* **73**, 3043 (C) (1994).

1.7 N. F. Mott and A. S. Alexandrov, *Sir Nevill Mott, 65 Years in Physics*, World Scientific Series in 20th Century Physics **12** (World Scientific, Singapore, 1995).

1.8 For an introduction, see H. Haug and S. W. Koch, *Quantum Theory of the Optical and Electronic Properties of Semiconductors* (World Scientific, Singapore, 1990).

1.9 R. Redmer in *Metal–Insulator Transitions Revisited*, ed. by P. P. Edwards and C. R. N. Rao (Taylor and Francis, London, 1995), p. 1.

1.10 W. Freyland and F. Hensel in *The Metallic and Non-Metallic States of Matter*, ed. by P. P. Edwards and C. R. N. Rao (Taylor and Francis, London, 1985), p. 93; J. C. Thompson, *ibid.*, p. 123; W. W. Warren Jr., *ibid.*, p. 139.

1.11 W. Freyland in *Metal–Insulator Transitions Revisited*, ed. by P. P. Edwards and C. R. N. Rao (Taylor and Francis, London, 1995), p. 167.

1.12 L. D. Landau and E. M. Lifshitz, *Quantum Mechanics*, Course of Theoretical Physics 3, § 33 (Pergamon Press, Oxford, 1981).

1.13 J. A. Hertz, Phys. Rev. B **14**, 1165 (1976).

1.14 M. Suzuki, Prog. Theor. Phys. **56**, 1454 (1976).

1.15 M. A. Continentino, Phys. Reports **239**, 179 (1994).

1.16 S. Sachdev, Phys. Rev. B **55**, 142 (1997).

1.17 F. Bloch, Z. Phys. **57**, 545 (1929); A. H. Wilson, Proc. Roy. Soc. London A **133**, 458 (1931); *ibid.* **134**, 277 (1931).

1.18 P. Fulde, *Electron Correlations in Molecules and Solids*, Springer Series in Solid-State Sciences **100** (Springer, Berlin, 1991).

1.19 D. B. McWhan, T. M. Rice, and P. H. Schmidt, Phys. Rev. **177**, 1063 (1969).

1.20 E. L. Itskevich and L. M. Fisher, Soviet Physics JETP **26**, 66 (1968).

1.21 M. Ross and T. W. Barbee III in *Metal–Insulator Transitions Revisited*, ed. by P. P. Edwards and C. R. N. Rao (Taylor and Francis, London, 1995), p. 43.

1.22 N. F. Mott, Phil. Mag. **6**, 287 (1961); L. V. Keldysh and Yu. V. Kupaev, Sov. Phys. Sol. State **6**, 2219 (1965); N. F. Mott, Rep. Prog. Phys. **47**, 909 (1984).

1.23 B. I. Halperin and T. M. Rice, *Solid-State Physics* **21**, ed. by F. Seitz, D. Turnbull, and H. Ehrenreich, p. 116 (1968); for an excitonic insulator in expanding liquid metals, see L. H. Turkevich and M. H. Cohen, Phys. Rev. Lett. **53**, 2323 (1984); W. Freyland and F. Hensel in *The Metallic and Non-Metallic States of Matter*, ed. by P. P. Edwards and C. R. N. Rao (Taylor and Francis, London, 1985), and [1.1, Chap. 10].

1.24 See, e.g., T. M. Rice, *Solid-State Physics* **32**, ed. by F. Seitz, D. Turnbull, and H. Ehrenreich, p. 1 (1977); D. W. Snoke and J. P. Wolfe, Phys. Rev. B **42**, 7876 (1990); J. L. Lin and J. P. Wolfe, Phys. Rev. Lett. **71**, 1222 (1993); J. P. Wolfe, J. L. Lin, and D. W. Snoke in *Bose–Einstein Condensation*, ed. by A. Griffin, D. W. Snoke, and S. Stringari (Cambridge University Press, Cambridge, 1995); P. Nozières, Ann. de Phys. **20**, colloque C2, 417 (1995).

1.25 See, e.g., J. C. Hensel, T. G. Phillips, and G. A. Thomas, *Solid-State Physics* **32**, ed. by F. Seitz, D. Turnbull, and H. Ehrenreich, p. 87 (1977); L. J. Schowalter, F. M. Sternanka, M. B. Salamon, and J. P. Wolfe, Solid-State Comm. **44**, 795 (1982); L. V. Butov, A. Zrenner, G. Abstreiter, G. Bohm, G. Weimann, Phys. Rev. Lett. **73**, 304 (1994); D. W. Snoke, Comm. on Cond. Matt. Physics **12**, 217 (1995).

1.26 J. Neuenschwander and P. Wachter, Phys. Rev. B **41**, 12693 (1990); B. Bucher, P. Steiner, and P. Wachter, Phys. Rev. B **67**, 2717 (1991); P. Wachter, J. Alloys and Compounds **225**, 133 (1995), in *Handbook on the Physics and Chemistry of Rare Earths* **19**, ed. by K. A. Gscheidner, Jr., L. Eyring, G. H. Lander, and G. R. Chopping (Elsevier, Amsterdam, 1994), p. 177.

1.27 R. E. Peierls, *Quantum Theory of Solids* (Clarendon Press, Oxford, 1955).

1.28 H. Fröhlich, Proc. Roy. Soc. London Ser. **223**, 296 (1954).

1.29 D. Baeriswyl, D. K. Campbell, and S. Mazumdar in *Conjugated Conducting Polymers*, ed. by H. Kiess, Springer Series in Solid-State Sciences **102** (Springer, Berlin, 1992), p. 7.

1.30 L. J. Sham in *Highly Conducting One-Dimensional Solids*, ed. by J. T. Devreese, R. P. Evrand, and V. E. van Doren (Plenum Press, New York, 1979), p. 227.

1.31 E. Lieb in *Interacting Electrons in Reduced Dimensions*, ed. by D. Baeriswyl and D. K. Campbell, NATO ASI Series B **213** (Plenum Press, New York, 1989), p. 1; B. Nachtergaele and E. H. Lieb, Phys. Rev. B **51**, 4777 (1995).

1.32 H. A. Jahn and E. Teller, Proc. Roy. Soc. London A **161**, 220 (1937); for a review, see G. A. Gehring and K. A. Gehring, Rep. Prog. Phys. **38**, p. 1 (1975).

1.33 A. J. Heeger, S. Kivelson, J. R. Schrieffer, and W.-P. Su, Rev. Mod. Phys. **60**, 781 (1988).

1.34 For a review, see *Highly Conducting One-Dimensional Solids*, ed. by J. Devreese, R. Evrand, and V. van Doren (Plenum Press, New York, 1979).

1.35 For a review, see *Low-Dimensional Electronic Properties of Molybdenum Bronzes and Oxides*, ed. by C. Schlenker (Kluwer Academic, Dordrecht, 1989).
1.36 C. N. R. Rao and B. Raveau, *Transition Metal Oxides* (VCH Publishers, New York, 1995).
1.37 P. A. Cox, *Transition Metal Oxides*, International Series of Monographs on Chemistry **27** (Clarendon Press, Oxford, 1992).
1.38 G. Grüner, Rev. Mod. Phys. **60**, 1129 (1988); *Density Waves in Solids* (Addison–Wesley, Reading, 1994).
1.39 See, e.g., J. P. Doumerc, M. Pouchard, and P. Hagenmuller in *The Metallic and Non-Metallic States of Matter*, ed. by P. P. Edwards and C. R. N. Rao (Taylor and Francis, London, 1985), p. 287.
1.40 R. J. Bouchard, J. H. Perlstein, and M. J. Sienko, Inorg. Chem. **6**, 1682 (1967); W. Fogle and J. H. Perlstein, Phys. Rev. B **6**, 1402 (1972).
1.41 R. M. Fleming, L. F. Schneemeyer, and D. E. Moncton, Phys. Rev. B **31**, 899 (1985); results from spectroscopy can be found in B. Dardel, D. Malterre, M. Grioni, P. Weibel, Y. Baer, C. Schlenker, and Y. Pétroff, Europhys. Lett. **19**, 525 (1992).
1.42 L. D. Landau, Sov. Phys. **3**, 664 (1933); H. Fröhlich, Proc. Phys. Soc. A **160**, 230 (1937).
1.43 W. Känzig, Phys. Rev. **99**, 1890 (1955); for a more recent experiment, see, e.g., S. Iwai, T. Tokizaki, A. Nakamura, K. Tanimura, N. Itoh, and A. Shluger, Phys. Rev. Lett. **76**, 1691 (1996).
1.44 M. Ueta, H. Kanzaki, K. Kobayashi, Y. Toyozawa, and E. Hanamura, *Excitonic Processes in Solids*, Springer Series in Solid-State Sciences **60** (Springer, Berlin, 1986).
1.45 B. Gerlach and H. Löwen, Rev. Mod. Phys. **63**, 63 (1991).
1.46 M. P. A. Fisher and W. Zwerger, Phys. Rev. B **34**, 5912 (1986).
1.47 E. H. Lieb and L. E. Thomas, to appear in Comm. Math. Phys. **183** (1997).
1.48 A. Rosch and T. Kopp, Phys. Rev. Lett. **75**, 1988 (1995), and references therein.
1.49 N. D. Mermin and H. Wagner, Phys. Rev. Lett. **17**, 1133 (1966); as a consequence of this theorem the Debye–Waller factor, a measure for the order in the system, vanishes at finite temperatures [1.5, Chap. 24].
1.50 P. W. Anderson, Phys. Rev. **109**, 1492 (1958).
1.51 G. Bergmann, Phys. Rep. **107**, 1 (1984).
1.52 J. M. Ziman, *Models of Disorder* (Cambridge University Press, Cambridge, 1979).
1.53 S. F. Edwards, Phil. Mag. **8**, 1020 (1958).
1.54 N. G. van Kampen, *Stochastic Processes in Physics and Chemistry* (North Holland, Amsterdam, 1981).
1.55 N. F. Mott and W. D. Twose, Adv. Phys. **10**, 107 (1961); R. E. Borland, Proc. Roy. Soc. London Ser. A **274**, 529 (1963); at the same time this was realized by Landauer, see R. Landauer, Phil. Mag. **21**, 863 (1970); Z. Phys. B **21**, 247 (1975).
1.56 D. J. Thouless, Phys. Rev. Lett. **39**, 1167 (1977).
1.57 N. Giordano, W. Gilson, and D. E. Prober, Phys. Rev. Lett. **43**, 725 (1979).
1.58 L. Bányai in *Physique de Semiconducteurs*, ed. by M. Hulin (Dunod, Paris, 1964).
1.59 N. F. Mott, Phil. Mag. **13**, 989 (1966); J. Phys. C **20**, 3075 (1987); see also M. H. Cohen, H. Fritzsche, and S. R. Ovshinsky, Phys. Rev. Lett. **22**, 1063 (1969).

1.60 H. Tokumoto, R. Mansfield, and M. J. Lea, Phil. Mag. B **46**, 93 (1982); S. von Molnar, A. Briggs, J. Flouquet, and G. Remenyi, Phys. Rev. Lett. **51**, 706 (1983).
1.61 M. Holthaus, G. H. Ristow, and D. W. Hone, Phys. Rev. Lett. **75**, 3914 (1995).
1.62 G. A. Thomas, Phil. Mag. B **52**, 479 (1985).
1.63 *Anderson Localization*, ed. by H. Nagaoka and H. Fukuyama, Springer Series in Solid-State Sciences **39** (Springer, Berlin, 1982).
1.64 P. A. Lee and T. V. Ramakrishnan, Rev. Mod. Phys. **57**, 287 (1985).
1.65 D. Belitz and T. R. Kirkpatrick, Rev. Mod. Phys. **66**, 261 (1994), and references therein.
1.66 H. Overhof and P. Thomas, *Electronic Transport in Hydrogenated Amorphous Semiconductors*, Springer Tracts in Modern Physics **114** (Springer, Berlin, 1989).
1.67 *Localization, Interactions and Transport Phenomena*, ed. by B. Kramer, G. Bergmann, and Y. Bruynseraede, Springer Series in Solid-State Sciences **61** (Springer, Berlin, 1985).
1.68 B. Kramer and A. MacKinnon, Reports on Progress in Physics **56**, 1469 (1993); I. Kh. Zharekeshev and B. Kramer, Phys. Rev. B **51**, 17239 (1995).
1.69 N. F. Mott, Phil. Mag. **26**, 1015 (1972).
1.70 For an alternative interpretation, see A. Moebius, J. Phys. C **18**, 4639 (1985).
1.71 E. Abrahams, P. W. Anderson, D. C. Licciardello, and T. V. Ramakrishnan, Phys. Rev. Lett. **42**, 673 (1979).
1.72 F. J. Wegner, Z. Phys. B **25**, 327 (1976); Z. Phys. B **35**, 207 (1979); L. Schäfer and F. J. Wegner, Z. Phys. B **38**, 113 (1980).
1.73 D. Vollhardt and P. Wölfle, Phys. Rev. Lett. **45**, 482 (1980); Phys. Rev. B **22**, 4666 (1980).
1.74 W. Götze, J. Phys. C **12**, 1279 (1979); D. Belitz, A. Gold, and W. Götze, Z. Phys. B **44**, 273 (1981); D. Belitz and W. Götze, Phys. Rev. B **28**, 5445 (1983).
1.75 F. J. Wegner, Nucl. Phys. B **316**, 663 (1989).
1.76 A. B. Harris, J. Phys. C **7**, 1671 (1974); a proof was given by J. T. Chayes, L. Chayes, D. S. Fisher, and T. Spencer, Phys. Rev. Lett. **57**, 2999 (1986).
1.77 M. Schreiber and H. Grussbach, Phys. Rev. Lett. **76**, 1687 (1996).
1.78 M. Pollak, Discuss. Faraday Soc. **50**, 11 (1970); G. Srinivasan, Phys. Rev. B **4**, 2581 (1971); A. L. Efros and B. I. Shklovskii, J. Phys. C **8**, L49 (1975); B. L. Altshuler and A. G. Aronov, Solid State Comm. **30**, 115 (1979); Sov. Phys. JETP **50**, 968 (1979).
1.79 For an introduction, see B. L. Shklovskii and A. L. Efros, *Electronic Processes in Doped Semiconductors*, Springer Series in Solid-State Sciences **45** (Springer, Berlin, 1984).
1.80 For a review, see *Electron-Electron Interactions in Disordered Systems*, ed. by M. Pollak and A. L. Efros (North Holland, Amsterdam, 1985); M. Pollak, Phil. Mag. B **65**, 657 (1992).
1.81 J. H. Davis, P. A. Lee, and T. M. Rice, Phys. Rev. B **29**, 4260 (1982); M. Grünewald, B. Pohlmann, L. Schweitzer, and D. Würtz, J. Phys. C **15**, L1153 (1982); M. Pollak and M. Ortuño, Sol. Energy Mater. **8**, 81 (1982).
1.82 S. D. Baranovski, A. L. Efros, and A. A. Uzakov, Sov. Phys. JETP **56**, 422 (1982); for more recent numerical studies, see, e.g., S. D. Baranovski, P. Thomas, and H. Vaupel, Phil. Mag. B **65**, 685 (1992).
1.83 D. E. Logan, Y. H. Szczech, and M. E. Tusch in *Metal–Insulator Transitions Revisited*, ed. by P. P. Edwards and C. R. N. Rao (Taylor and Francis, London, 1995), p. 395.

1.84 H. Stupp, M. Hornung, M. Lakner, O. Madel, and H. v. Löhneysen, Phys. Rev. Lett. **71**, 2634 (1993).

1.85 T. F. Rosenbaum, G. A. Thomas, and M. A. Paalanen, Phys. Rev. Lett. **72**, 2121 (C) (1994); H. Stupp, M. Hornung, M. Lakner, O. Madel, and H. v. Löhneysen, *ibid.*, 2122 (C) (1994); T. G. Castner, Phys. Rev. Lett. **73**, 3600 (C) (1994); H. Stupp, M. Hornung, M. Lakner, O. Madel, and H. v. Löhneysen, *ibid.*, 3601 (C) (1994).

1.86 L. D. Landau, Sov. Phys. JETP **3**, 920 (1957); *ibid.* **5**, 101 (1957).

1.87 D. Pines and P. Nozières, *The Theory of Quantum Liquids*, 2nd edition (Addison–Wesley, New York, 1988).

1.88 A. B. Migdal, Sov. Phys. JETP **5**, 333 (1957); J. M. Luttinger and J. C. Ward, Phys. Rev. **118**, 1417 (1960); J. M. Luttinger, Phys. Rev. **119**, 1153 (1960); *ibid.* **121**, 942 (1961); for details, see P. Nozières, *Theory of Interacting Fermi Systems* (Addison–Wesley, New York, 1964).

1.89 A. L. Fetter and J. D. Walecka, *Quantum Theory of Many-Particle Systems* (McGraw–Hill, New York, 1971).

1.90 G. Baym, *Lectures on Quantum Mechanics* (Benjamin/Cummings, Menlo Park, 1973).

1.91 O. Schaerpf, H. Capellmann, T. Brueckel, A. Comberg, and H. Passing, Z. Phys. B **78**, 345 (1990).

1.92 J. C. Slater, Phys. Rev. **82**, 538 (1951).

1.93 For a historical overview, see B. H. Brandow, Adv. Phys. **26**, 651 (1977).

1.94 H. Capellmann, J. Phys. F **4**, 1466 (1974); V. Korenmann, J. L. Murray, and R. E. Prange, Phys. Rev. B **16**, 4032, 4048, 4058 (1977); H. Capellmann and V. Vieira, Solid State Comm. **43**, 747 (1982); for a review, see V. Korenmann in *Metallic Magnetism*, ed. by H. Capellmann, Springer Topics in Current Physics **42** (Springer, Berlin, 1987), p. 109.

1.95 N. F. Mott, Proc. Phys. Soc. A **62**, 416 (1949).

1.96 N. W. Ashcroft, Phys. Rev. B. **41**, 10963 (1990); H. Chacham, X. Zhu, and S. G. Louie, Phys. Rev. B **46**, 6688 (1992).

1.97 P. W. Anderson, *Solid-State Physics* **14**, ed. by F. Seitz and D. Turnbull, p. 99 (1963).

1.98 E. Wigner, Phys. Rev. **46**, 1002 (1934).

1.99 A. A. Kugler, Annals of Physics **53**, 133 (1969); D. Ceperley, Phys. Rev. B **18**, 3126 (1978).

1.100 B. Sutherland, Phys. Rev. Lett. **34**, 1083 (1975); *ibid.* **35**, 185 (1975).

1.101 C. C. Grimes and G. Adams, Phys. Rev. Lett. **42**, 795 (1979); for a review, see F. M. Peeters in *The Physics of the Two-Dimensional Electron Gas*, ed. by J. T. Devreese and F. M. Peeters, NATO ASI Series B **157** (Plenum Press, New York, 1987), p. 393.

1.102 E. Y. Andrei, G. Deville, D. C. Glattli, F. I. B. Williams, E. Paris, and B. Etienne, Phys. Rev. Lett. **62**, 2765 (1988); Surf. Sci. **229**, 344 (1990).

1.103 H. L. Stormer and R. L. Willett, Phys. Rev. Lett. **62**, 972 (C) (1992); E. Y. Andrei, G. Deville, D. C. Glattli, F. I. B. Williams, E. Paris, and B. Etienne, Phys. Rev. Lett. **62**, 973 (C) (1992), *ibid.*, 1926 (E) (1992).

1.104 V. J. Goldmann, M. Santos, M. Shayegan, and J. E. Cunningham, Phys. Rev. Lett. **65**, 2189 (1990).

1.105 I. V. Kukushkin, N. J. Pulsford, K. v. Klitzing, R. J. Haug, K. Ploog, and V. B. Timofeev, Europhys. Lett. **23**, 211 (1993).

1.106 R. Willett in *Phase Transitions and Relaxation in Systems with Competing Energy Scales*, ed. by T. Riste and D. Sherrington, NATO ASI Series C **415** (Kluwer Academic, Dordrecht, 1993), p. 367;

1.107 X. Zhu and S. G. Louie, Phys. Rev. Lett. **70**, 335 (1993); R. Price, P. M. Platzmann, and S. He, Phys. Rev. Lett. **70**, 339 (1993).
1.108 P. A. Miles, W. B. Westphal, and A. von Hippel, Rev. Mod. Phys. **29**, 279 (1957).
1.109 E. J. W. Verwey, Z. Kristallogr. **91**, 65 (1935); Nature **144**, 327 (1939).
1.110 A review on recent work is given by J. M. Honig in *The Metallic and Non-Metallic States of Matter*, ed. by P. P. Edwards and C. R. N. Rao (Taylor and Francis, London, 1985), p. 261; see also R. D. Barlow, M. Jones, P. J. Phillips, A. J. Poynton, and T. E. Whall, J. Appl. Phys. **61**, 3546 (1987).
1.111 J. R. Cullen and E. Callen, J. Appl. Phys. **41**, 879 (1970); Phys. Rev. Lett. **26**, 236 (1971); D. Ihle and B. Lorenz, J. Phys. C **19**, 5239 (1986).
1.112 C. Herring in *Magnetism, Vol. IV*, ed. by G. T. Rado and H. Suhl (Academic Press, New York, 1966), p. 1.
1.113 J. Friedel in *The Physics of Metals*, ed. by J. M. Ziman (Cambridge University Press, Cambridge, 1969), p. 340.
1.114 C. M. Varma and T. Giamarchi in *Strongly Interacting Fermions and High-T_c Superconductivity*, ed. by B. Douçot and J. Zinn-Justin, Les Houches Session LVI 1991 (North Holland, Amsterdam, 1995), p. 149.
1.115 V. Heine and L. F. Mattheiss, J. Phys. C **4**, L191 (1971); L. F. Mattheiss, Phys. Rev. B **5**, 290 (1972).
1.116 J. B. Goodenough, Prog. Sol. State Chem. **5**, 145 (1972); see also [1.110].
1.117 K. E. Smith and V. E. Henrich, Phys. Rev. B **38**, 5965 (1988), and references therein.
1.118 L. F. Mattheiss, J. Phys. Cond. Matt. **6**, 6477 (1994).
1.119 K. I. Kugel and D. I. Khomskii, Sov. Phys. JETP **37**, 725 (1973); Sov. Phys. Usp. **25**, 231 (1982).
1.120 C. Castellani, C. R. Natoli, and J. Ranninger, Phys. Rev. B **18**, 4945, 4967, 5001 (1978).
1.121 T. M. Rice in *Spectroscopy of Mott Insulators and Correlated Metals*, ed. by A. Fujimori and Y. Tokura, Springer Series in Solid-State Sciences **119** (Springer, Berlin, 1995), p. 221.
1.122 D. B. McWhan, J. P. Remeika, T. M. Rice, W. F. Brinkman, J. P. Maita, and A. Menth, Phys. Rev. Lett. **27**, 941 (1971); D. B. McWhan, A. Menth, J. P. Remeika, W. F. Brinkman, and T. M. Rice, Phys. Rev. B **7**, 1920 (1973).
1.123 S. A. Carter, T. F. Rosenbaum, M. Lu, H. M. Jaeger, P. Metcalf, J. M. Honig, and J. Spałek, Phys. Rev. B **49**, 7898 (1994).
1.124 A. Fujimori, I. Hase, H. Namatame, Y. Fujishima, Y. Tokura, H. Eisaki, S. Uchida, K. Takegahara, and F. M. F. de Groot, Phys. Rev. Lett. **69**, 1796 (1992); A. Fujimori, I. Hase, H. Namatame, Y. Fujishima, Y. Tokura, M. Nakamura, A. Misu, M. Abbate, F. M. F. de Groot, M. Czyzyk, J. C. Fuggle, O. Strebel, F. Lopez, M. Domke, and G. Kaindl, Phys. Rev. B **46**, 9841 (1992).
1.125 G. A. Thomas, D. H. Rapkine, S. A. Carter, A. J. Millis, T. F. Rosenbaum, P. Metcalf, and D. F. Honig, Phys. Rev. Lett. **73**, 1529 (1994); M. J. Rozenberg, G. Kotliar, H. Kajueter, G. A. Thomas, D. H. Rapkine, J. M. Honig, and P. Metcalf, Phys. Rev. Lett. **75**, 105 (1995).
1.126 A. Menth and J. P. Remeika, Phys. Rev. B **2**, 3756 (1970); S. Shin, Y. Tezuka, T. Kinoshita, A. Kakizaki, T. Ishii, Y. Ueda, W. Jang, H. Takei, Y. Chiba, and M. Ishigame, Phys. Rev. B **46**, 9224 (1992).
1.127 R. B. Moon, Phys. Rev. Lett. **25**, 527 (1970).
1.128 P. Majumdar and H. R. Krishnamurthy, Phys. Rev. Lett. **73**, 1525 (1994); Phys. Rev. B **52**, 5479 (1995).

1.129 S. A. Carter, T. F. Rosenbaum, J. M. Honig, and J. Spałek, Phys. Rev. Lett. **67**, 3440 (1991); S. A. Carter, T. F. Rosenbaum, P. Metcalf, J. M. Honig, and J. Spałek, Phys. Rev. B **48**, 16841 (1993); for a review, see J. M. Honig, Acta Phys. Polon. **85**, 102 (1994).
1.130 J. Zaanen, G. A. Sawatzky, and J. W. Allen, Phys. Rev. Lett. **55**, 418 (1985); J. Zaanen and G. A. Sawatzky, Can. J. Phys. **65**, 1262 (1987).
1.131 K. Terakura, A. R. Williams, T. Oguchi, and J. Kübler, Phys. Rev. Lett. **52**, 1830 (1984); Phys. Rev. B **30**, 4734 (1984).
1.132 A thorough discussion of the electronic structure of NiO and results of angle-resolved photoemission spectroscopy can be found in Z.-X. Shen, R. S. List, D. S. Dessau, B. O. Wells, O. Jepsen, A. J. Arko, R. Barttlet, C. K. Shih, F. Parmigiani, J. C. Huang, and P. A. P. Lindberg, Phys. Rev. B **44**, 3604 (1991), and references therein.
1.133 J. Bała, A. M. Oleś, and J. Zaanen, Phys. Rev. Lett. **72**, 2600 (1994).
1.134 J. A. Wilson in *The Metallic and Non-Metallic States of Matter*, ed. by P. P. Edwards and C. R. N. Rao (Taylor and Francis, London, 1985), p. 215.
1.135 M. P. Pasternak, R. D. Taylor, A. Chen, C. Maede, L. M. Falicov, A. Giesekus, R. Jeanloz, and P. Y. Yu, Phys. Rev. Lett. **65**, 790 (1990); for a theoretical treatment, see A. Giesekus and L. M. Falicov, Phys. Rev. B **44**, 19449 (1991); J. K. Freericks and L. M. Falicov, Phys. Rev. B **45**, 1896 (1992).

Chapter 2

2.1 N. W. Ashcroft and N. D. Mermin, *Solid-State Physics* (Holt–Saunders, Philadelphia, 1976).
2.2 P. W. Anderson, *Solid-State Physics* **14**, ed. by F. Seitz and D. Turnbull, p. 99 (1963). Here, the Fock contributions are included in the band structure calculations; for the technical and conceptional reasons mentioned in the text we will not include the Fock terms.
2.3 P. Hohenberg and W. Kohn, Phys. Rev. **136**, B864 (1964); see also W. Kohn and L. J. Sham, Phys. Rev. **140**, A1133 (1964), and [2.8].
2.4 P. Fulde, *Electron Correlations in Molecules and Solids*, Springer Series in Solid-State Sciences **100** (Springer, Berlin, 1991).
2.5 O. Gunnarsson, O. K. Andersen, and A. Svane in *Interacting Electrons in Reduced Dimensions*, ed. by D. Baeriswyl and D. K. Campbell, NATO ASI Series B **213** (Plenum Press, New York, 1989), p. 139.
2.6 V. L. Moruzzi, J. F. Janak, and A. R. Williams, *Calculated Electronic Properties of Metals* (Pergamon Press, New York, 1978).
2.7 R. M. Dreizler and E. K. U. Gross, *Density-Functional Theory* (Springer, Berlin, 1990).
2.8 R. O. Jones and O. Gunnarsson, Rev. Mod. Phys. **61**, 689 (1989).
2.9 G. D. Mahan, *Many-Particle Physics*, 2nd edition (Plenum Press, New York, 1990).
2.10 J. C. Slater, *The Self-Consistent Field for Molecules and Solids*, Quantum Theory of Molecules and Solids, Vol. IV (McGraw–Hill, New York, 1974).
2.11 W. E. Pickett, Rev. Mod. Phys. **61**, 433 (1989).
2.12 K. Terakura, A. R. Williams, T. Oguchi, and J. Kübler, Phys. Rev. Lett. **52**, 1830 (1984); Phys. Rev. B **30**, 4734 (1984).
2.13 G. A. Sawatzky and J. W. Allen, Phys. Rev. Lett. **53**, 2339 (1984).
2.14 A. Svane and O. Gunnarsson, Phys. Rev. B **37**, 9919 (1988); Europhys. Lett. **7**, 171 (1988); for recent applications of the SIC-LDA, see A. Svane, Phys. Rev. Lett. **72**, 1248 (1994), and references therein.

2.15 A. Svane and O. Gunnarsson, Phys. Rev. Lett. **65**, 1148 (1990); ideas in this direction were also formulated by R. D. Cowan, Phys. Rev. **163**, 54 (1967); I. Lindgren, Int. J. Quantum Chem. **5**, 411 (1971); A. Zunger, J. P. Perdew, and G. L. Oliver, Solid-State Comm. **34**, 933 (1980).

2.16 A. Fujimori, F. Minami, and S. Sugano, Phys. Rev. B **29**, 5225 (1984); *ibid.* **30**, 957 (1984).

2.17 V. I. Anisimov, J. Zaanen, and O. K. Andersen, Phys. Rev. B **44**, 943 (1991); for an application of this concept, see, e.g., P. Wei and Z. Q. Qi, Phys. Rev. B **49**, 10864 (1994).

2.18 An extensive and correct introduction to second quantization can be found in A. L. Fetter and J. D. Walecka, *Quantum Theory of Many-Particle Systems* (McGraw–Hill, New York, 1971).

2.19 P. W. Anderson, Phys. Rev. **115**, 2 (1959); see also [2.2].

2.20 J. Hubbard, Proc. Roy. Soc. London Ser. A **276**, 238 (1963); *ibid.* **277**, 237 (1963).

2.21 J. Hubbard, Proc. Roy. Soc. London **281**, 401 (1964).

2.22 M. Cyrot, Physica B **91**, 141 (1977).

2.23 M. W. Long, Int. J. Mod. Phys. B **5**, 865 (1991).

2.24 M. C. Gutzwiller, Phys. Rev. Lett. **10**, 159 (1963).

2.25 J. Kanamori, Prog. Theor. Phys. **30**, 275 (1963).

2.26 N. F. Mott, Proc. Phys. Soc. A **62**, 416 (1949).

2.27 See the contributions by J. C. Slater, E. P. Wohlfarth, and J. H. van Vleck on a conference on magnetism in 1952, printed as Rev. Mod. Phys. **25**, pp. 199–226 (1953); see also J. C. Slater, *Quantum Theory of Molecules and Solids*, Vol. III, Chap. 11 (McGraw–Hill, New York, 1963), and D. E. Logan and P. P. Edwards in *The Metallic and Non-Metallic States of Matter*, ed. by P. P. Edwards and C. R. N. Rao (Taylor and Francis, London, 1985), p. 65.

2.28 R. G. Parr, J. Chem. Phys. **20**, 1499 (1952); R. Pariser and R. G. Parr, J. Chem. Phys. **21**, 466, 767 (1953); J. A. Pople, Trans. Faraday Soc. **49**, 1375 (1953); Proc. Phys. Soc. London A **68**, 81 (1955).

2.29 Reviews on various organic conductors and semiconductors are presented in *Organic Conductors*, ed. by J.-P. Farges (Marcel Dekker, New York, 1994).

2.30 D. Baeriswyl, D. K. Campbell, and S. Mazumdar in *Conjugated Conducting Polymers*, ed. by H. Kiess, Springer Series in Solid-State Sciences **102** (Springer, Berlin, 1992), p. 7.

2.31 V. Y. Krivnov and A. A. Ovchinnikov, Sov. Phys. JETP Lett. **39**, 159 (1984); Sov. Phys. JETP **63**, 414 (1986).

2.32 G. König and G. Stollhoff, Phys. Rev. Lett. **65**, 1239 (1990).

2.33 S. Kivelson, W.-P. Su, J. R. Schrieffer, and A. J. Heeger, Phys. Rev. Lett. **58**, 1899 (1987); C.-Q. Wu, X. Sun, and K. Nasu, Phys. Rev. Lett. **59**, 831 (1987); D. Baeriswyl, P. Horsch, and K. Maki, Phys. Rev. Lett. **60**, 70 (C) (1988); J. T. Gammel and D. K. Campbell, *ibid.*, 71 (1988); S. Kivelson, W.-P. Su, J. R. Schrieffer, and A. J. Heeger, *ibid.*, 72 (1988); D. K. Campbell, J. T. Gammel, and E. Y. Loh, Phys. Rev. B **38**, 12043 (1988); *ibid.*, **42**, 475 (1990).

2.34 A. Painelli and A. Girlando, Synth. Met. **27**, A15 (1988); in *Interacting Electrons in Reduced Dimensions*, ed. by D. Baeriswyl and D. K. Campbell, NATO ASI Series B **213** (Plenum Press, New York, 1989), p. 165; Phys. Rev. B**39**, 2830 (1989).

2.35 D. K. Campbell, J. T. Gammel, and E. Y. Loh, Phys. Rev. B **38**, 12043 (1988), *ibid.* **42**, 475 (1990), and references therein.

2.36 V. J. Emery in *Highly Conducting One-Dimensional Solids*, ed. by J. Devreese, R. Evrand, and V. van Doren (Plenum Press, New York, 1979), p. 247.

2.37 C. Herring in *Magnetism, Vol. IV*, ed. by G. T. Rado and H. Suhl (Academic Press, New York, 1966), p. 1.
2.38 J. Friedel in *The Physics of Metals*, ed. by J. M. Ziman (Cambridge University Press, Cambridge, 1969), p. 340.
2.39 *Metallic Magnetism*, ed. by H. Capellmann, Springer Topics in Current Physics **42** (Springer, Berlin, 1987).
2.40 T. Moriya, *Spin Fluctuations in Itinerant Electron Magnetism*, Springer Series in Solid-State Sciences **56** (Springer, Berlin, 1985).
2.41 C. N. R. Rao and B. Raveau, *Transition Metal Oxides* (VCH Publishers, New York, 1995).
2.42 P. A. Cox, *Transition Metal Oxides*, International Series of Monographs on Chemistry **27** (Clarendon Press, Oxford 1992).
2.43 K. I. Kugel and D. I. Khomskii, Sov. Phys. JETP **37**, 725 (1973); Sov. Phys. Usp. **25**, 231 (1982).
2.44 C. Castellani, C. R. Natoli, and J. Ranninger, Phys. Rev. B **18**, 4945, 4967, 5001 (1978).
2.45 L. Roth, Phys. Rev. **149**, 306 (1966); H. Capellmann, Phys. Rev. B **7**, 4153 (1973); M. Cyrot and C. Lyon-Caen, J. Phys. C **6**, L274 (1975); Journ. de Physique **36**, 253 (1975), and references therein.
2.46 W. Nolting, G. Greipel, and K. Ertl, Phys. Rev. B **45**, 5790 (1992).
2.47 For an overview, see, e.g., *Electronic Properties of Fullerenes*, ed. by H. Kuzmany, J. Fink, M. Mehring, and S. Roth, Springer Series in Solid-State Sciences **117** (Springer, Berlin, 1993).
2.48 S. Chakravarty, M. P. Gelfand, and S. Kivelson, Science **254**, 970 (1991); S. Chakravarty and S. Kivelson, Europhys. Lett. **16**, 751 (1991).
2.49 R. W. Lof, M. A. van Veenendaal, B. Koopmans, H. T. Jonkman, and G. A. Sawatzky, Phys. Rev. Lett. **68**, 3924 (1992).
2.50 V. P. Antropov, O. Gunnarsson, and O. Jepsen, Phys. Rev. B **46**, 13647 (1992); O. Gunnarsson, E. Koch, and R. M. Martin, Phys. Rev. B **54**, 11026 (1996).
2.51 S. L. Sondhi, M. P. Gelfand, H. Q. Lin, and D. K. Campbell, Phys. Rev. B **51**, 5943 (1995).
2.52 E. L. Shirley and S. G. Louie, Phys. Rev. Lett. **71**, 133 (1993).
2.53 C. H. Pennington, V. A. Stenger, C. H. Recchia, C. Hahm, K. Gorny, V. Nandor, D. R. Buffinger, S. M. Lee, and R. P. Ziebarth, Phys. Rev. B **53**, 2967 (1996).
2.54 C. M. Varma, S. Schmitt-Rink, and E. Abrahams, Sol. State Comm. **62**, 681 (1987); V. J. Emery, Phys. Rev. Lett. **58**, 2794 (1987).
2.55 M. S. Hybertsen, M. Schlüter, and N. E. Christensen, Phys. Rev. B **39**, 9028 (1989).
2.56 G. Dopf, A. Muramatsu, and W. Hanke, Phys. Rev. B **41**, 9264 (1990); Phys. Rev. Lett. **68**, 353 (1992); G. Dopf, J. Wagner, P. Dieterich, A. Muramatsu, and W. Hanke, Phys. Rev. Lett. **68**, 2082 (1992).
2.57 J. C. Fuggle, P. J. W. Weijs, R. Schoorl, G. A. Sawatzky, J. Fink, N. Nücker, P. J. Durham, and W. M. Temmermann, Phys. Rev. B **37**, 123 (1988); N. Nücker, J. Fink, J. C. Fuggle, P. J. Durham, and W. M. Temmermann, Phys. Rev. B **37**, 5158 (1988); P. Kuiper, G. Kruizinga, J. Ghijsen, M. Grioni, P. J. W. Eijs, F. M. F. de Groot, G. A. Sawatzky, H. Verweij, L. F. Feiner, and H. Petersen, Phys. Rev. B **38**, 6483 (1988).
2.58 P. W. Anderson, Phys. Rev. Lett. **64**, 1839 (1990); *ibid.*, **65**, 2306 (1990).
2.59 P. W. Anderson, G. Baskaran, Z. Zou, and T. Hsu, Phys. Rev. Lett. **58**, 2790 (1987); F. C. Zhang and T. M. Rice, Phys. Rev. B **37**, 3759 (1988).

2.60 G. R. Stewart, Rev. Mod. Phys. **56**, 755 (1984); Z. Fisk, D. W. Hess, C. J. Pethick, D. Pines, J. L. Smith, J. D. Thompson, and J. O. Willis, Science **239**, 33 (1988).
2.61 P. Fulde, J. Keller, and G. Zwicknagel, *Solid-State Physics* **41**, ed. by F. Seitz and D. Turnbull, p. 1 (1988); G. Czycholl, Phys. Reports **143**, 277 (1986).
2.62 C. Varma and Y. Yafet, Phys. Rev. B **13**, 2950 (1976).
2.63 P. W. Anderson, Phys. Rev. **124**, 41 (1961).
2.64 J. R. Schrieffer and P. A. Wolff, Phys. Rev. **149**, 491 (1966).
2.65 J. Kondo, Prog. Theor. Phys. **32**, 37 (1964); B. Coqblin and J. R. Schrieffer, Phys. Rev. **185**, 847 (1969).
2.66 D. Vollhardt and P. Wölfle, *The Suprafluid Phases of Helium-3* (Taylor and Francis, London, 1990).
2.67 D. Vollhardt, Rev. Mod. Phys. **56**, 99 (1984); D. Vollhardt, P. Wölfle, and P. W. Anderson, Phys. Rev. B **35**, 6703 (1987).
2.68 An introduction can be found in E. Feenberg, *Theory of Quantum Fluids* (Academic Press, New York, 1969); C. E. Campbell in *Progress in Liquid Physics*, ed. by C. A. Croxton (Wiley, New York, 1978); see also B. E. Clements, E. Krotscheck, J. A. Smith, and C. E. Campbell, Phys. Rev. B **47**, 5239 (1993), and references therein.
2.69 M. P. A. Fisher, P. B. Weichman, G. Grinstein, and D. S. Fisher, Phys. Rev. B **40**, 546 (1989); N. Trivedi, D. M. Ceperley, and W. Krauth, Phys. Rev. Lett. **67**, 2307 (1991); K. J. Runge, Phys. Rev. B **45**, 13136 (1992); A. Kampf and G. T. Zimanyi, Phys. Rev. B **47**, 279 (1993); C. Bruder, R. Fazio, and G. Schön, Phys. Rev. B **47**, 342 (1993); J. K. Freericks and H. Monien, Europhys. Lett. **26**, 545 (1994);
2.70 L. M. Falicov and J. C. Kimball, Phys. Rev. Lett. **22**, 997 (1969); this simplification is also the basis of the "alloy-analogy approximation" in the third of Hubbard's papers on his model [2.21].
2.71 B. S. Shastry, Phys. Rev. Lett. **56**, 1529 (1986); *ibid.*, 2334 (E) (1986); *ibid.*, 2453 (1986); J. Stat. Phys. **50**, 57 (1988).
2.72 D. B. Uglov and V. E. Korepin, Phys. Lett. A **190**, 238 (1994).
2.73 This symmetry was discovered several times; see, e.g., E. Nowak, Z. Phys. B **45**, 173 (1981); the idea acquired new momentum by the work of C. N. Yang, Phys. Rev. Lett. **63**, 2144 (1989); see also C. N. Yang and F. C. Zhang, Mod. Phys. B **4**, 759 (1990); F. C. Zhang, Mod. Phys. B **5**, 153 (1990); M. Pernici, Europhys. Lett. **12**, 75 (1990); for a "no-go" theorem, see G.-S. Tian, J. Phys. A **26**, 2325 (1993); Sol. Stat Comm. **90**, 837 (1994).
2.74 F. H. L. Eßler, V. E. Korepin, and K. Schoutens, Phys. Rev. Lett. **67**, 3848 (1992); Nucl. Phys. B **372**, 559 (1992); *ibid.* **384**, 431 (1992); Phys. Rev. Lett. **72**, 908 (1994); the states which were missing in the Bethe ansatz have an energy gap and are thus irrelevant for the physics at low temperatures.
2.75 E. H. Lieb and F. Y. Wu, Phys. Rev. Lett. **20**, 1445 (1968).
2.76 B. Sutherland in *Exactly Solvable Problems in Condensed Matter and Relativistic Field Theory*, ed. by B. S. Shastry, S. S. Jha, and V. Singh, Lecture Notes in Physics **242** (Springer, Berlin, 1985), p. 1.
2.77 A. Messiah, *Quantum Mechanics*, Vol. 2 (North-Holland, Amsterdam, 1962).
2.78 G. Beni, P. Pincus, and T. Holstein, Phys. Rev. B **8**, 312 (1973); D. J. Klein, *ibid.*, 3452 (1973).
2.79 Y. Nagaoka, Phys. Rev. **147**, 392 (1966).
2.80 T. Hanisch and E. Müller-Hartmann, Annalen der Physik **2**, 8. Serie, 381 (1993); E. Müller-Hartmann, T. Hanisch, and R. Hirsch, Physica B **186–188**, 834 (1993); F. Gebhard and X. Zotos, Phys. Rev. B **43**, 1176 (1991).

2.81 One might think about other scenarios, e.g., phase separation; see, P. B. Visscher, Phys. Rev. B **10**, 943 (1974); M. Grilli, C. Castellani and G. Kotliar, Phys. Rev. B **45**, 10805 (1992); P. G. J. van Dongen, Phys. Rev. Lett. **74**, 182 (1995), and references therein.
2.82 A. B. Harris and R. V. Lange, Phys. Rev. **157**, 295 (1967).
2.83 P. G. J. van Dongen, Phys. Rev. Lett. **67**, 757 (1993); Phys. Rev. B **49**, 7904 (1994); Phys. Rev. B **50**, 14016 (1994).
2.84 J. H. van Vleck, Rev. Mod. Phys. **25**, 220 (1955).
2.85 P, G. J. van Dongen and V. Janiš, Phys. Rev. Lett. **72**, 3258 (1994).
2.86 E. H. Lieb in *Advances in Dynamic Systems and Quantum Physics*, ed. by V. Figari et al. (World Scientific, Singapore, 1994), p. 173; in *The Hubbard Model: Its Physics and Mathematical Physics*, ed. by D. Baeriswyl, D. K. Campbell, J. M. P. Carmelo, F. Guinea, and E. Louis, NATO ASI Series B **343** (Plenum Press, New York, 1995), p. 1.
2.87 H. Tasaki, Phys. Rev. Lett. **75**, 4678 (1995); for a review, see H. Tasaki, J. Stat. Phys. **84**, 535 (1996).
2.88 R. J. Baxter, *Exactly Solved Models in Statistical Mechanics*, 3rd printing (Academic Press, New York, 1989).
2.89 K. Kubo and T. Kishi, Phys. Rev. Lett. **61**, 2585 (1988); T. Kennedy, E. H. Lieb, and B. S. Shastry, J. Stat. Phys. **53**, 1019 (1988).
2.90 M. B. Walker and T. W. Ruijgrok, Phys. Rev. **171**, 513 (1968); D. K. Gosh, Phys. Rev. Lett. **27**, 1584 (1971), *ibid.* **28**, 330 (E) (1972); G. Uhrig, Phys. Rev. B **45**, 4738 (1992).
2.91 E. H. Lieb and D. C. Mattis, Phys. Rev. **125**, 164 (1962).
2.92 D. C. Mattis, *The Theory of Magnetism I+II*, Springer Series in Solid-State Sciences **17**, **55** (Springer, Berlin, 1981, 1985); see also K. Yosida, *Theory of Magnetism*, Springer Series in Solid-State Sciences **122** (Springer, Berlin, 1996).
2.93 E. H. Lieb, Phys. Rev. Lett. **62**, 1201 (1989); *ibid.* **62**, 1927 (E) (1989).
2.94 S.-Q. Shen and Z.-M. Qiu, Phys. Rev. Lett. **71**, 4238 (1993); an extensive review on the Hubbard model with attractive interaction is given by R. Micnas, J. Ranninger, and S. Robaskiewicz, Rev. Mod. Phys. **62**, 113 (1990).
2.95 H. Tasaki, Phys. Rev. B **40**, 9192 (1989); G. S. Tian, J. Phys. A **23**, 2231 (1990).
2.96 G. S. Tian, Phys. Rev. B **44**, 4444 (1991); for finite lattices the fully polarized state is unstable for two holes; see, e.g., B. Doucot and X. G. Wen, Phys. Rev. B **40**, 2719 (1989).
2.97 A. Mielke, J. Phys. A **24**, 3311 (1991); H. Tasaki, Phys. Rev. Lett. **69** , 1608 (1992); A. Mielke and H. Tasaki, Comm. Math. Phys. **158**, 341 (1993).
2.98 E. Müller-Hartmann, J. Low Temp. Phys. **99**, 349 (1995).
2.99 K. Penc, H. Shiba, F. Mila, and T. Tsukagoshi, Phys. Rev. B **54**, 4056 (1996); P. Pieri, S. Daul, D. Baeriswyl, M. Dzierzawa, and P. Fazekas, *ibid.*, 9250 (1996).
2.100 M. W. Long, C. W. M. Castleton, and C. A. Hayward, J. Phys. Cond. Matt. **6**, 481 (1994); K. Ueda, T. Nishino, and H. Tsunetsugu, Phys. Rev. B **50**, 612 (1994).
2.101 P. Fazekas, to appear in Phil. Mag. B **75** (1997).
2.102 For a brief review, see R. Strack and D. Vollhardt, J. Low Temp. Phys. **99**, 385 (1995).
2.103 H. J. Schulz, Phys. Rev. B **39**, 2940 (1989); Phys. Rev. Lett. **64**, 1445 (1990).
2.104 M. Inui and P. B. Littlewood, Phys. Rev. B **44**, 4415 (1991), and references therein.

Chapter 3

3.1 V. Bach, E. H. Lieb, and J. P. Solovej, J. Stat. Phys. **76**, 3 (1994).

3.2 W. Langer, M. Plischke, and D. Mattis, Phys. Rev. Lett. **23**, 1448 (1969).

3.3 K. Dichtel, R. J. Jelitto, and H. Koppe, Z. Phys. **246**, 248 (1971); *ibid.* **251**, 173 (1972).

3.4 P. Kopietz, Phys. Rev. B **48**, 13789 (1993).

3.5 P. G. J. van Dongen, Phys. Rev. Lett. **67**, 757 (1991); Phys. Rev. B **49**, 7904 (1994); Phys. Rev. B **50**, 14016 (1994).

3.6 E. C. Stoner, Journ. de Phys. et Rad. **12**, 372 (1951).

3.7 D. R. Penn, Phys. Rev. **142**, 350 (1966).

3.8 H. J. Schulz, Phys. Rev. B **39**, 2940 (1989); Phys. Rev. Lett. **64**, 1445 (1990).

3.9 M. Inui and P. B. Littlewood, Phys. Rev. B **44**, 4415 (1991), and references therein.

3.10 See, e.g., R. P. Feynman, *Statistical Mechanics*, Frontiers in Physics **36**, 7th printing (Addison–Wesley, New York, 1982).

3.11 Lower bounds for the ground-state energy of correlated electron systems can be obtained from cluster partitions of the corresponding Hamiltonians; see, e.g., R. Valentí, J. Stolze, and P. J. Hirschfeld, Phys. Rev. B **43**, 13743 (1991); R. Valentí, P. J. Hirschfeld, and J.-C. Anglés d'Auriac, Phys. Rev. B **44**, 3995 (1991).

3.12 G. Baym and L. P. Kadanoff, Phys. Rev. **124**, 287 (1961); G. Baym, Phys. Rev. **127**, 1391 (1962); see also L. P. Kadanoff and G. Baym, *Quantum Statistical Mechanics* (Benjamin, New York, 1962).

3.13 Further results from perturbation theory for the ground state energy can be found, e.g., in W. Metzner and D. Vollhardt, Phys. Rev. B **39**, 4462 (1989); H. Schweitzer and G. Czycholl, Z. Phys. B **83**, 93 (1991); V. Zlatić, B. Horvatić, G. Schliecker, and D. Schotte, Phil. Mag. B **65**, 1255 (1992).

3.14 A. L. Fetter and J. D. Walecka, *Quantum Theory of Many-Particle Systems* (McGraw–Hill, New York, 1971).

3.15 J. Kanamori, Prog. Theor. Phys. **30**, 275 (1963).

3.16 A. Georges and J. S. Yedidia, Phys. Rev. B **43**, 3475 (1991).

3.17 P. Kopietz, J. Phys. Cond. Matt. **6**, 4885 (1994).

3.18 A. Martín-Rodero and F. Flores, Phys. Rev. B **45**, 13008 (1992); J. K. Freericks, Phys. Rev. B **50**, 403 (1994).

3.19 M. Jarrell, Phys. Rev. Lett. **69**, 168 (1992).

3.20 D. E. Logan, Y. H. Szczech, and M. E. Tusch in *Metal–Insulator Transitions Revisited*, ed. by P. P. Edwards and C. R. N. Rao (Taylor and Francis, London, 1995), p. 395.

3.21 J. R. Schrieffer, X.-G. Wen, and S. C. Zhang, Phys. Rev. B **39**, 11663 (1989).

3.22 N. E. Bickers and D. J. Scalapino, Annals of Physics **193**, 206 (1989); N. E. Bickers and S. R. White, Phys. Rev. B **43**, 8044 (1991).

3.23 K. Schönhammer, O. Gunnarsson, and R. M. Noack, Phys. Rev. B **52**, 2504 (1995).

3.24 M. Levy, Proc. Nat. Acad. Sci. (USA) **76**, 6062 (1979).

3.25 A. Messiah, *Quantum Mechanics*, Vol. 2 (North-Holland, Amsterdam, 1962).

3.26 P. Hohenberg and W. Kohn, Phys. Rev. **136**, B864 (1964); see also W. Kohn and L. J. Sham, Phys. Rev. **140**, A1133 (1964).

3.27 A. Schindlmayr and R. W. Godby, Phys. Rev. B **51**, 10427 (1995).

3.28 G. B. Bachelet, M. Schlüter, and G. A. Baraff, Phys. Rev. B **27**, 2545 (1983); G. A. Baraff and M. Schlüter, Phys. Rev. B **30**, 1853, 3460 (1984); Phys. Rev. Lett. **55**, 1327 (1985), and references therein.

3.29 J. P. Perdew, R. G. Parr, M. Levy, and J. L. Baduz, Phys. Rev. Lett. **49**, 1691 (1982); J. P. Perdew and M. Levy, Phys. Rev. Lett. **51**, 1884 (1983).
3.30 L. J. Sham and M. Schlüter, J. P. Perdew and M. Levy, Phys. Rev. Lett. **51**, 1888 (1983).
3.31 O. Gunnarsson and K. Schönhammer, Phys. Rev. Lett. **56**, 1968 (1986); L. J. Sham and M. Schlüter, Phys. Rev. Lett. **60**, 1582 (C) (1988); O. Gunnarsson and K. Schönhammer, *ibid.*, 1583 (C) (1988); K. Schönhammer and O. Gunnarsson, J. Phys. C **20**, 3675 (1987).
3.32 V. I. Anisimov, J. Zaanen, and O. K. Andersen, Phys. Rev. B **44**, 943 (1991); for an application of this concept, see, e.g., P. Wei and Z. Q. Qi, Phys. Rev. B **49**, 10864 (1994).
3.33 See, for example, G. Strinati, H. J. Mattausch, and W. Hanke, Phys. Rev. B **25**, 2867 (1982); W. E. Pickett and C. S. Wang, *ibid.* **30**, 4719 (1984); M. Hybertson and S. G. Louie, *ibid.* **34**, 5390 (1986); R. W. Godby, M. Schlüter, and L. J. Sham, Phys. Rev. Lett. **56**, 2415 (1986); Phys. Rev. B **35**, 4170 (1987).
3.34 A. Svane and O. Gunnarsson, Phys. Rev. B **37**, 9919 (1988); Europhys. Lett. **7**, 171 (1988); D. M. Edwards, S. M. Manning, and J. Rettke-Grover, Phil. Mag. B **69**, 849 (1994); for recent applications of the SIC-LDA, see A. Svane, Phys. Rev. Lett. **72**, 1248 (1994), and references therein.
3.35 E. H. Lieb and F. Y. Wu, Phys. Rev. Lett. **20**, 1445 (1968).
3.36 S. M. Manning and D. M. Edwards, J. Phys. Cond. Matt. **5**, 6203 (1993).
3.37 O. Gunnarsson, B. I. Lundqvist, and J. W. Wilkins, Phys. Rev. B **10**, 1319 (1974).
3.38 O. Gunnarsson, O. K. Andersen, and A. Svane, in *Interacting Electrons in Reduced Dimensions*, ed. by D. Baeriswyl and D. K. Campbell, NATO ASI Series B **213** (Plenum Press, New York, 1989), p. 139.
3.39 R. O. Jones and O. Gunnarsson, Rev. Mod. Phys. **61**, 689 (1989).
3.40 Y. Ishii and K. Terakura, Phys. Rev. B **42**, 10924 (1990); T. Miyazaki, Y. Ishii, and K. Terakura, Prog. Theor. Phys. Supp. **106**, 173 (1991).
3.41 J. Hubbard, Proc. Roy. Soc. London Ser. A **276**, 238 (1963); *ibid.* **277**, 237 (1963).
3.42 R. A. Beni and T. A. Kaplan, Phys. Lett. A **33**, 400 (1970).
3.43 A. Kawabata, Prog. Theor. Phys. **48**, 1793 (1972).
3.44 Similar structures appear in the context of the single-impurity Anderson model; see A. Theumann, Phys. Rev. **178**, 978 (1969); C. Lacroix, J. Phys. F: Metal Phys. **11**, 2389; compare Sect. VI B 4 of A. Georges, G. Kotliar, W. Krauth, and M. J. Rozenberg, Rev. Mod. Phys. **68**, 13 (1996); in fact, any expansion around the atomic limit requires four self-energies; see, e.g., T. Pruschke and N. Grewe, Z. Phys. B **74**, 439 (1989).
3.45 J. Hubbard, Proc. Roy. Soc. London **281**, 401 (1964).
3.46 M. Cyrot, Physica B **91**, 141 (1977).
3.47 L. M. Falicov and J. C. Kimball, Phys. Rev. Lett. **22**, 997 (1969).
3.48 P. Soven, Phys. Rev. **156**, 809 (1967); D. W. Taylor, Phys. Rev. **156**, 1017 (1967); see also R. J. Elliott, J. A. Krumhansl, and P. L. Leath, Rev. Mod. Phys. **46**, 465 (1974).
3.49 J. M. Ziman, *Models of Disorder* (Cambridge University Press, Cambridge, 1979).
3.50 L. Schwartz and E. Siggia, Phys. Rev. B **5**, 383 (1972).
3.51 R. Vlaming and D. Vollhardt, Phys. Rev. B **45**, 4637 (1992).
3.52 A. O. Anokhin, V. Y. Irkhin, and M. I. Katsnelson, J. Phys. Cond. Matt. **3**, 1475 (1991).

3.53 V. Janiš and D. Vollhardt, Int. J. Mod. Phys. **B 6**, 283 (1992); V. Janiš and D. Vollhardt, Z. Phys. B **91**, 317 (1993); V. Janiš, J. Mašek, and D. Vollhardt, Z. Phys. B **91**, 325 (1993).
3.54 D. M. Edwards and A. C. Hewson, Rev. Mod. Phys. **40**, 810 (1967).
3.55 A. Kawabata, Prog. Theor. Phys. **54**, 45 (1975).
3.56 A decoupling in higher order of the Green-function hierarchy does not lead to any qualitative improvements; see C. Gros, Phys. Rev. B **50**, 7295 (1994).
3.57 A. B. Harris and R. V. Lange, Phys. Rev. **157**, 295 (1967).
3.58 R. A. Tahir-Kheli and H. S. Jarrett, Phys. Rev. **180**, 544 (1969); W. Nolting, Z. Phys. B **255**, 25 (1972); *ibid.* **265**, 173 (1973); for recent applications, see also W. Nolting and W. Borgieł, Phys. Rev. B **39**, 6962 (1989).
3.59 E. G. Goryachev, E. V. Kuzmin, and S. G. Ovchinnikov, J. Phys. C **15**, 1481 (1982).
3.60 M. Cyrot, Phys. Rev. Lett. **25**, 871 (1970); Phil. Mag. **25**, 1031 (1972).
3.61 M. Cyrot, Journ. de Physique **33**, 125 (1972).
3.62 See also Chap. 8 in T. Moriya, *Spin Fluctuations in Itinerant Electron Magnetism*, Springer Series in Solid-State Sciences **56** (Springer, Berlin, 1985).
3.63 U. Behn and W. Weller, phys. stat. sol. (b) **99**, 71 (1980); U. Behn, phys. stat. sol. (b) **100**, 129 (1980).
3.64 D. M. Edwards and J. A. Hertz, Physica B **163**, 527 (1990); D. M. Edwards, J. Phys. Cond. Matt. **5**, 161 (1993).
3.65 P. M. Mors, J. R. Iglesias, and E. V. Anda, Solid-State Comm. **55**, 827 (1985).
3.66 A. P. Kampf, Phys. Rev. B **44**, 2637 (1991).
3.67 An introduction can be found in E. Feenberg, *Theory of Quantum Fluids* (Academic Press, New York, 1969); C. E. Campbell in *Progress in Liquid Physics*, ed. by C. A. Croxton (Wiley, New York, 1978); see also B. E. Clements, E. Krotscheck, J. A. Smith, and C. E. Campbell, Phys. Rev. B **47**, 5239 (1993), and references therein.
3.68 M. C. Gutzwiller, Phys. Rev. Lett. **10**, 159 (1963).
3.69 M. C. Gutzwiller, Phys. Rev. **134**, A923 (1964); *ibid.* **137**, A1726 (1965).
3.70 A. J. Millis and S. N. Coppersmith, Phys. Rev. B **43**, 13770 (1991).
3.71 P. Nozières, *Theory of Interacting Fermi Systems* (Addison–Wesley, New York, 1964).
3.72 P. Fazekas, B. Menge, and E. Müller-Hartmann, Z. Phys. B **78**, 69 (1990).
3.73 D. Baeriswyl and K. Maki, Phys. Rev. B **31**, 6633 (1985); D. Baeriswyl, J. Carmelo, and K. Maki, Synth. Met. **21**, 271 (1987).
3.74 P. G. J. van Dongen, F. Gebhard, and D. Vollhardt, Z. Phys. B **76**, 199 (1989).
3.75 W. F. Brinkman and T. M. Rice, Phys. Rev. B **2**, 4302 (1970).
3.76 W. Metzner and D. Vollhardt, Phys. Rev. Lett. **59**, 121 (1987); Phys. Rev. B **37**, 7382 (1988).
3.77 F. Gebhard and D. Vollhardt, Phys. Rev. Lett. **59**, 1472 (1987); Phys. Rev. B **38**, 6911 (1988).
3.78 A brief review can be found in D. Vollhardt, P. G. J. van Dongen, F. Gebhard, and W. Metzner, Mod. Phys. Lett. B **4**, 499 (1990).
3.79 T. Ogawa, K. Kanda, and T. Matsubara, Prog. Theor. Phys. **53**, 614 (1975).
3.80 W. Metzner and D. Vollhardt, Phys. Rev. Lett. **62**, 324 (1989).
3.81 F. Gebhard, Phys. Rev. B **41**, 9452 (1990).
3.82 W. Metzner, Z. Phys. B **77**, 253 (1989); *ibid.* **82**, 183 (1991).
3.83 F. Gebhard, Phys. Rev. B **44**, 992 (1991).

3.84 R. Strack and D. Vollhardt, Mod. Phys. Lett. B **5**, 1377 (1991); R. Strack and D. Vollhardt, J. Low Temp. Phys. **84**, 357 (1991); Z. Gulási, R. Strack, and D. Vollhardt, Phys. Rev. B **47**, 8594 (1993).
3.85 D. Vollhardt, Rev. Mod. Phys. **56**, 99 (1984); D. Vollhardt, P. Wölfle, and P. W. Anderson, Phys. Rev. B **35**, 6703 (1987).
3.86 For an application to lattice models for Helium-3, see [3.85].
3.87 For an application to models for valence fluctuations and heavy fermions (periodic Anderson model), see C. M. Varma, W. Weber, and L. J. Randall, Phys. Rev. B **33**, 1015 (1985); T. M. Rice and K. Ueda, Phys. Rev. Lett. **55**, 995 (1985); *ibid.*, 2093 (E) (1985); Phys. Rev. B **34**, 6420 (1986); P. Fazekas, Sol. State Comm. **60**, 431 (1986); B. H. Brandow, Phys. Rev. B **33**, 215 (1986); P. Fazekas and B. H. Brandow, Phys. Scr. **36**, 809 (1987); V. Z. Vulović and E. Abrahams, Phys. Rev. B **36**, 2614 (1987).
3.88 For an application to models for high-temperature superconductivity (e.g., tJ model, Emery model), see F. C. Zhang, C. Gros, T. M. Rice, and H. Shiba, Supercond. Sci. Techn. **1**, 36 (1988); A. M. Oleś and J. Zaanen, Phys. Rev. B **39**, 9175 (1989); S. K. Sarker, Phys. Rev. B **39**, 2155 (1989).
3.89 J. Spałek and P. Gopalan, Phys. Rev. Lett. **64**, 2823 (1990); P. Korbel, J. Spałek, W. Wójcik, and M. Acquarone, Phys. Rev. B **52**, 2213 (1995); J. Spałek and W. Wójcik in *Spectroscopy of the Mott Insulators and Correlated Metals*, Springer Series in Solid-State Sciences **119**, (Springer, Berlin, 1995), p. 41.
3.90 Supporting arguments are given by P. B. Visscher, Phys. Rev. B **10**, 943 (1974); M. Grilli, C. Castellani, and G. Kotliar, Phys. Rev. B **45**, 10805 (1992); P. G. J. van Dongen, Phys. Rev. Lett. **74**, 182 (1995), and references therein.
3.91 T. A. Kaplan, P. Horsch, and P. Fulde, Phys. Rev. Lett. **49**, 889 (1982).
3.92 D. Baeriswyl in *Nonlinearity in Condensed Matter*, ed. by R. Bishop, D. K. Campbell, P. Kumar and S. E. Trullinger, Springer Series in Solid-State Sciences **69** (Springer, Berlin, 1987), p. 183.
3.93 H. Otsuka, J. Phys. Soc. Jpn. **61**, 1645 (1992).
3.94 M. Dzierzawa, D. Baeriswyl, and M. di Stasio, Phys. Rev. B **51**, 1993 (1995).
3.95 F. Gebhard and A. Girndt, Z. Phys. B **93**, 455 (1994).
3.96 F. Gebhard and A. E. Ruckenstein, Phys. Rev. Lett. **68**, 244 (1992); F. Gebhard, A. Girndt, and A. E. Ruckenstein, Phys. Rev. B **49**, 10926 (1994).
3.97 P.-A. Bares and F. Gebhard, Europhys. Lett. **29**, 573 (1995).
3.98 For an introduction, see J. W. Negele and H. Orland, *Quantum Many-Particle Systems*, Frontiers in Physics **68** (Addison–Wesley, New York, 1987); E. Fradkin, *Field Theories of Condensed Matter Systems*, Frontiers in Physics **82** (Addison–Wesley, New York, 1991).
3.99 H. Hasegawa, J. Phys. Cond. Matt. **1**, 9325 (1989).
3.100 E. O. Tüngler and T. Kopp, Nucl. Phys. B **443**, 516 (1995).
3.101 J. Hubbard, Phys. Rev. Lett. **3**, 77 (1959); R. L. Stratonovich, Sov. Phys. Dokl. **2**, 416 (1957).
3.102 H. J. Schulz, Phys. Rev. Lett. **65**, 2462 (1990); in *The Hubbard Model*, ed. by D. Baeriswyl, D. K. Campbell, J. M. P. Carmelo, F. Guinea, and E. Louis, NATO ASI Series **343** (Plenum Press, New York, 1995), p. 89; J. Low Temp. Phys. **99**, 615 (1995).
3.103 See, e.g., J. E. Hirsch, Phys. Rev. B **28**, 4059 (1983); *ibid.* **31**, 4403 (1985); E. Y. Loh, J. E. Gubernatis, R. T. Scalettar, S. R. White, D. J. Scalapino, and R. L. Sugar, Phys. Rev. B **41**, 9301 (1990), and references therein.

3.104 For recent applications, see, e.g., X. G. Wen, Phys. Rev. B **39**, 7223 (1989); W. Weller, phys. stat. sol. (b) **162**, 251 (1990); N. Nagaosa and P. A. Lee, Phys. Rev. Lett. **64**, 2450 (1990); M. U. Ubbens and P. A. Lee, Phys. Rev. B **46**, 8434 (1992); C. Kübert and A. Muramatsu, Phys. Rev. B **47**, 787 (1993), and in *The Hubbard Model*, ed. by D. Baeriswyl, D. K. Campbell, J. M. P. Carmelo, F. Guinea, and E. Louis, NATO ASI Series **343** (Plenum Press, New York, 1995), p. 103; X.-G. Wen and P. A. Lee, Phys. Rev. Lett. **76**, 503 (1996).
3.105 G. Kotliar and A. E. Ruckenstein, Phys. Rev. Lett. **57**, 1362 (1986).
3.106 L. Lilly, A. Muramatsu, and W. Hanke, Phys. Rev. Lett. **65**, 1379 (1990); E. Arrigoni and G. C. Strinati, Phys. Rev. B **52**, 13707 (1995).
3.107 M. Lavagna, Phys. Rev. B **41**, 142 (1990); Int. J. Mod. Phys. B **5**, 885 (1991).
3.108 A. J. Leggett, Rev. Mod. Phys. **47**, 331 (1975).
3.109 T. Li, P. Wölfle, and P. J. Hirschfeld, Phys. Rev. B **40**, 6817 (1989); P. Wölfle and T. Li, Z. Phys. B **78**, 45 (1990); T. Li, Y. S. Sun, and P. Wölfle, Z. Phys. B **82**, 369 (1991); P. Wölfle, J. Low Temp. Phys. **99**, 625 (1995).
3.110 For example by C. A. Balseiro, M. Avignon, A. G. Rojo, and B. Alascio, Phys. Rev. Lett. **62**, 2624 (1989); L. Lilly, A. Muramatsu, and W. Hanke, Phys. Rev. Lett. **65**, 1379 (1990); E. Arrigoni, G. C. Strinati, and C. Castellani, Phys. Rev. B **41**, 4838 (1990); J. Schmalian, G. Baumgärtel, and K. H. Bennemann, Phys. Rev. Lett. **68**, 1406 (1992).
3.111 M. Deeg, H. Fehske, and H. Büttner, Z. Phys. B **91**, 31 (1993); Z. Phys. B **88**, 283 (1992); D. Schmeltzer and A. R. Bishop, Europhys. Lett. **12**, 369 (1990).
3.112 K. Schönhammer, Phys. Rev. B **42**, 2591 (1990).
3.113 R. Frésard, M. Dzierzawa, and P. Wölfle, Europhys. Lett. **15**, 325 (1991); R. Frésard and P. Wölfle, J. Phys. Cond. Matt. **4**, 3625 (1992); R. Frésard and P. Wölfle, Int. J. Mod. Phys. B **6**, 685, 3087 (E) (1992); K. Doll, M. Dzierzawa, R. Frésard, and P. Wölfle, Z. Phys. B **90**, 297 (1993).
3.114 An early example is given by R. Bari, Phys. Rev. B **5**, 2736 (1972).
3.115 R. Raimondi and C. Castellani, Phys. Rev. B **48**, 11453 (1993); as in [3.107] the limits $\delta \to 0$ and $\overline{d} \to 0$ were exchanged.
3.116 E. Arrigoni and G. C. Strinati, Phys. Rev. Lett. **71**, 3178 (1993); Phys. Rev. B **52**, 2428 (1995); in *The Hubbard Model*, ed. by D. Baeriswyl, D. K. Campbell, J. M. P. Carmelo, F. Guinea, and E. Louis, NATO ASI Series **343** (Plenum Press, New York, 1995), p. 209; J. Low Temp. Phys. **99**, 599 (1995).
3.117 T. Jolicoeur and J. C. Guillon, Phys. Rev. B **44**, 2403 (1991).
3.118 S. E. Barnes, J. Phys. F **6**, 1375 (1976); *ibid.* **7**, 2637 (1977).
3.119 P. Coleman, Phys. Rev. B **28**, 5255 (1983); *ibid.* **29**, 3035 (1984); *ibid.* **35**, 5072 (1987).
3.120 N. Read and D. M. Newns, J. Phys. C **16**, 3273 (1983); Adv. Phys. **36**, 799 (1987).
3.121 N. Read, D. M. Newns, and S. Doniach, Phys. Rev. B **30**, 3841 (1984).
3.122 A. Auerbach and K. Levin, Phys. Rev. Lett. **57**, 877 (1986); Phys. Rev. B **34**, 3524 (1986); A. Auerbach, J. H. Kim, K. Levin, and M. R. Norman, Phys. Rev. Lett. **60**, 623 (1988).
3.123 M. Lavagna, A. J. Millis, and P. A. Lee, Phys. Rev. Lett. **58**, 266 (1987); A. J. Millis and P. A. Lee, Phys. Rev. B **35**, 3394 (1987).
3.124 M. Grilli, G. Kotliar, and A. J. Millis, Phys. Rev. B **42**, 329 (1990); M. Grilli, R. Raimondi, C. Castellani, C. di Castro, and G. Kotliar, Phys. Rev. Lett. **67**, 259 (1991).
3.125 P. Fulde, *Electron Correlations in Molecules and Solids*, Springer Series in Solid-State Sciences **100** (Springer, Berlin, 1991).

3.126 P. Fulde, J. Keller, and G. Zwicknagel, *Solid-State Physics* **41**, ed. by F. Seitz and D. Turnbull, p. 1 (1988); G. Czycholl, Phys. Reports **143**, 277 (1986).
3.127 P. G. J. van Dongen, Phys. Rev. Lett. **69**, 2878 (C) (1993); G. Czycholl and Q. Qin, Phys. Rev. Lett. **69**, 2879 (C) (1993).
3.128 An excellent review on $1/N_f$ techniques for dilute magnetic alloys is given by N. E. Bickers, Rev. Mod. Phys. **59**, 845 (1987); see also N. E. Bickers, D. L. Cox, and J. W. Wilkins, Phys. Rev. B **36**, 2036 (1987).
3.129 O. Gunnarsson and K. Schönhammer, Phys. Rev. Lett. **50**, 604 (1983), Phys. Rev. B **28**, 4315 (1983); in *Handbook on the Physics and Chemistry of Rare Earths* **10**, ed. by K. A. Gscheidner, Jr., L. Eyring, and S. Hüfner (Elsevier, Amsterdam, 1987), p. 103; the Gutzwiller-correlated wave function agrees with the Gunnarsson–Schönhammer wave function to lowest order in $(1/N_f)$ and yields the same results for the single-site problem [3.83].
3.130 N. Read, J. Phys. C **18**, 2651 (1987).
3.131 E. Arrigoni, C. Castellani, R. Raimondi, and G. C. Strinati, Phys. Rev. B **50**, 2700 (1994).
3.132 F. Mila, D. Poilblanc, and C. Bruder, Phys. Rev. B **43**, 7891 (1991).
3.133 J. Ferrer, Phys. Rev. B **47**, 8769 (1993).
3.134 P. Coleman, R. Miranda, and A. Tsvelik, Physica B **186–188**, 362 (1993); Phys. Rev. B **49**, 8955 (1994).
3.135 Z. Y. Weng, D. N. Sheng, C. S. Ting, and Z. B. Su, Phys. Rev. B **45**, 7850 (1992); Phys. Rev. Lett. **67**, 3318 (1991); S. Feng, Z. B. Su, and L. Yu, Phys. Rev. B **49**, 2368 (1994).

Chapter 4

4.1 A. B. Harris and R. V. Lange, Phys. Rev. **157**, 295 (1967).
4.2 M. W. Long and R. Fehrenbacher, J. Phys. Cond. Matt. **2**, 10343 (1990).
4.3 N. Andrei in *Low-Dimensional Quantum Field Theories for Condensed Matter Physicists*, Series on Modern Condensed Matter Physics **6**, ed. by S. Lundquist, G. Morandi, and Yu Lu (World Scientific, Singapore, 1995), p. 457; this article also contains a review on the standard approach to elementary excitations of the Kondo and the Hubbard model.
4.4 *Exactly Solvable Models of Strongly Correlated Electrons*, ed. by V. E. Korepin and F. H. L. Eßler (World Scientific, Singapore, 1995).
4.5 F. H. Eßler, V. E. Korepin, and K. Schoutens, Phys. Rev. Lett. **67**, 3848 (1991); Nucl. Phys. B **372**, 559 (1992); *ibid.* **384**, 431 (1992).
4.6 F. H. L. Eßler and V. E. Korepin, Phys. Rev. Lett. **72**, 908 (1994); Nucl. Phys. B **426**, 505 (1994).
4.7 B. S. Shastry, Phys. Rev. Lett. **56**, 1529 (1986); *ibid.*, 2334 (E) (1986); *ibid.*, 2453 (1986); J. Stat. Phys. **50**, 57 (1988).
4.8 C. N. Yang, Phys. Rev. Lett. **19**, 1312 (1967).
4.9 M. Gaudin, *La Fonction d'Onde de Bethe pour les Modèles Exacts de la Méchanique Statistique* (Commisariat à l'Energie Atomique, Paris, 1983).
4.10 J. Zinn-Justin and E. Brezin, C. R. Acad. Sci. (Paris) **263**, 670 (1966).
4.11 R. J. Baxter, *Exactly Solved Models in Statistical Mechanics*, 3rd printing (Academic Press, New York, 1989).
4.12 B. Sutherland in *Exactly Solvable Problems in Condensed Matter and Relativistic Field Theory*, ed. by B. S. Shastry, S. S. Jha, and V. Singh, Lecture Notes in Physics **242** (Springer, Berlin, 1985), p. 1. Sutherland uses the definition $\theta_{\text{Suth.}}(k_1, k_2) = -\theta(k_1, k_2)$.

4.13 An introduction is given by P. P. Kulish and E. K. Sklyanin in *Integrable Quantum Field Theories*, ed. by J. Hietarinta and C. Montonen, Lecture Notes in Physics **151** (Springer, Berlin, 1982); L. A. Takhtajan, and N. M. Bogoliubov, A. G. Izergin, and V. E. Korepin in *Exactly Solvable Problems in Condensed Matter and Relativistic Field Theory*, ed. by B. S. Shastry, S. S. Jha, and V. Singh, Lecture Notes in Physics **242** (Springer, Berlin, 1985), p. 220; V. E. Korepin, A. G. Izergin, and N. M. Bogoliubov, *Quantum Inverse Scattering Method and Correlation Functions* (Cambridge University Press, New York, 1993).
4.14 B. S. Shastry, Phys. Rev. Lett. **56**, 2453 (1986); J. Stat. Phys. **50**, 57 (1988).
4.15 P. B. Ramos and M. J. Martins, available on the e-print archive `xxx.lanl.gov` as hep-th/9605141, to appear as letter in J. Phys. A **30** (1997).
4.16 E. H. Lieb and F. Y. Wu, Phys. Rev. Lett. **20**, 1445 (1968); the equation for the jump in the chemical potential lacks a minus sign.
4.17 A. Klümper, A. Schadschneider, and J. Zittartz, Z. Phys. B. **78**, 99 (1990); A. Schadschneider and J. Zittartz, *ibid.* **82**, 387 (1991).
4.18 H. Shiba, Phys. Rev. B **6**, 930 (1972); analytical results for large U/t or small densities are also presented by J. Carmelo and D. Baeriswyl, Phys. Rev. B **37**, 7541 (1987).
4.19 Results from perturbation theory for the ground state energy are found in W. Metzner and D. Vollhardt, Phys. Rev. B **39**, 4462 (1989), and in H. Schweitzer and G. Czycholl, Z. Phys. B **83**, 93 (1991); an analysis of the Lieb–Wu equations for small interactions has been carried out by H. Frahm and M. P. Pfannmüller, Phys. Lett. A **204**, 347 (1995).
4.20 E. N. Economou and P. N. Poulopoulos, Phys. Rev. B **20**, 4756 (1979).
4.21 F. Woynarovich, J. Phys. C **15**, 85, 97 (1982); *ibid.* **16**, 5293 (1983); J. Phys. A **22**, 4243 (1989).
4.22 H. Frahm and V. E. Korepin, Phys. Rev. B **42**, 10553 (1990); *ibid.* **43**, 5653 (1991); they use $\mu^{\mathrm{FK}} = -\mu$.
4.23 C. N. Yang and C. P. Yang, J. Math. Phys. **10**, 1115 (1969).
4.24 N. Kawakami and S.-K. Yang, Phys. Lett. **A 148**, 359 (1990); Phys. Rev. Lett. **65**, 3063 (1990); Phys. Rcv. B **44**, 7844 (1991).
4.25 J. Carmelo and A. A. Ovchinnikov, J. Phys. Cond. Matt. **3**, 757 (1991); J. Carmelo, P. Horsch, P. A. Bares, and A. A. Ovchinnikov, Phys. Rev. B **44**, 9967 (1991); J. M. P. Carmelo, P. Horsch, and A. A. Ovchinnikov, Phys. Rev. B **45**, 7899 (1992); P.-A. Bares, J. M. P. Carmelo, J. Ferrer, and P. Horsch, Phys. Rev. B **46**, 14624 (1992); J. M. P. Carmelo, P. Horsch, and A. A. Ovchinnikov, Phys. Rev. B **46**, 14728 (1992).
4.26 A. A. Ovchinnikov, Zh. Eksp. Teor. Fiz. **57**, 2137 (1969) [Sov. Phys. JETP **30**, 1160 (1970)].
4.27 K. Schönhammer, O. Gunnarsson, and R. M. Noack, Phys. Rev. B **52**, 2504 (1995).
4.28 C. A. Stafford and A. J. Millis, Phys. Rev. B **48**, 1409 (1993).
4.29 M. Imada, J. Phys. Soc. Jpn **62**, 1105 (1993); *ibid.* **63**, 851, 3059, 4294 (1994); *ibid.* **64**, 2954 (1995); in *Spectroscopy of Mott Insulators and Correlated Metals*, ed. by A. Fujimori and Y. Tokura, Springer Series in Solid-State Sciences **119** (Springer, Berlin, 1995), p. 3; J. Low Temp. Phys. **99**, 437 (1995); F. F. Assaad and M. Imada, Phys. Rev. Lett. **76**, 3176 (1996).
4.30 E. Melzer, Nucl. Phys. B **443**, 553 (1995).
4.31 W. Kohn, Phys. Rev. **133**, A171 (1964); recent applications can be found in C. A. Stafford, A. J. Millis and B. S. Shastry, Phys. Rev. B **43**, 13660 (1991) and in R. M. Fye, M. J. Martins, D. J. Scalapino, J. Wagner, and W. Hanke, Phys. Rev. B **45**, 7311 (1992).

4.32 B. S. Shastry and B. Sutherland, Phys. Rev. Lett. **65**, 243 (1990).
4.33 N. Yu and M. Fowler, Phys. Rev. B **45**, 11795 (1992).
4.34 H. J. Schulz, Phys. Rev. Lett. **64**, 2831 (1990); H. J. Schulz in *Strongly Correlated Electron Systems II*, Proceedings of the Adriatico Research Conference and Miniworkshop, Trieste, Italy, 18 June–27 July 1990, ed. by G. Baskaran, A. E. Ruckenstein, E. Tosatti, and Y. Lu (World Scientific, Singapore, 1991), p. 57.
4.35 N. W. Ashcroft and N. D. Mermin, *Solid-State Physics* (Holt–Saunders, Philadelphia, 1976).
4.36 F. D. M. Haldane, Phys. Rev. Lett. **45**, 1358 (1980); J. Phys. C **14**, 2585 (1981); Phys. Lett. A **81**, 153 (1981).
4.37 K. Schönhammer and V. Meden, Phys. Rev. B **46**, 15753 (1992); J. Elec. Spectr. Relat. Phenom. **62**, 225 (1993); Phys. Rev. B **47**, 16205 (1993).
4.38 J. Voit, Phys. Rev. B **45**, 4027 (1992); Phys. Rev. B **47**, 6740 (1993); J. Phys. Cond. Matt. **5**, 8305 (1993).
4.39 T. Giamarchi, Phys. Rev. B **44**, 2905 (1991); *ibid.* **46**, 342 (1992); T. Giamarchi and A. J. Millis, Phys. Rev. B **46**, 9325 (1992).
4.40 F. Gebhard, K. Bott, M. Scheidler, P. Thomas, S. W. Koch, Phil. Mag. B **75**, pp. 1, 13, 47 (1997), and references therein.
4.41 M. Gulásci and K. S. Bedell, Phys. Rev. Lett. **72**, 2765 (1994).
4.42 H. Eskes and A. M. Oleś, Phys. Rev. Lett. **73**, 1279 (1994).
4.43 H. J. Schulz in *Strongly Correlated Electronic Materials: The Los Alamos Symposium 1993*, ed. by K. S. Bedell, Z. Wang, D. E. Meltzer, A. V. Balatsky, and E. Abrahams (Addison–Wesley, Reading, 1994), p. 187.
4.44 M. A. Olshanetsky and A. M. Perelomov, Phys. Rep. **94**, 313 (1983).
4.45 F. Calogero, J. Math. Phys. **10**, 2191, 2197 (1969); *ibid.* **12**, 419 (1971); B. Sutherland, *ibid.* **12**, 246, 251 (1971); Phys. Rev. A **4**, 2019 (1971); *ibid.* **5**, 1372 (1972); Phys. Rev. B **45**, 907 (1994); see also [4.12].
4.46 F. J. Dyson and M. L. Mehta, J. Math. Phys. **4**, 701, 714 (1963); see also M. L. Mehta, *Random Matrices*, 2nd edition (Academic Press, Boston, 1991).
4.47 B. S. Shastry and B. Sutherland, Phys. Rev. Lett. **71**, 5 (1993).
4.48 A. Polychronakos, Phys. Rev. Lett. **70**, 2329 (1993).
4.49 F. Gebhard and D. Vollhardt, Phys. Rev. Lett. **59**, 1472 (1987); Phys. Rev. B **38**, 6911 (1988).
4.50 F. D. M. Haldane, Phys. Rev. Lett. **60**, 635 (1988).
4.51 B. S. Shastry, Phys. Rev. Lett. **60**, 639 (1988).
4.52 F. D. M. Haldane, Z. Ha, J. C. Talstra, D. Bernard, V. Pasquier, Phys. Rev. Lett. **69**, 2021 (1992); J. C. Talstra and F. D. M. Haldane, J. Phys. A **28**, 2369 (1995).
4.53 F. D. M. Haldane, Phys. Rev. Lett. **66**, 1529 (1991).
4.54 F. D. M. Haldane and M. R. Zirnbauer, Phys. Rev. Lett. **71**, 4055 (1993); Phys. Rev. B **52**, 8729 (1995); J. C. Talstra and F. D. M. Haldane, Phys. Rev. B **50**, 6889 (1994).
4.55 J. Talstra, PhD. thesis (Princeton University, 1995), available on the e-print archive `xxx.lanl.gov` as cond-mat/9509178.
4.56 F. Gebhard and A. E. Ruckenstein, Phys. Rev. Lett. **68**, 244 (1992); here, $-\pi \leq Q < \pi$ for the total momentum was used.
4.57 F. Gebhard and A. Girndt, Z. Phys. B **93**, 455 (1994).
4.58 D. F. Wang, Q. F. Zhong, and P. Coleman, Phys. Rev. B **48**, 8476 (1993); D. F. Wang, Phys. Rev. B **48**, 10556 (1993); D. F. Wang and C. Gruber, J. Stat. Phys. **82**, 421 (1996).

4.59 F. Gebhard, A. Girndt, and A. E. Ruckenstein, Phys. Rev. B **49**, 10926 (1994); F. Gebhard in *The Hubbard Model: Its Physics and Mathematical Physics*, ed. by D. Baeriswyl, D. K. Campbell, J. M. P. Carmelo, F. Guinea, and E. Louis, NATO ASI Series B **343** (Plenum Press, New York, 1995), p. 29.
4.60 J. Voit, Rep. Prog. Phys. **58**, 977 (1995); H. J. Schulz in *Fermi Liquids and Non-Fermi Liquids*, ed. by E. Akkermans, G. Montambaux, J. Picard, and J. Zinn-Justin, Les Houches Session LXI 1994 (Elsevier, Amsterdam, 1995), p. 533.
4.61 M. Takahashi, Prog. Theor. Phys. **47**; T. Koma, Prog. Theor. Phys. **83**, 655 (1990).
4.62 M. Takahashi, Prog. Theor. Phys. **52**, 103 (1974).
4.63 N. Kawakami, T. Usuki, and A. Okiji, Phys. Lett. A **137**, 287 (1989); T. Usuki, N. Kawakami, and A. Okiji, J. Phys. Soc. Jpn **59**, 1357 (1990); for some early exact diagonalization studies, see H. Shiba and P. A. Pincus, Phys. Rev. B **5**, 1966 (1972).
4.64 A. Klümper and R. Z. Bariev, Nucl. Phys. B **458** [FS], 623 (1996).
4.65 M. Suzuki, Phys. Rev. B **31**, 2957 (1985).
4.66 J. Ashkin and E. Teller, Phys. Rev. **64**, 178 (1943).
4.67 I. Affleck, Phys. Rev. Lett. **56**, 746 (1986).
4.68 V. I. Inozemtsev, Journ. Stat. Phys. **59**, 1143 (1990); Comm. Math. Phys. **148**, 359 (1992).
4.69 A. Polychronakos, Phys. Rev. Lett. **70**, 2329 (1993); B. Sutherland and B. S. Shastry, Phys. Rev. Lett. **71**, 5 (1993).
4.70 B. Sutherland, R. A. Römer, and B. S. Shastry, Phys. Rev. Lett. **73**, 2154 (1994).
4.71 V. I. Inozemtsev and A. L. Kuzemsky, Phys. Rev. B **43**, 1090 (1991).
4.72 E. R. Hansen, *A Table of Series and Products* (Prentice–Hall, Englewood Cliffs, 1975), (14.17.17.).
4.73 M. Abramovitz and I. A. Stegun, *Handbook of Mathematical Functions*, 9th edition (Dover Publications, New York, 1970), (16.29.4).
4.74 P.-A. Bares and F. Gebhard, Europhys. Lett. **29**, 573 (1995); J. Low Temp. Phys. **99**, 565 (1995).
4.75 B. Sutherland, J. Math. Phys. **12**, 251 (1971); Sutherland uses the definition $\theta_{\mathrm{Suth.}}(k_1, k_2) = -\theta(k_1, k_2)$.
4.76 J. Sólyom, Adv. Phys. **28**, 201 (1979).
4.77 I. E. Dzyaloshinskii and A. I. Larkin, Sov. Phys. JETP **34**, 422 (1972); *ibid.* **38**, 202 (1974); see also C. di Castro and W. Metzner, Phys. Rev. Lett. **67**, 3852 (1991); W. Metzner and C. di Castro, Phys. Rev. B **47**, 16107 (1993).
4.78 J. M. Luttinger, J. Math. Phys. **4**, 1154 (1963).
4.79 D. C. Mattis and E. H. Lieb, J. Math. Phys. **6**, 304 (1965).
4.80 A. Luther and V. J. Emery, Phys. Rev. Lett. **33**, 589 (1974); A. Luther and I. Peschel, Phys. Rev. B **9**, 2911 (1974); V. J. Emery, A. Luther, and I. Peschel, Phys. Rev. B **13**, 1272 (1976).
4.81 A thorough review is presented by R. Shankar, Rev. Mod. Phys. **66**, 129 (1994).
4.82 J. des Cloizeaux and M. Gaudin, J. Math. Phys. **7**, 1384 (1966); J. des Cloizeaux, *ibid.*, 2136 (1966); C. N. Yang and C. P. Yang, Phys. Rev. **150**, 321, 327 (1966); see also [4.11].
4.83 For an instructive example, see V. Meden and K. Schönhammer, J. Am. Phys. **64**, 1168 (1996).

Chapter 5

5.1 D. Vollhardt in *Correlated Electron Systems*, Vol. 9, ed. by V. J. Emery (World Scientific, Singapore, 1993), p. 57; Int. J. Mod. Phys. B **3**, 2189 (1989).
5.2 E. Müller-Hartmann, Int. J. Mod. Phys. B **3**, 2169 (1989).
5.3 T. Pruschke, M. Jarrell, and J. K. Freericks, Adv. Phys. **44**, 187 (1995).
5.4 A. Georges, G. Kotliar, W. Krauth, and M. J. Rozenberg, Rev. Mod. Phys. **68**, 13 (1996).
5.5 R. Brout, Phys. Rev. **118**, 1009 (1960).
5.6 C. J. Thompson, Comm. Math. Phys. **36**, 255 (1974).
5.7 R. J. Baxter, *Exactly Solved Models in Statistical Mechanics*, 3rd printing (Academic Press, New York, 1989).
5.8 D. C. Mattis, *The Theory of Magnetism I+II*, Springer Series in Solid-State Sciences **17**, **55** (Springer, Berlin, 1981, 1985).
5.9 R. O. Zaitsev and M. I. Dushenat, Fiz. Tverd. Tela **25**, 3440 (1983) [Sov. Phys. Solid-State **25**, 1979 (1984)].
5.10 W. Metzner and D. Vollhardt, Phys. Rev. Lett. **62**, 324 (1989).
5.11 E. Müller-Hartmann, Z. Phys. B **74**, 507 (1989); *ibid.* **76**, 211 (1989).
5.12 P. G. J. van Dongen, F. Gebhard, and D. Vollhardt, Z. Phys. B **76**, 199 (1989).
5.13 Y. L. Wang, S. Shtrikman, and H. Callen, Phys. Rev. **148**, 419 (1966).
5.14 V. Dobrosavljevič and G. Kotliar, Phys. Rev. Lett. **71**, 3218 (1993); Phys. Rev. B **50**, 1430 (1994).
5.15 W. Metzner, Z. Phys. B **77**, 253 (1989).
5.16 A. L. Fetter and J. D. Walecka, *Quantum Theory of Many-Particle Systems* (McGraw–Hill, New York, 1971).
5.17 H. Schweitzer and G. Czycholl, Z. Phys. B **83**, 93 (1991), and references therein.
5.18 P. A. Wolff, Phys. Rev. **124**, 1030 (1961); A. M. Clogston, Phys. Rev. **125**, 439 (1961); the Hartree–Fock treatment of the model is contained in T. Moriya, *Spin Fluctuations in Itinerant Electron Magnetism*, Springer Series in Solid-State Sciences **56** (Springer, Berlin, 1985).
5.19 A. Khurana, Phys. Rev. Lett. **64**, 1990 (1990).
5.20 H. Schweitzer and G. Czycholl, Phys. Rev. Lett. **67**, 3724 (1991).
5.21 T. Pruschke, D. L. Cox, and M. Jarrell, Europhys. Lett. **21**, 593 (1993); Phys. Rev. B **47**, 3553 (1993); M. Jarrell, J. K. Freericks, and T. Pruschke, Phys. Rev. B **51**, 11704 (1995).
5.22 G. Uhrig and D. Vollhardt, Phys. Rev. B **52**, 5617 (1995).
5.23 R. Vlaming and D. Vollhardt, Phys. Rev. B **45**, 4637 (1992); V. Janiš and D. Vollhardt, Phys. Rev. B **46**, 15712 (1992).
5.24 F. J. Ohkawa, J. Phys. Soc. Jpn **60**, 3218 (1991); Prog. Theor. Phys. (Supp.) **106**, 95 (1991).
5.25 V. Janiš and D. Vollhardt, Int. J. Mod. Phys. **B 6**, 283 (1992); V. Janiš and D. Vollhardt, Z. Phys. B **91**, 317 (1993); V. Janiš, J. Mašek, and D. Vollhardt, Z. Phys. B **91**, 325 (1993).
5.26 M. Jarrell, Phys. Rev. Lett. **69**, 168 (1992).
5.27 A. Georges and G. Kotliar, Phys. Rev. B **45**, 6479 (1992).
5.28 D. M. Edwards and J. A. Hertz, Physica B **163**, 527 (1990); D. M. Edwards, J. Phys. Cond. Matt. **5**, 161 (1993).
5.29 M. J. Rozenberg, X. Y. Zhang, and G. Kotliar, Phys. Rev. Lett. **69**, 1236 (1992).

5.30 G. D. Mahan, *Many-Particle Physics*, 2nd edition (Plenum Press, New York, 1990).
5.31 A. C. Hewson, *The Kondo Problem to Heavy Fermions*, Cambridge Studies in Magnetism **2** (Cambridge University Press, Cambridge, 1993); K. Yosida, *Theory of Magnetism*, Springer Series in Solid-State Sciences **122** (Springer, Berlin, 1996).
5.32 K. Yosida and K. Yamada, Prog. Theor. Phys. Suppl. **46**, 244 (1970); Prog. Theor. Phys. **53**, 1286 (1975); K. Yamada, *ibid.*, 970 (1975).
5.33 V. Zlatić and B. Horvatić, Phys. Rev. B **28**, 6904 (1983); *ibid.* **30**, 6717 (1984).
5.34 E. Feenberg, Phys. Rev. **74**, 206 (1948); a clear description can be found in [5.58].
5.35 P. B. Wiegmann and A. M. Tsvelik, J. Phys. C **16**, 2281 (1983); see also the references in [5.33].
5.36 A. B. Migdal, Sov. Phys. JETP **5**, 333 (1957); J. M. Luttinger and J. C. Ward, Phys. Rev. **118**, 1417 (1960); J. M. Luttinger, Phys. Rev. **119**, 1153 (1960); *ibid.* **121**, 942 (1961); for details, see P. Nozières, *Theory of Interacting Fermi Systems* (Addison–Wesley, New York, 1964).
5.37 E. Müller-Hartmann, Int. J. Mod. Phys. B **3**, 2169 (1989); J. R. Engelbrecht and K. S. Bedell, Phys. Rev. Lett. **74**, 4265 (1995).
5.38 B. Menge and E. Müller-Hartmann, Z. Phys. B **82**, 237 (1991).
5.39 J. K. Freericks, Phys. Rev. B **50**, 403 (1994); see also H. R. Krishnamurthy, D. M. Newns, P. C. Pattnaik, C. C. Tsuei, and C. C. Chi, Phys. Rev. B **49**, 3520 (1994).
5.40 V. Zlatić, B. Horvatić, G. Schliecker, and D. Schotte, Phil. Mag. B **65**, 1255 (1992).
5.41 W. Metzner, Phys. Rev. B **43**, 8549 (1991).
5.42 J. K. Freericks, Phys. Rev. B **48**, 3881 (1993).
5.43 W. F. Brinkman and T. M. Rice, Phys. Rev. B **2**, 1324 (1970).
5.44 W. Metzner, P. Schmit, and D. Vollhardt, Phys. Rev. B **45**, 2237 (1992); R. Strack and D. Vollhardt, Phys. Rev. B **46**, 13852 (1992); A. Schiller, P. Kumar, R. Strack, and D. Vollhardt, Phys. Rev. B **51**, 8337 (1995).
5.45 Y. Nagaoka, Phys. Rev. **147**, 392 (1966).
5.46 F. Gebhard, Phys. Rev. B **41**, 9452 (1990).
5.47 F. Gebhard, Phys. Rev. B **44**, 992 (1991).
5.48 J. Hubbard, Proc. Roy. Soc. London Ser. A **276**, 238 (1963); *ibid.* **277**, 237 (1963).
5.49 M. C. Gutzwiller, Phys. Rev. **134**, A923 (1964); *ibid.* **137**, A1726 (1965).
5.50 L. M. Falicov and J. C. Kimball, Phys. Rev. Lett. **22**, 997 (1969).
5.51 T. Kennedy and E. H. Lieb, Physica **138A**, 320 (1986); E. H. Lieb, Physica **140A**, 240 (1986).
5.52 E.-N. Foo and H. Amar, Phys. Rev. Lett. **25**, 1748 (1970); M. Plischke and D. Mattis, Phys. Rev. Lett. **27**, 42 (1971); for more recent results, see C. Gruber, D. Ueltschi, and J. Jedrzejewski, J. Stat. Phys. **76**, 125 (1994), and references therein.
5.53 U. Brandt and C. Mielsch, Z. Phys. B **75**, 365 (1989), *ibid.* **79**, 295 (1990); *ibid.* **82**, 37 (1991); U. Brandt and M. Urbanek, *ibid.* **89**, 297 (1992).
5.54 V. Janiš, Z. Phys. B **83**, 227 (1991).
5.55 P. G. J. van Dongen and D. Vollhardt, Phys. Rev. Lett. **65**, 1663 (1990); P. G. J. van Dongen, Phys. Rev. B **45**, 2267 (1992).
5.56 Q. Si and G. Kotliar, Phys. Rev. B **46**, 1261 (1992); J. K. Freericks, Phys. Rev. B **48**, 14797 (1994).
5.57 Y. M. Li and N. d'Ambrumenil, Phys. Rev. B **49**, 6058 (1994).

5.58 E. N. Economou, *Green's Functions in Quantum Physics*, Springer Series in Solid-State Sciences **7** (Springer, Berlin, 1979).

5.59 A renormalization group analysis on the question of the existence of a Fermi liquid in infinite dimensions was carried out by Q. Si and G. Kotliar, Phys. Rev. Lett. **70**, 3143 (1993); Phys. Rev. B **48**, 13881 (1993).

5.60 N. F. Mott, *Metal–Insulator Transitions*, 2nd edition (Taylor and Francis, London, 1990).

5.61 J. R. Cullen and E. Callen, J. Appl. Phys. **41**, 879 (1970); Phys. Rev. Lett. **26**, 236 (1971); D. Ihle and B. Lorenz, J. Phys. C **19**, 5239 (1986).

5.62 J. des Cloizeaux and M. Gaudin, J. Math. Phys. **7**, 1384 (1966); J. des Cloizeaux, *ibid.*, 2136 (1966); C. N. Yang and C. P. Yang, Phys. Rev. **150**, 321, 327 (1966); see also R. J. Baxter, *Exactly Solved Models in Statistical Mechanics*, 3rd printing (Academic Press, New York, 1989).

5.63 A. A. Ovchinnikov, Soviet Physics JETP **37**, 176 (1973); M. Fowler and M. W. Puga, Phys. Rev. B **18**, 1 (1978).

5.64 P. Lemberger and N. Macris, Lett. Math. Phys. **28**, 295 (1993).

5.65 A thorough review is presented by R. Shankar, Rev. Mod. Phys. **66**, 129 (1994).

5.66 G. S. Uhrig and R. Vlaming, Phys. Rev. Lett. **71**, 271 (1993); Annalen der Physik **4** (8. Serie), 778 (1995).

5.67 E. Halvorsen, G. S. Uhrig, and G. Czycholl, Z. Phys. B **94**, 291 (1994).

5.68 R. Valming, G. S. Uhrig, and D. Vollhardt, J. Phys. Cond. Matt. **4**, 7774, 10103 (E) (1992); G. S. Uhrig and R. Vlaming, J. Phys. Cond. Matt. **5**, 2561 (1993).

5.69 L. Laloux, A. Georges, and W. Krauth, Phys. Rev. B **50**, 3092 (1994).

5.70 J. K. Freericks and M. Jarrell, Phys. Rev. Lett. **74**, 186 (1995).

5.71 A. Giesekus and U. Brandt Phys. Rev. B **48**, 10311 (1993).

5.72 F. J. Ohkawa, J. Phys. Soc. Jpn **61**, 4521 (1992).

5.73 A. Georges, G. Kotliar, and W. Krauth, Z. Phys. B **92**, 313 (1993).

5.74 M. Jarrell, H. Akhlagpour, and T. Pruschke, Phys. Rev. Lett. **70**, 1670 (1993).

5.75 Q. Si and G. Kotliar, Phys. Rev. Lett. **70**, 3143 (1993); Phys. Rev. B **48**, 13881 (1993); Q. Si, M. J. Rozenberg, G. Kotliar, and A. E. Ruckenstein, Phys. Rev. Lett. **72**, 2761 (1994).

5.76 J. K. Freericks, M. Jarrell, and D. J. Scalapino, Phys. Rev. B **48**, 6302 (1993).

5.77 V. Janiš, M. Ulmke, and D. Vollhardt, Europhys. Lett. **24**, 287 (1993); Phys. Rev. B **51**, 10411 (1995).

5.78 X. Y. Zhang and G. M. Zhang, Phys. Rev. B **49**, 7929 (1994).

5.79 G. Toulouse, C. R. Acad. Sc. (Paris) B **268**, 1200 (1969).

5.80 J. E. Hirsch and R. M. Fye, Phys. Rev. Lett. **56**, 2521 (1986); J. E. Gubernatis, J. E. Hirsch, and D. J. Scalapino, Phys. Rev. B **35**, 8478 (1987).

5.81 X. Y. Zhang, M. J. Rozenberg, and G. Kotliar, Phys. Rev. Lett. **70**, 1666 (1993); M. J. Rozenberg, G. Kotliar, and X. Y. Zhang, Phys. Rev. B **49**, 10181 (1994).

5.82 A. Georges and W. Krauth, Phys. Rev. Lett. **69**, 1240 (1992); Phys. Rev. B **48**, 7167 (1993).

5.83 M. Jarrell and T. Pruschke, Z. Phys. B **90**, 187 (1993).

5.84 G. Santoro, M. Airoldi, S. Sorella, and E. Tosatti, Phys. Rev. B. **47**, 16216 (1993).

5.85 J. Hubbard, Phys. Rev. Lett. **3**, 77 (1959); R. L. Stratonovich, Sov. Phys. Dokl. **2**, 416 (1957).

5.86 H.-B. Schüttler and D. J. Scalapino, Phys. Rev. Lett. **55**, 1204 (1985); Phys. Rev. B **34**, 4744 (1986).

5.87 J. E. Gubernatis, M. Jarrell, R. N. Silver, and D. S. Sivia, Phys. Rev. B **44**, 6011 (1991).
5.88 M. Caffarel and W. Krauth, Phys. Rev. Lett. **72**, 1545 (1994).
5.89 M. J. Rozenberg, G. Moeller, and G. Kotliar, Mod. Phys. Lett. B **8**, 535 (1994).
5.90 D. Förster and N. Schopohl, Phys. Rev. B **48**, 993 (1993).
5.91 Q. Si, M. J. Rozenberg, G. Kotliar, and A. E. Ruckenstein, Phys. Rev. Lett. **72**, 2761 (1994).
5.92 C. Gros, Phys. Rev. B **50**, 7295 (1994); C. Gros, W. Wenzel, R. Valentí, G. Hülsenbeck, and J. Stolze, Europhys. Lett. **27**, 299 (1994); C. Gros, W. Wenzel, R. Valentí, and J. Stolze, J. Low Temp. Phys. **99**, 603 (1995).
5.93 H. Kajueter and G. Kotliar, Phys. Rev. Lett. **77**, 131 (1996).
5.94 N. E. Bickers, Rev. Mod. Phys. **59**, 845 (1987); see also N. E. Bickers, D. L. Cox, and J. W. Wilkins, Phys. Rev. B **36**, 2036 (1987).
5.95 T. Pruschke and N. Grewe, Z. Phys. B **74**, 439 (1989).
5.96 J. Hubbard, Proc. Roy. Soc. London **281**, 401 (1964).
5.97 A. Theumann, Phys. Rev. **178**, 978 (1969); C. Lacroix, J. Phys. F: Metal Phys. **11**, 2389.
5.98 D. E. Logan, M. P. Eastwood, and M. A. Tusch, Phys. Rev. Lett. **76**, 4785 (1996).
5.99 D. E. Logan, M. P. Eastwood, and M. Tusch, J. Phys. Cond. Matt. (1997).
5.100 J. Ferrer, A. Martín-Rodero, and F. Flores, Phys. Rev. B **36**, 6149 (1987).
5.101 L. S. Macarie and N. d'Ambrumenil, J. Phys. Cond. Matt. **7**, 3237 (1995).
5.102 J. K. Freericks and M. Jarrell, Phys. Rev. B **50**, 6939 (1994).
5.103 G. Moeller, Q. Si, G. Kotliar, M. J. Rozenberg, and D. S. Fisher, Phys. Rev. Lett. **74**, 2082 (1995).
5.104 P. Majumdar and H. R. Krishnamurthy, Phys. Rev. Lett. **73**, 1525 (1994); Phys. Rev. B **52**, 5479 (1995).

Chapter 6

6.1 H. Frahm and A. Schadschneider, J. Phys. A **26**, 1463 (1993); in *The Hubbard Model: Its Physics and Mathematical Physics*, ed. by D. Baeriswyl, D. K. Campbell, J. M. P. Carmelo, F. Guinea, and E. Louis, NATO ASI Series B **343** (Plenum Press, New York, 1995), p. 21.
6.2 F. D. M. Haldane, Phys. Lett. **80A**, 281 (1980); *ibid.* **81A**, 545 (E) (1981); T. C. Choy, Phys. Lett. **80A**, 49 (1980).
6.3 K. Lee and P. Schlottmann, Phys. Rev. Lett. **63**, 2299 (1989); Physica B **163**, 398 (1990).
6.4 P. Schlottmann, Phys. Rev. B **45**, 5784 (1992).
6.5 T. C. Choy and F. D. M. Haldane, Phys. Lett. **90A**, 83 (1982).
6.6 P. Schlottmann, Phys. Rev. B **43**, 3101 (1991).
6.7 For a recent overview on the one-dimensional extended Hubbard model, see H. Q. Lin, E. R. Gagliano, D. K. Campbell, E. H. Fradkin, and J. E. Gubernatis in in *The Hubbard Model: Its Physics and Mathematical Physics*, ed. by D. Baeriswyl, D. K. Campbell, J. M. P. Carmelo, F. Guinea, and E. Louis, NATO ASI Series B **343** (Plenum Press, New York, 1995), p. 315.
6.8 J. des Cloizeaux and M. Gaudin, J. Math. Phys. **7**, 1384 (1966); J. des Cloizeaux, *ibid.*, 2136 (1966); C. N. Yang and C. P. Yang, Phys. Rev. **150**, 321, 327 (1966); see also R. J. Baxter, *Exactly Solved Models in Statistical Mechanics*, 3rd printing (Academic Press, New York, 1989).

6.9 F. Mila and X. Zotos, Europhys. Lett. **24**, 133 (1993); K. Penc and F. Mila, Phys. Rev. B **49**, 9670 (1994).
6.10 J. Voit, Phys. Rev. B **45**, 4027 (1992); Rep. Prog. Phys. **58**, 977 (1995).
6.11 R. A. Bari, Phys. Rev. B **3**, 2662 (1971); C. Borgs, J. Jedrzejewski, and R. Kotecky, J. Phys. A **29**, 733 (1996).
6.12 P. G. J. van Dongen, Phys. Rev. Lett. **67**, 757 (1993); Phys. Rev. B **49**, 7904 (1994); Phys. Rev. B **50**, 14016 (1994); Phys. Rev. Lett. **74**, 182 (1995); in *The Hubbard Model: Its Physics and Mathematical Physics*, ed. by D. Baeriswyl, D. K. Campbell, J. M. P. Carmelo, F. Guinea, and E. Louis, NATO ASI Series B **343** (Plenum Press, New York, 1995), p. 185.
6.13 P. G. J. van Dongen and D. Vollhardt, Phys. Rev. Lett. **65**, 1663 (1990); P. G. J. van Dongen, Phys. Rev. B **45**, 2267 (1992).
6.14 C. Castellani, C. di Castro, D. Feinberg, and J. Ranninger, Phys. Rev. Lett. **43**, 1957 (1979); C. Castellani, C. di Castro, and M. Grilli, Phys. Rev. Lett. **72**, 3626 (C) (1994).
6.15 J. de Boer and A. Schadschneider, Phys. Rev. Lett. **75**, 4298 (1995); for applications to spin systems, see A. Klümper, A. Schadschneider, and J. Zittartz, J. Phys. A **24**, L955 (1991); Z. Phys. B **87**, 281 (1992); Europhys. Lett. **24**, 293 (1993); G. Su, Phys. Lett. A **213**, 93 (1996).
6.16 U. Brandt and A. Giesekus, Phys. Rev. Lett. **68**, 2648 (1992); R. Strack and D. Vollhardt, Phys. Rev. Lett. **70**, 2637 (1993); *ibid.* **72**, 3425 (1994); J. Low Temp. Phys. **99**, 385 (1995); in *The Hubbard Model: Its Physics and Mathematical Physics*, ed. by D. Baeriswyl, D. K. Campbell, J. M. P. Carmelo, F. Guinea, and E. Louis, NATO ASI Series B **343** (Plenum Press, New York, 1995), p. 145.
6.17 A. A. Ovchinnikov, Mod. Phys. Lett. B **7**, 1397 (1993); J. Phys. Cond. Matt. **6**, 11957 (1994).
6.18 See, e.g., P. W. Anderson, Phys. Rev. B **83**, 1260 (1951); W. D. Langer and D. C. Mattis, Phys. Lett. A **36**, 139 (1971); for recent applications to the Hubbard model, see R. Valentí, J. Stolze, and P. J. Hirschfeld, Phys. Rev. B **43**, 13743 (1991); R. Valentí, P. J. Hirschfeld, and J.-C. Anglés d'Auriac, Phys. Rev. B **44**, 3995 (1991).
6.19 A. B. Harris and R. V. Lange, Phys. Rev. **157**, 295 (1967).
6.20 A. Messiah, *Quantum Mechanics*, Vol. 2 (North-Holland, Amsterdam, 1962).
6.21 J. E. Hirsch, Physica C **158**, 326 (1989); J. E. Hirsch and F. Marsiglio, Phys. Rev. B **39**, 11515 (1989); J. E. Hirsch, Phys. Rev. B **43**, 11400 (1991).
6.22 J. de Boer, V. E. Korepin, and A. Schadschneider, Phys. Rev. Lett. **74**, 789 (1995); A. Schadschneider, Phys. Rev. B **51**, 10386 (1995).
6.23 A. A. Aligia and L. Arrachea, Phys. Rev. Lett. **73**, 2240 (1994).
6.24 R. Z. Bariev, J. Phys. A **24**, L549 (1991); R. Z. Bariev, A. Klümper, A. Schadschneider, and J. Zittartz, J. Phys. A **26**, 1249, 4863 (1993); M. Quaisser, A. Schadschneider, and J. Zittartz, Z. Phys. B **95**, 427 (1994).
6.25 I. N. Karnaukhov, Phys. Rev. Lett. **73**, 1130 (1994); Phys. Rev. B **51**, 7858 (1995); G. Bedürftig and H. Frahm, Phys. Rev. Lett. **74**, 5284 (C) (1995); I. N. Karnaukhov, *ibid.*, 5285 (1995); G. Bedürftig and H. Frahm, J. Phys. A **28**, 4453 (1995).
6.26 R. Z. Bariev, A. Klümper, and J. Zittartz, Europhys. Lett. **32**, 85 (1995); see also A. J. Bracken, M. D. Gould, J. R. Links, and Y.-Z. Zhang, Phys. Rev. Lett. **74**, 2768 (1995).
6.27 F. Gebhard, K. Bott, M. Scheidler, P. Thomas, S. W. Koch, Phil. Mag. B **75**, pp. 1, 13, 47 (1997).

6.28 A. Montorsi and M. Rasetti, Phys. Rev. Lett. **66**, 1383 (1991); the model naturally appeared in the discussion about the relevance of Hubbard X, V, and W terms for polymers in [2.33].
6.29 V. Janiš, F. Gebhard, R. Strack, and D. Vollhardt, Phys. Rev. Lett. **69**, 2443 (C) (1992).
6.30 K. Michielsen, H. de Raedt, and T. Schneider, Phys. Rev. Lett. **68**, 1410 (1992); P. de Vries, K. Michielsen, and H. De Raedt, Phys. Rev. Lett. **70**, 2463 (1993); Z. Phys. B **92**, 353 (1993); K. Michielsen, Int. J. Mod. Phys. B **7**, 2571 (1993); Z. Phys. B **95**, 475 (1994).
6.31 C. K. Lai and C. N. Yang, Phys. Rev. A **3**, 393 (1971); C. K. Lai, J. Math. Phys. **15**, 954, 1675 (1974); B. Sutherland, Phys. Rev. B **12**, 3795 (1975); P. Schlottmann, Phys. Rev. B **36**, 5177 (1987); P.-A. Bares and G. Blatter, Phys. Rev. Lett. **64**, 2567 (1990); P.-A. Bares, G. Blatter, and M. Ogata, Phys. Rev. B **44**, 130 (1991).
6.32 F. H. L. Eßler, V. E. Korepin, and K. Schoutens, Phys. Rev. Lett. **68**, 2960 (1992); Phys. Rev. Lett. **70**, 73 (1994); F. H. L. Eßler and V. E. Korepin, in *The Hubbard Model: Its Physics and Mathematical Physics*, ed. by D. Baeriswyl, D. K. Campbell, J. M. P. Carmelo, F. Guinea, and E. Louis, NATO ASI Series B **343** (Plenum Press, New York, 1995), p. 81; F. H. L. Eßler, J. Low Temp. Phys. **99**, 415 (1995).
6.33 D. F. Wang and C. Gruber, Phys. Rev. B **51**, 7476 (1995).
6.34 P. van Dongen and D. Vollhardt, Phys. Rev. B **40**, 7252 (1989).
6.35 K. Seiler, C. Gros, T. M. Rice, K. Ueda, and D. Vollhardt, Journ. Low Temp. Phys. **64**, 195 (1986).
6.36 D. C. Mattis and E. H. Lieb, J. Math. Phys. **6**, 304 (1965).
6.37 D. Pines and P. Nozières, *The Theory of Quantum Liquids*, 2nd edition (Addison–Wesley, New York, 1988).
6.38 J. Spałek and W. Wójcik, Phys. Rev. B **37**, 1532 (1988); K. Byczuk and J. Spałek, Phys. Rev. B **51**, 7934 (1995); G. S. Joyce, S. Sarkar, J. Spałek, and K. Byczuk, Phys. Rev. B **53**, 990 (1996).
6.39 G. Baskaran, Mod. Phys. Lett. B **5**, 643 (1991); see also M. A. Continentino and M. D. Coutinho-Filho, Solid State Comm. **90**, 619 (1994).
6.40 Y. Hatsugai and M. Kohmoto, J. Phys. Soc. Jpn **61**, 2056 (1992).
6.41 Y. Hatsugai, M. Kohmoto, T. Koma, and Y.-S. Wu, Phys. Rev. B **54**, 5358 (1996).
6.42 F. D. M. Haldane, Phys. Rev. Lett. **67**, 937 (1991); Y.-S. Wu, Phys. Rev. Lett. **73**, 922 (1994); see also [6.38].

Chapter 7

7.1 N. F. Mott, Proc. Phys. Soc. A **62**, 416 (1949).
7.2 N. F. Mott, *Metal–Insulator Transitions*, 2nd edition (Taylor and Francis, London, 1990).
7.3 V. Gates, E. Kangaroo, M. Roachcook, and W. C. Gall, Physica D **15**, 289 (1985).

Appendix

A.1 P. G. J. van Dongen, Phys. Rev. Lett. **67**, 757 (1993); Phys. Rev. B **49**, 7904 (1994); Phys. Rev. B **50**, 14016 (1994).

A.2 K. Dichtel, R. J. Jelitto, and H. Koppe, Z. Phys. **246**, 248 (1971); *ibid.* **251**, 173 (1972).

A.3 P. Kopietz, Phys. Rev. B **48**, 13789 (1993).

A.4 F. Gebhard and A. Girndt, Z. Phys. B **93**, 455 (1994).

A.5 G. D. Mahan, *Many-Particle Physics*, 2nd edition (Plenum Press, New York, 1990).

A.6 A. L. Fetter and J. D. Walecka, *Quantum Theory of Many-Particle Systems* (McGraw–Hill, New York, 1971).

A.7 For an introduction, see J. W. Negele and H. Orland, *Quantum Many-Particle Systems*, Frontiers in Physics **68** (Addison–Wesley, New York, 1987).

A.8 J. Hubbard, Proc. Roy. Soc. London Ser. A **276**, 238 (1963); *ibid.* **277**, 237 (1963).

A.9 D. N. Zubarev, Sov. Phys. Usp. **3**, 320 (1960).

A.10 N. N. Bogoliubov and S. V. Tjablikov, Dokl. Akad. Nauk USSR **126**, 53 (1959); see also S. V. Tjablikov, *Methods in the Quantum Theory of Magnetism* (Plenum Press, New York, 1967).

A.11 H. Shiba, Phys. Rev. B **6**, 930 (1972); analytical results for large U/t or small densities are presented by J. Carmelo and D. Baeriswyl, Phys. Rev. B **37**, 7541 (1987).

A.12 E. H. Lieb and F. Y. Wu, Phys. Rev. Lett. **20**, 1445 (1968).

A.13 H. Frahm and V. E. Korepin, Phys. Rev. B **42**, 10553 (1990); *ibid.* **43**, 5653 (1991).

Index

Springer Tracts in Modern Physics

121 **Current-Induced Nonequilibrium Phenomena in Quasi-One-Dimensional Superconductors**
By R. Tidecks 1990. 109 figs. IX, 341 pages

122 **Particle Induced Electron Emission I**
With contributions by M. Rösler, W. Brauer, and J. Devooght, J.-C. Dehaes, A. Dubus, M. Cailler, J.-P. Ganachaud 1991. 64 figs. X, 130 pages

123 **Particle Induced Electron Emission II**
With contributions by D. Hasselkamp and H. Rothard, K.-O. Groeneveld, J. Kemmler and P. Varga, H. Winter 1992. 90 figs. IX, 220 pages

124 **Ionization Measurements in High Energy Physics**
By B. Sitar, G. I. Merson, V. A. Chechin, and Yu. A. Budagov 1993. 184 figs. X, 337 pages

125 **Inelastic Scattering of X-Rays with Very High Energy Resolution**
By E. Burkel 1991. 70 figs. XV, 112 pages

126 **Critical Phenomena at Surfaces and Interfaces**
Evanescent X-Ray and Neutron Scattering
By H. Dosch 1992. 69 figs. X, 145 pages

127 **Critical Behavior at Surfaces and Interfaces**
Roughening and Wetting Phenomena
By R. Lipowsky 1999. 80 figs. X, Approx. 180 pages

128 **Surface Scattering Experiments with Conduction Electrons**
By D. Schumacher 1993. 55 figs. IX, 95 pages

129 **Dynamics of Topological Magnetic Solitons**
By V. G. Bar'yakhtar, M. V. Chetkin, B. A. Ivanov, and S. N. Gadetskii 1994. 78 figs. VIII, 179 pages

130 **Time-Resolved Light Scattering from Excitons**
By H. Stolz 1994. 87 figs. XI, 210 pages

131 **Ultrathin Metal Films**
Magnetic and Structural Properties
By M. Wuttig 1998. 103 figs. X, Approx. 180 pages

132 **Interaction of Hydrogen Isotopes with Transition-Metals and Intermetallic Compounds**
By B. M. Andreev, E. P. Magomedbekov, G. Sicking 1996. 72 figs. VIII, 168 pages

133 **Matter at High Densities in Astrophysics**
Compact Stars and the Equation of State
In Honor of Friedrich Hund's 100th Birthday
By H. Riffert, H. Müther, H. Herold, and H. Ruder 1996. 86 figs. XIV, 278 pages

134 **Fermi Surfaces of Low-Dimensional Organic Metals and Superconductors**
By J. Wosnitza 1996. 88 figs. VIII, 172 pages

135 **From Coherent Tunneling to Relaxation**
Dissipative Quantum Dynamics of Interacting Defects
By A. Würger 1996. 51 figs. VIII, 216 pages

136 **Optical Properties of Semiconductor Quantum Dots**
By U. Woggon 1997. 126 figs. VIII, 252 pages

137 **The Mott Metal-Insulator Transition**
Models and Methods
By F. Gebhard 1997. 38 figs. XVI, 320 pages

138 **The Partonic Structure of the Photon**
Photoproduction of the Lepton-Proton Collider HERA
By M. Erdmann 1997. 54 figs. VIII, XXX pages

Printing: Mercedesdruck, Berlin
Binding: Buchbinderei Lüderitz & Bauer, Berlin